Systematics

Systematics: A Course of Lectures

Ward C. Wheeler

WILEY-BLACKWELL

A John Wiley & Sons, Ltd., Publication

For

Kurt Milton Pickett
(1972–2011)
Ave atque vale

Contents

Color plate section between pp. 76 and 77

Preface

These notes are intended for use in an advanced undergraduate or introductory level graduate course in systematics. As such, the goal of the materials is to encourage knowledge of core systematic literature (*e.g.* works of Aristotle, Linné, Mayr, Hennig, Sokal, Farris, Kluge, Felsenstein) and concepts (*e.g.* Classification, Optimality, Optimization, Trees, Diagnosis, Medians, Computational Hardness). A component of this goal is specific understanding of methodologies and theory (*e.g.* Cluster Analysis, Parsimony, Likelihood, String Match, Tree Search). Exercises are provided to enhance familiarity with concepts and common analytical tools. These notes are focused on the study of pattern in biodiversity; notions of process receive limited attention and are better discussed elsewhere.

Each chapter covers a topic that could easily be the subject of an entire book-length treatment and many have. As a result, the coverage of large literatures is confined to what I think could be covered in a lecture or two, but may seem brief, idiosyncratic, but hopefully not too superficial. These notes are not meant to be the last word in systematics, but the first.

Students should have basic knowledge of biology and diversity including anatomy and molecular genetics. Some knowledge of computation, statistics, and linear algebra would be nice but not required. Relevant highlights of these fields are covered where necessary.

Using these notes

This is not a fugue. In most cases, sections can be rearranged, or separated entirely without loss of intelligibility. Several sections do build on others (*e.g.* sections on tree searching and support), while others can be deleted entirely if students have the background (*e.g.* sections on computational and statistical basics). The book was developed for a single semester course and, in general, each chapter is designed to be covered in a single 90 minute class period. The chapters on Parsimony, Likelihood, Posterior Probability, and Tree Searching are exceptions, spanning two such classes.

Exercises are of three types: those that can be worked by hand, those that require computational aids, and lastly those that are more suited to larger projects or group work. Hopefully, they are useful.

Acknowledgments

I would like to thank the support of the National Science Foundation (NSF) as well as the National Aeronautics and Space Administration (NASA) and the Defense Advanced Research Projects Agency (DARPA) for supporting the research that has gone into many sections of these notes.

John Wenzel and Adam Kashuba were supportive and persistent in urging me to publish these notes.

All errors, polemics, and disturbing asides are of course my own.

I thank the following people for offering expert advice in reviewing text, suggesting improvements, and identifying errors:

- Benjamin de Bivort, Ph.D.
  Rowland Institute at Harvard

- James M. Carpenter, Ph.D.
  Division of Invertebrate Zoology
  American Museum of Natural History

- Megan Cevasco, Ph.D.
  Coastal Carolina University

- Ronald M. Clouse, Ph.D.
  Division of Invertebrate Zoology
  American Museum of Natural History

- Louise M. Crowley, Ph.D.
  Division of Invertebrate Zoology
  American Museum of Natural History

- John Denton, M.A.
  Division of Vertebrate Zoology
  American Museum of Natural History

- James S. Farris, Ph.D.
  Molekylärsystematiska laboratoriet
  Naturhistoriska riksmuseet

- John V. Freudenstein, Ph.D.
  Director of the Herbarium and Museum of Zoology
  Herbarium, 1350 Museum of Biological Diversity

- Darrel Frost, Ph.D.
  Division of Vertebrate Zoology
  American Museum of Natural History

- Taran Grant, Ph.D.
  Departamento de Zoologia
  Instituto de Biociências
  Universidade de São Paulo

- Gonzalo Giribet, Ph.D.
  Department of Organismic and Evolutionary Biology
  Museum of Comparative Zoology
  Harvard University

- Pablo Goloboff, Ph.D.
  Instituto Superior de Entomologa
  Miguel Lillo

- Lin Hong, M.S.
  Division of Invertebrate Zoology
  American Museum of Natural History

- Jaakko Hyvönen, Ph.D.
  Plant Biology (Biocenter 3)
  University of Helsinki

- Daniel Janies, Ph.D.
  Biomedical Informatics
  The Ohio State University

- Isabella Kappner, Ph.D.
  Division of Invertebrate Zoology
  American Museum of Natural History

- Lavanya Kannan, Ph.D.
  Division of Invertebrate Zoology
  American Museum of Natural History

- Arnold G. Kluge, Ph.D.
  Museum of Zoology
  The University of Michigan

- Nicolas Lucaroni, B.S.
  Division of Invertebrate Zoology
  American Museum of Natural History

- Brent D. Mishler, Ph.D.
  Department of Integrative Biology
  and Jepson Herbaria
  University of California, Berkeley

- Jyrki Muona, Ph.D.
  Division of Entomology
  Zoological Museum
  University of Helsinki

- Paola Pedraza, Ph.D.
  Institute of Systematic Botany
  New York Botanical Garden

- Norman Platnick, Ph.D.
  Division of Invertebrate Zoology
  American Museum of Natural History

- Christopher P. Randle, Ph.D.
  Department of Biological Sciences
  Sam Houston State University

- Randall T. Schuh, Ph.D.
  Division of Invertebrate Zoology
  American Museum of Natural History

- William Leo Smith, Ph.D.
  Department of Zoology
  The Field Museum of Natural History

- Katherine St. John, Ph.D.
  Dept. of Mathematics & Computer Science
  Lehman College, City University of New York

- Alexandros Stamatakis, Ph.D.
  Heidelberg Institute for Theoretical Studies (HITS GmbH)

- Andrés Varón, Ph.D.
  Jane Street Capital

- Peter Whiteley, Ph.D.
  Division of Anthropology
  American Museum of Natural History

List of Algorithms

Part I

Fundamentals

Chapter 1

History

Roman bust of Aristotle
(384–322 BCE)

Ibn Rushd (Averroes)
(1126–1198)

Systematics has its origins in two threads of biological science: classification and evolution. The organization of natural variation into sets, groups, and hierarchies traces its roots to Aristotle and evolution to Darwin. Put simply, systematization of nature can and has progressed in absence of causative theories relying on ideas of "plan of nature," divine or otherwise. Evolutionists (Darwin, Wallace, and others) proposed a rationale for these patterns. This mixture is the foundation of modern systematics.

Originally, systematics was natural history. Today we think of systematics as being a more inclusive term, encompassing field collection, empirical comparative biology, and theory. To begin with, however, taxonomy, now known as the process of naming species and higher taxa in a coherent, hypothesis-based, and regular way, and systematics were equivalent.

1.1 Aristotle

Systematics as classification (or taxonomy) draws its Western origins from Aristotle[1]. A student of Plato at the Academy and reputed teacher of Alexander the Great, Aristotle founded the Lyceum in Athens, writing on a broad variety of topics including what we now call biology. To Aristotle, living things (*species*) came from nature as did other physical classes (*e.g.* gold or lead). Today, we refer to his classification of living things (Aristotle, 350 BCE) that show similarities with the sorts of classifications we create now. In short, there are three features of his methodology that we recognize immediately: it was functional, binary, and empirical.

Aristotle's classification divided animals (his work on plants is lost) using functional features as opposed to those of habitat or anatomical differences: "Of land animals some are furnished with wings, such as birds and bees." Although he recognized these features as different in aspect, they are identical in use.

---

[1]Largely through translation and commentary by Ibn Rushd (Averroes).

Systematics: A Course of Lectures, First Edition. Ward C. Wheeler.
© 2012 Ward C. Wheeler. Published 2012 by Blackwell Publishing Ltd.

Features were also described in binary terms: "Some are nocturnal, as the owl and the bat; others live in the daylight." These included egg- or live-bearing, blooded or non-blooded, and wet or dry respiration.

An additional feature of Aristotle's work was its empirical content. Aspects of creatures were based on observation rather than ideal forms. In this, he recognized that some creatures did not fit into his binary classification scheme: "The above-mentioned organs, then, are the most indispensable parts of animals; and with some of them all animals without exception, and with others animals for the most part, must needs be provided." Sober (1980) argued that these departures from Aristotle's expectations (Natural State Model) were brought about (in Aristotle's mind) by errors due to some perturbations (hybridization, developmental trauma) resulting in "terata" or monsters. These forms could be novel and helped to explain natural variation within his scheme.

- Blooded Animals
    - Live-bearing animals
        - humans
        - other mammals
    - Egg-laying animals
        - birds
        - fish

- Non-Blooded Animals
    - Hard-shelled sea animals: Testacea
    - Soft-shelled sea animals: Crustacea
    - Non-shelled sea animals: Cephalopods
    - Insects
    - Bees

- Dualizing species (potential "terata," errors in nature)
    - Whales, seals and porpoises—in water, but bear live young
    - Bats—have wings and can walk
    - Sponges—like plants and like animals.

Aristotle clearly had notions of biological progression (*scala naturae*) from lower (plant) to higher (animals through humans) forms that others later seized upon as being evolutionary and we reject today. Aristotle's classification of animals was neither comprehensive nor entirely consistent, but was hierarchical, predictive (in some sense), and formed the beginning of modern classification.

1.2 Theophrastus

Theophrastus succeeded Aristotle and is best known in biology for his *Enquiry into Plants* and *On the Causes of Plants*. As a study of classification, his work

Theophrastus (*c*.371–*c*.287 BCE)

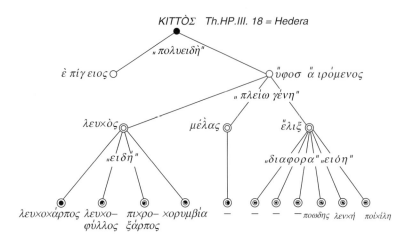

Figure 1.1: Branching diagram after Theophrastus (Vácsy, 1971).

Pierre Belon
(1517–1564)

on ivy (κιττός) discussed extensively by Nelson and Platnick (1981), has been held to be a foundational work in taxonomy based (in part at least) on dichotomous distinctions (*e.g.* growing on ground versus upright) of a few essential features.

Theophrastus distinguished ivies based on growth form and color of leaves and fruit. Although he never presented a branching diagram, later workers (including Nelson and Platnick) have summarized these observations in a variety of branching diagrams (Vácsy, 1971) (Fig. 1.1).

1.3 Pierre Belon

Trained as a physician, Pierre Belon, studied botany and traveled widely in southern Europe and the Middle East. He published a number of works based on these travels and is best known for his comparative anatomical representation of the skeletons of humans and birds (Belon, 1555) (Fig. 1.2).

1.4 Carolus Linnaeus

Carl von Linné
(1707–1778)

Carolus Linnaeus (Carl von Linné) built on Aristotle and created a classification system that has been the basis for biological nomenclature and communication for over 250 years. Through its descendants, the current codes of zoological, botanical, and other nomenclature, his influence is still felt today. Linnaeus was interested in both classification and identification (animal, plant, and mineral species), hence his system included descriptions and diagnoses for the creatures he included. He formalized the custom of binomial nomenclature, genus and species we use today.

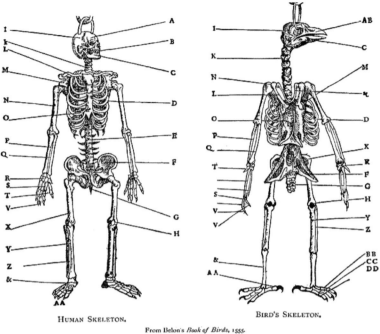

HUMAN SKELETON.　　　BIRD'S SKELETON.

From Belon's *Book of Birds*, 1555.

Figure 1.2: Belon's funky chicken (Belon, 1555).

Linnaeus was known, somewhat scandalously in his day, for his sexual system of classification (Fig. 1.3). This was most extensively applied to plants, but was also employed in the classification of minerals and fossils. Flowers were described using such terms as visible (public marriage) or clandestine, and single or multiple husbands or wives (stamens and pistils). Floral parts were even analogized to the foreskin and labia.

Nomenclature for many fungal, plant, and other eukaryote groups[2] is founded on the *Species Plantarum* (Linnaeus, 1753), and that for animals the 10th Edition of *Systema Naturae* (Linnaeus, 1758). The system is hierarchical with seven levels reflecting order in nature (as opposed to the views of Georges Louis Leclerc, 1778 [Buffon], who believed the construct arbitrary and natural variation a result of the combinatorics of components).

- Imperium (Empire)—everything

- Regnum (Kingdom)—animal, vegetable, or mineral

- Classis (Class)—in the animal kingdom there were six (mammals, birds, amphibians, fish, insects, and worms)

- Ordo (Order)—subdivisions of Class

- Genus—subdivisions of Order

---

[2]For the current code of botanical nomenclature see `http://ibot.sav.sk/icbn/main.htm`.

- Species—subdivisions of Genus

- Varietas (Variety)—species varieties or "sub-species."

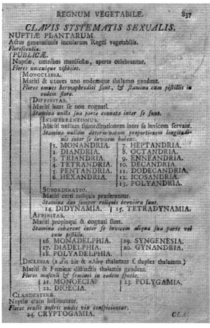

(a) Sexual system for plants (Linnaeus, 1758).   (b) English translation.

Figure 1.3: Linnaeus' sexual system for classification (a) with English translation (b) (Linnaeus, 1758).

The contemporary standard hierarchy includes seven levels: Kingdom, Phylum, Class, Order, Family, Genus, and Species, although other levels are often created as needed to describe diversity conveniently (*e.g.* McKenna and Bell, 1997).

1.5   Georges Louis Leclerc, Comte de Buffon

Georges Louis Leclerc, Comte de Buffon, began his scientific career in mathematics and probability theory[3]. He was appointed director of the *Jardin du Roi* (later *Jardin des Plantes*), making it into a research center.

Buffon is best known for the encyclopedic and massive *Histoire naturelle, générale et particulière* (1749–1788). He was an ardent anti-Linnean, believing taxa arbitrary, hence there could be no preferred classification. He later thought, however, that species were real (due to the *moule intérieur*—a concept at the

Georges Louis Leclerc, Comte de Buffon (1707–1788)

---

[3]Buffon's Needle: Given a needle of length l dropped on a plane with a series of parallel lines d apart, what is the probability that the needle will cross a line? The solution, $\frac{2l}{d\pi}$ can be used to estimate π.

foundation of comparative biology). Furthermore, Buffon believed that species could "improve" or "degenerate" into others, (*e.g.* humans to apes) changing in response to their environment. Some (*e.g.* Mayr, 1982) have argued that Buffon was among the first evolutionary thinkers with mutable species. His observation that the mammalian species of tropical old and new world, though living in similar environments, share not one taxon, went completely against then-current thought and is seen as the foundation of biogeography as a discipline (Nelson and Platnick, 1981).

Jean-Baptiste Lamarck
(1744–1829)

1.6   Jean-Baptiste Lamarck

Jean-Baptiste Lamarck (who coined the word "Biologie" in 1802) believed that classifications were entirely artificial, but still useful (especially if dichotomous). His notion of classification is closer to our modern keys (Nelson and Platnick, 1981). An example of this comes from his *Philosophie zoologique* (Lamarck, 1809), with the division of animal life into vertebrates and invertebrates on the presence or absence of "blood" (Fig. 1.4(a)).

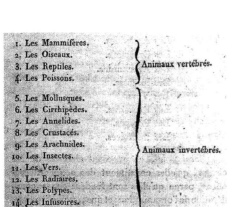

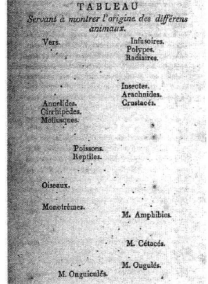

(a) Lamarck's classification of animals.   (b) Lamarck's transmutational tree.

Figure 1.4: Lamarck's division of animal life (a) and transmutational tree (b) (Lamarck, 1809).

Lamarck is best known for his theory of Transmutation (Fig. 1.4(b))—where species are immutable, but creatures may move through one species to another based on a motivating force to perfection and complexity, as well as the familiar "use and dis-use." Not only are new species created in this manner, but species can "re-evolve" in different places or times as environment and innate drive allow.

1.7   Georges Cuvier

Georges Cuvier
(1769–1832)

The hugely influential Léopold Chrétien Frédéric Dagobert "Georges" Cuvier divided animal life not into the *Scala Naturae* of Aristotle, or two-class Vertebrate/ Invertebrate divide of Lamarck, but into four "embranchements": Vertebrata, Articulata, Mollusca, and Radiata (Cuvier, 1812). These branches were representative of basic body plans or "archetypes" derived (in Cuvier's view) from functional requirements as opposed to common genealogical origin of structure. Based on his comparative anatomical work with living and fossil taxa, Cuvier believed that species were immutable but could go extinct, ("catastrophism") leaving an unfillable hole. New species, then, only appeared to be new, and were really migrants not seen before. Cuvier established the process of extinction as fact, a revolutionary idea in its day.

1.8   Étienne Geoffroy Saint-Hilaire

Étienne Geoffroy Saint-Hilaire
(1772–1844)

Although (like Lamarck), the comparative anatomist Étienne Geoffroy Saint-Hilaire is remembered for his later evolutionary views[4], Geoffroy believed that there were ideal types in nature and that species might transform among these immutable forms. Unlike Lamarck, who believed that the actions of creatures motivated transmutation, Geoffroy believed environmental conditions motivated change. This environmental effect was mediated during the development of the organism. He also believed in a fundamental unity of form for all animals (both living and extinct), with homologous structures performing similar tasks. In this, he disagreed sharply with Cuvier and his four archetypes (embranchements), not with the existence of archetypes, but with their number.

1.9   Johann Wolfgang von Goethe

Johann Wolfgang von Goethe
(1749–1832)

With Oken and Owen, Goethe was one of the foremost "ideal morphologists" of the 19th century in that he saw universal patterns underlying the forms of organisms. He coined the term "Morphology" to signify the entirety of an organism's form through development to adult as opposed to "gestalt" (or type— which was inadequate in his view). This is similar to Hennig's concept of the "semaphoront" to represent the totality of characters expressed by an organism over its entire life cycle.

Goethe applied these ideas to the comparative morphology and development of plants (von Goethe, 1790)[5] as Geoffroy did to animals, creating morphological ideals to which all plants ascribed. He claimed, based on observation, that

---

[4] "The external world is all-powerful in alteration of the form of organized bodies... these are inherited, and they influence all the rest of the organization of the animal, because if these modifications lead to injurious effects, the animals which exhibit them perish and are replaced by others of a somewhat different form, a form changed so as to be adapted to the new environment" (Saint-Hilaire, 1833).

[5] In his spare time, he wrote a book called *Faust*.

archetypes contained the inherent nature of a taxon, such as "bird-ness" or "mammal-ness." This ideal was not thought to be ancestral or primitive in any way, but embodied the morphological relationships of the members of the group.

1.10 Lorenz Oken

Oken was a leader in the "Naturphilosophie" (Oken, 1802) and an ideal morphologist. In this, he sought general laws to describe the diversity in nature through the identification of ideal forms. One of the central tenets of the *Naturphilosophie* was that there were aspects of natural law and organization that would be perceived by all observers. He applied this to his classification of animal life, and created five groups based on his perception of sense organs.

Lorenz Oken
(1779–1851)

1. Dermatazoa—invertebrates

2. Glossozoa—fish (with tongue)

3. Rhinozoa—reptiles (with nose opening)

4. Otozoa—birds (with external ear)

5. Ophthalmozoa—mammals (nose, ears, and eyes).

Oken is also known for his attempts to serially homologize vertebral elements with the vertebrate skull, suggesting fusion of separate elements as the main developmental mechanism. Although falsified for vertebrates, the idea found ground in discussions of the development of the arthropod head.

1.11 Richard Owen

Richard Owen was a vertebrate comparative anatomist known for his role in founding the British Museum (Natural History), the definitions of homology and analogy, and his opposition (after initial favor) to Darwinian evolution. Owen (1847) defined a homologue as "The same organ in different animals under every variety of form and function." Analogy was, in his view, based on function, "A part or organ in one animal which has the same function as another part or organ in a different animal."

Richard Owen
(1804–1892)

Owen derived the general archetype for vertebrates based (as in Oken) on the serial homology of vertebral elements (Fig. 1.5).

Owen's notion of homology and archetype was tightly connected with the component parts that made up the archetype—the homologues.

A system based on concentric groupings of creatures in sets of five, "Quinarianism" (Macleay, 1819), was briefly popular in early 19th century Britain.

1.12 Charles Darwin

To Aristotle, biological "species" were a component of nature in the same way that rocks, sky, and the moon were. Linnaeus held that the order of natural variation was evidence of divine plan. Darwin (1859b) brought the causative theory

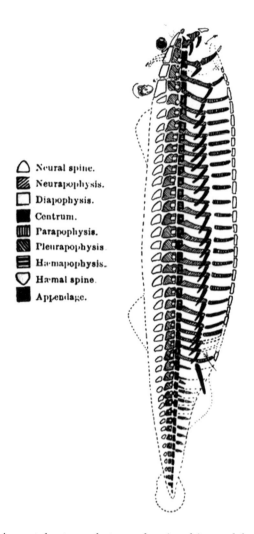

Figure 1.5: Owen's vertebrate archetype showing his model of a series of unmodified vertebral elements (Russell, 1916; after Owen, 1847).

Charles Darwin
(1809–1882)

of evolution to generate and explain the hierarchical distribution of biological variation. This had a huge intellectual impact in justifying classification as a reflection of genealogy for the first time, and bringing intellectual order (however reluctantly) to a variety of conflicting, if reasonable, classificatory schemes.

The genealogical implications of Darwin's work led him to think in terms of evolutionary "trees," (Fig. 1.6), the ubiquitous metaphor we use today. The relationship between classification and evolutionary genealogy, however, was not particularly clarified (Hull, 1988). Although the similarities between genealogy and classification were ineluctable, Darwin was concerned (as were many who followed) with representing both degree of genealogical relationship and degree of evolutionary modification in a single object. He felt quite clearly that

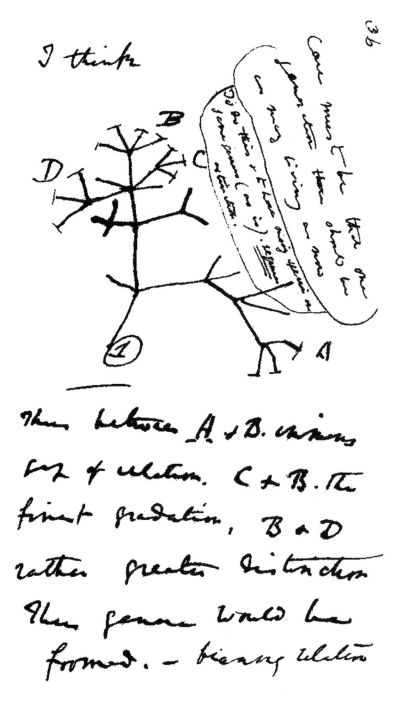

Figure 1.6: Darwin's famous "I think..." tree depiction.

classifications were more than evolutionary trees, writing that "genealogy by itself does not give classification" (Darwin, 1859a).

How to classify even a hypothetical case of genealogy (Fig. 1.7)? Darwin's Figure presents many issues—ancestral species, extinction, different "degrees of modification," different ages of taxa. As discussed by Hull (1988), Darwin gave no clear answer. He provided an intellectual framework, but no guide to actually determining phylogenetic relationships or constructing classifications based on this knowledge.

Darwin transformed Owen's archetype into an ancestor. Cladistics further transformed the ancestor into a median.

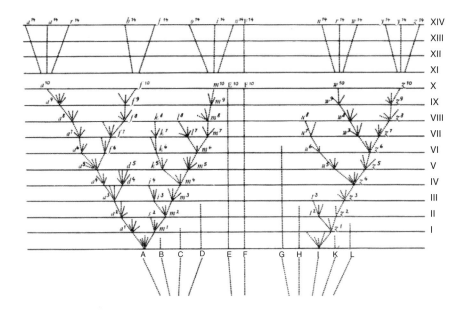

Figure 1.7: Darwin's hypothetical phylogeny from the *Origin*.

1.13   Stammbäume

Haeckel (1866) presented the situation in a graphical form (Fig. 1.8), including both genealogical relationships (as branches), degrees of modification (distance from root), and even Aristotle's *Scala Naturae* beginning with Monera at the root and progressing through worms, mollusks, echinoderms, tetrapods, mammals, and primates before crowning with humans. In his 1863 lecture, Haeckel divided the scientific community into Darwinians (progressives) and traditionalists (conservatives): "Development and progress!" ("Entwicklung und Fortschritt!") versus "Creation and species!" ("Schöpfung und Species!"). He even coined the word "Phylogeny" (Haeckel, 1866) to describe the scheme of genealogical relationships[6]. Haeckel felt that paleontology and development were the primary

---

[6] And the term "First World War" in 1914.

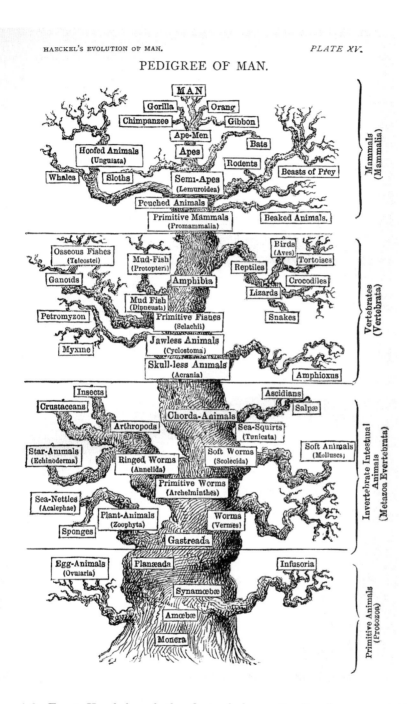

Figure 1.8: Ernst Haeckel and the first phylogenetic "tree" representation (Haeckel, 1866).

ways to discover phylogeny (Haeckel, 1876). Morphology was a third leg, but of lesser importance. Bronn (1858, 1861) also had a tree like representation and was the translator of Darwin into the German version that Haeckel read (Richards, 2005). Bronn found Darwin's ideas untested, while Haeckel did not.

August Schleicher constructed linguistic trees as Darwin had biological. A friend of Haeckel, Schleicher "tested" Darwin with language (Schleicher, 1869). Interestingly, he thought there were better linguistic fossils than biological, and hence they could form a strong test of Darwin's ideas.

1.14 Evolutionary Taxonomy

Ernst Mayr
(1904–2005)

George Gaylord Simpson
(1902–1984)

After publication of the *Origin*, evolution, genetics, and paleontology went their own ways. In the middle of the 20th century, these were brought together in what became known as the "New Synthesis." Among many, Dobzhansky (1937), Mayr (1942), Simpson (1944)[7], and Wright (1931) were most prominent. The New Synthesis brought together these strands of biology creating a satisfyingly complete (to them) Darwinian theory encompassing these formerly disparate fields (Provine, 1986; Hull, 1988). The New Synthesis begat the "New Systematics" (Huxley, 1940), which grew to become known as Evolutionary Taxonomy. Evolutionary Taxonomy competed with Phenetics (sometimes referred to as Numerical Taxonomy) and Phylogenetic Systematics (Cladistics) in the Cladistics Wars of the 1970s and 1980s, transforming systematics and classification and forming the basis for contemporary systematic research.

Here, we are limited to a brief precis of the scientific positions and differences among these three schools of systematics. Hull (1988) recounts, in great detail, the progress of the debate beginning in the late 1960s. They were amazing and frequently bitter times. As Hull writes, "Perhaps the seminar rooms of the American Museum of Natural History are not as perilous as Wallace's upper Amazon, but they come close."

Evolutionary Taxonomy as promulgated by Simpson (1961) and Mayr (1969) reached its apex in the late 1960s. This branch of systematics seized on the problem Darwin had seen in classification in that he felt that genealogy alone was not sufficient to create a classification—that systematics needed to include information on ancestors, processes, and degrees of evolutionary difference (similarity) as well as strict genealogy of taxa. There was also a great emphasis on species concepts that will be discussed later (Chapter 3).

At its heart (and the cause of its eventual downfall), Evolutionary Taxonomy was imprecise, authoritarian, and unable to articulate a specific goal other than ill-defined "naturalness." The only rule, *per se*, was that all the members of a taxonomic group should be descended from a single common ancestor. These groups were called "monophyletic" in a sense attributed to Haeckel (1866). This is in contrast to the Hennigian (Hennig, 1950, 1966) notion of monophyly that required a monophyletic group to contain *all* descendants of a common ancestor. Hennig would have called some of the "monophyletic" groups of Evolutionary

---

[7]Whose AMNH office I occupy.

Taxonomy paraphyletic (*e.g.* "Reptilia"), while Hennig's monophyly was referred to as "holophyly" by Mayr (Fig. 1.9). We now follow Hennig's concepts and their strict definitions (Farris, 1974). According to Simpson (1961), even the "monophyly" rule could be relaxed in order to maintain cherished group definitions (*e.g.* Simpson's Mammalia).

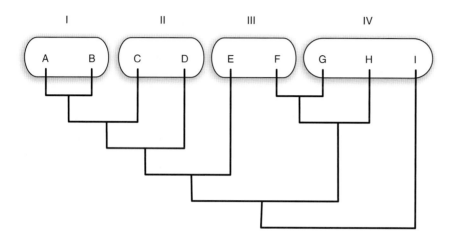

Figure 1.9: Alternate valid groups. Evolutionary Taxonomy would allow groups I, II, and IV; Hennigian Phylogenetic Systematics only I; Phenetics would allow III (as well as the others depending on degree of similarity).

In applying this rule, there were no specific criteria. Since Evolutionary Taxonomy strove to include evolutionary level (grade) information, individual investigators had to judge the relative importance of different features themselves. This weighting of information relied on the expert or authority status of the proponents of a given scenario. Great weight was given to the identification of fossil ancestors and their inclusion in systematic discussions because they were links in the Darwinian chain.

Furthermore, given that genealogy was only one element of a classification, a single genealogy could yield multiple, contradictory classifications. As stated by Mayr (1969), "Even if we had perfect understanding of phylogeny, it would be possible to convert it into many different classifications."

The lack of rules, authoritarian basis for interpretation of evidence, and inherent imprecision in the meaning of classifications produced doomed Evolutionary Taxonomy. Little remains today that is recognizably derived from this research program other than, ironically enough, the term "Cladistics."

1.15 Phenetics

Phenetics, or as it was once referred to, Numerical Taxonomy (as with Cladistics, Mayr, 1965, was the origin of the name), arose through criticisms of Evolutionary

Robert Sokal

Taxonomy. As articulated by Charles Michener, Robert Sokal, Peter Sneath, and others (Michener and Sokal, 1957; Sokal and Sneath, 1963; Sneath and Sokal, 1973), Phenetics had many features lacking in Evolutionary Taxonomy, and was free of some of its more obvious problems. Phenetic classification was based on overall similarity and required an explicit matrix of features, equally weighted. The idea was that the observations of creatures should be explicit and open to objective criticisms by other workers. The equal weighting was specified to avoid the authoritarian arguments about the relative importance of features and to produce generally useful classifications. Similarity was expressed in a phenogram, a branching tree diagram representing levels of similarity among taxa.

The method was explicit, rules-based, and objective. It also made no reference to, and had no necessary relationship with, genealogy or evolutionary trees at all. In fact, phenetic classifications could include groups of genealogically unrelated, but similar, taxa in groups termed "polyphyletic" by both Evolutionary Taxonomy and Cladistics (Fig. 1.9). This was an unavoidable consequence of lumping all similarity in the same basket, a fault found as well (if to a lesser extent) in Evolutionary Taxonomy (see Schuh and Brower, 2009, for more discussion). The specifics of phenetic (and distance methods in general) tree building are discussed later (Chapter 9).

There are few advocates of phenetic classification in contemporary science. Several contributions, however, remain. The ideas of objectivity and explicitness of evidence, specificity of rule-based tree construction, and liberation from authoritarianism all helped systematics move from art to science. Phenetics was mistaken in several major aspects, but its influence can be seen in modern, computational systematic analysis.

1.16   Phylogenetic Systematics

Willi Hennig
(1913–1976)

Phylogenetic Systematics, or as it is more commonly known, Cladistics, has its foundation in the work of Hennig (1950)[8]. Although known and read by German speakers (*e.g.* Mayr and Sokal), Hennig's work did not become widely known until later publications (Hennig, 1965, 1966). The presentation of the work (in German as well as in English) was regarded as difficult, even though the concepts were few, simple, and clear. As promulgated by Nelson (1972) and Brundin (1966), Hennig's ideas became more broadly known following the path of Nelson from Stockholm to London to New York (Schuh and Brower, 2009).

1.16.1   Hennig's Three Questions

English uses the term "sister-group" because *Gruppe* is feminine in German. Those systematists in romance-language speaking countries use "brother" group.

Hennig proposed three questions: "what is a phylogenetic relationship, how is it established, and how is knowledge of it expressed so that misunderstandings are

---

[8]The concept of what constitutes phylogenetic relationship and has come to be known as the "sister-group" was discussed both by Rosa (1918) and more prominently by Zimmermann (1931). Hennig (1950) cited Zimmerman as important to the development of his ideas (Nelson and Platnick, 1981; Donoghue and Kadereit, 1992; Williams and Ebach, 2008).

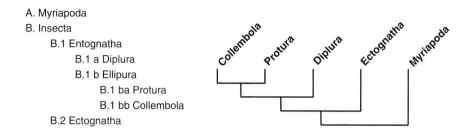

A. Myriapoda
B. Insecta
    B.1 Entognatha
        B.1 a Diplura
        B.1 b Ellipura
            B.1 ba Protura
            B.1 bb Collembola
    B.2 Ectognatha

Figure 1.10: Isomorphism between Hennigian classification (left) and genealogy (right).

excluded" (Hennig, 1965). His answers were equally precise. Phylogenetic relationship meant genealogical relationship, expressed as a series of nested sister-group relationships where two taxa are more closely related (in terms of recency of common ancestry) to each other than they are to a third[9]. These sister-group relationships are established by "special" similarity or synapomorphy—a derived (= advanced) feature present in the sister taxa and absent in others. The expression of these relationships is presented in a branching diagram summarizing the sister-group relationships termed a "cladogram" (Fig. 1.10). In Hennig's sense, a cladogram was not an evolutionary tree since it did not contain ancestor–descendant relationships, but was built on sister-group statements only.

Although Hennig had a view of species very close to that of Mayr and the evolutionary taxonomists (and the pheneticists as well), the answers to Hennig's three questions set his framework apart. In the first place, he defined phylogenetic relationship strictly in terms of recency of common ancestry. His emphasis was entirely on the "clade" as opposed to the "grade" (terms coined by Huxley, 1959) as Mayr (1965) would say. This was a definition that removed the uncertainties that existed in nearly all (phenetics aside) classification schemes.

The rules of evidence he proposed also set him apart from others in that he limited evidence of relationship to aspects that were shared and derived

Plato (360 BCE) was also the originator of $\log n$ binary search—"To separate off at once the subject of investigation, is a most excellent plan, if only the separation be rightly made. . . But you should not chip off too small a piece, my friend; the safer way is to cut through the middle; which is also the more likely way of finding classes. Attention to this principle makes all the difference in a process of enquiry."

---

[9]As Platnick (1989) has pointed out, the distinction between those groups that positively share features and those that are united only by their absence was known to the ancient Greeks. Plato (360 BCE): "The error was just as if some one who wanted to divide the human race, were to divide them after the fashion which prevails in this part of the world; here they cut off the Hellenes as one species, and all the other species of mankind, which are innumerable, and have no ties or common language, they include under the single name of 'barbarians,' and because they have one name they are supposed to be of one species also. Or suppose that in dividing numbers you were to cut off ten thousand from all the rest, and make of it one species, comprehending the first under another separate name, you might say that here too was a single class, because you had given it a single name. Whereas you would make a much better and more equal and logical classification of numbers, if you divided them into odd and even; or of the human species, if you divided them into male and female; and only separated off Lydians or Phrygians, or any other tribe, and arrayed them against the rest of the world, when you could no longer make a division into parts which were also classes."

AMNH *circa* 1910

(synapomorphy). Phenetics made no distinction between similarity that was primitive or general (symplesiomorphy), and that which was restricted or derived (synapomorphy). Furthermore, unique features of a lineage or group played no role in their placement. An evolutionary taxonomist might place a group as distinct from its relatives purely on the basis of how different its features were from other creatures (autapomorphy) such as Mayr's rejection of Archosauria (Aves + Crocodilia). The patristic (amount of change) distinctions were irrelevant to their cladistic relationships. [These terms will be discussed in later sections.] This would all have been fine if all evidence agreed, but that is not the case. Alternate statements of synapomorphy or homoplasy (convergence or parallelism) confused this issue.

Hennig annoyed many in that his cladograms made no reference to ancestors. His methodology required that ancestral species went extinct as splitting (cladogenetic) events occurred. Species only existed between splitting events, hence ancestors were difficult if not impossible to recognize (Chapter 3). This seemed anti-evolutionary, even heretical and won no friends among paleontologists. Extinct taxa could be accorded no special status—they were to be treated as any extant taxon (Chapter 2).

1.17   Molecules and Morphology

The 1980s saw tremendous technological improvement in molecular data gathering techniques. By the end of the decade, DNA sequence data were becoming available in sufficient quantity to play a role in supporting and challenging phylogenetic hypotheses, an activity that had previously been the sole province of anatomical (including developmental) data. Many meeting symposia and papers were produced agonizing over the issue (*e.g.* Patterson, 1987). In the intervening years, the topic has become something of a non-issue. Molecular sequence data are ubiquitous and easily garnered (for living taxa), forming a component of nearly all modern analyses. Anatomical information is a direct link to the world in which creatures live and is the only route to analysis of extinct taxa. Data are data and all are qualified to participate in systematic hypothesis testing.

A current descendant of this argument is that over the analysis of combined or partitioned data sets. This plays out in the debates over "Total Evidence" (Chapter 2) and, to some extent, over supertree consensus techniques (Chapter 16).

1.18   We are all Cladists

Today we struggle with different criteria to distinguish between competing and disagreeing evidence. In contemporary systematics, several methods are used to make these judgements based on Ockham's razor (parsimony) or stochastic evolutionary models (likelihood and Bayesian techniques). Although they differ in their criteria, they all agree that groups must be monophyletic in the

Hennigian sense, that classifications must match genealogy exactly, and that evidence must rely on special similarity (if differently weighted). All systematists today, whether they like it or not, are Hennigian cladists.

1.19 Exercises

1. Were the pre-Darwinians Cladists?

2. What remains of Phenetics?

3. Do we read what we want into the older literature?

4. What about the "original intent" of terms (*e.g.* monophyly)? Does it matter? Can we know? Is definitional consistency important?

5. What are the relationships among the following terms: archetype, bauplan, semaphoront, ancestor, and hypothetical ancestor?

6. What constitutes "reality" and "natural-ness" in a taxon?

Chapter 2

Fundamental Concepts

This section is a bit of a grab-bag. These are the fundamental concepts upon which systematics discussions are based. They include concepts and definitions of characters, taxa, trees, and optimality. From these, definitions of higher level concepts such as homology, polarity, and ancestors are built.

2.1   Characters

Karl Popper
(1902–1994)

Characters are the basis of systematic analysis. In principle, any variant in an organism (at any life stage) could be used for comparison, but we usually limit ourselves to those features that are intrinsic and heritable. However impartially observations can be made, characters are theory-laden objects (Popper, 1934, 1959). By this, we mean that characters are not unorganized observations, but ones that convey notions of relevance, comparability, and correspondence. It is important to keep this in mind as we attempt to test hypotheses of character evolution and relationship in as rigorous a manner as possible (Popper, 1959). Patterson (1982) and DePinna (1991) regarded the establishment of the characters themselves as the primary step (or test in the case of Patterson) of establishing homology.

Biological variants may be intrinsic or extrinsic to an organism. Intrinsic features would include the familiar character types of morphology, behavior, and biochemistry. Extrinsic features are a diverse lot, including variation in population size, geographic location, or environmental conditions. Such external features are not usually a component of phylogenetic analysis (at least in the construction and testing of hypotheses) due to the difficulty in establishing homology relationships and the absence of a direct connection to the organism itself. There is a gradation here, however, from aspects that are clearly properties of an organism itself (*e.g.* obligate feeding on a specific host) to those that are not (*e.g.* annual mean temperature).

Intrinsic features are the more frequent sources of systematic information. These may be divided further into genotypic and phenotypic aspects. Genotypic

Systematics: A Course of Lectures, First Edition. Ward C. Wheeler.
© 2012 Ward C. Wheeler. Published 2012 by Blackwell Publishing Ltd.

information is the most obviously appropriate source of comparative variation since its genomic origin requires that variation be passed through nucleic acids from generation to generation. All changes are *inheritable*. Aspects of the phenotype, such as anatomy, behavior, overall shape and size, are clearly more similar in parents and offspring than they are to other creatures (*heritable*; Fig. 2.1) even if their specific genetic basis is unknown, and hence have utility as characters. For example, the precise genomic origins of the collum in Diplopoda (millipedes) are unknown, but its strict passage from parent to offspring and restricted variation show that it is clearly appropriate for systematic study. Similarly, behavioral features such as stridulation in Orthoptera—whose genetics are also unknown—show intrinsic variation useful to systematics (Fig. 2.2). Developmental features may straddle this division, but are technically phenotypic.

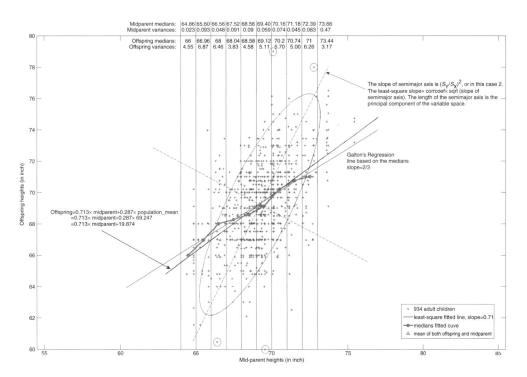

Figure 2.1: Heritability *sensu* Galton (1889). See Plate 2.1 for the color figure.

There are many heritable features that are not intrinsic to organisms, hence, they are not usually employed as grouping information. Examples of these would be location (offspring usually live near parents), mean rainfall, and population size. There are gray areas, however. A larval lepidopteran species may be found exclusively on a particular plant taxon and may eat only the leaves of that species. The notion that the metabolism of the caterpillar (and its genetics) is specific to this habitat suggest that this is an intrinsic, even inherited feature, hence of comparative use (Freudenstein et al., 2003).

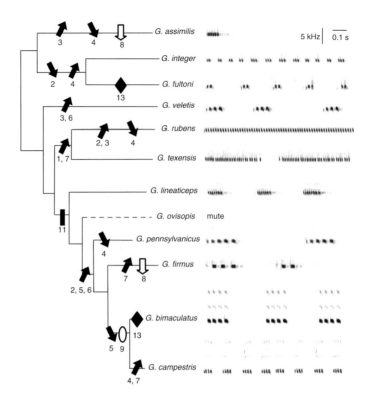

Figure 2.2: Cladogram of orthopteran stridulation (Robillard et al., 2006).

In general, we would like to include as broad and large a collection of characters as possible. This may include molecular sequence, developmental expression, anatomical information, and behavioral observations. With burgeoning molecular genetic and developmental data, situations are rapidly approaching where an observed variation may be present in multiple data types. Clearly, if we "know" the genetic origins of an anatomical variant, we cannot code a single feature in both data sets [contra Freudenstein et al. (2003)]. The issue, in this case, is independent information. Can the transformations be traced back to a single change or multiple? If single, only one variant can be coded, if multiple (or unknown), the changes are potentially independent and should all be used.

2.1.1   Classes of Characters and Total Evidence

Systematists, being classifiers, typically divide characters into classes: morphological, molecular, behavioral, developmental and so forth. Although these classes can have descriptive meaning, they do not require that their variation be valued differentially. The observations "compound eyes" and "adenine at position 234," although helpful in understanding where these characters come from (anatomy and molecular sequence) and how they were observed, do not convey any inherent strength or weakness in their ability to participate in hypothesis testing. In short, there may be descriptive character classes, but analytical classes do not necessarily follow.

The argument over whether to evaluate all characters simultaneously ("Total Evidence" Kluge, 1989; or "Simultaneous Analysis" Nixon and Carpenter, 1996b) or separately (partitioned analysis) has focused on two very different ideas of the determination of the "best" phylogenetic hypothesis. The concept behind partitioned analysis is one of robustness (*i.e.* how do different data sources agree or disagree) as opposed to one of optimality and quantity (which is the best hypothesis given all the data). These will be discussed later in more detail (Chapter 16).

Arnold Kluge

2.1.2 Ontogeny, Tokogeny, and Phylogeny

Characters and character states can have three types of relationship: "ontogenetic," "tokogenetic," and "phylogenetic" (Hennig, 1950, 1966). When states transform into one another during development, they have an ontogenetic relationship. An example is the imaginal disks of holometabolous insects, which transform into adult features such as eyes, genitalia, and wings. Although ontogenetic relationships have been used as indicators of character polarity (primitive versus derived) since Haeckel (1868), the transformations in this type of relationship are within a single organism.

Tokogeny is the relationship among features that vary within a sexually reproducing species[1] (Fig. 2.3). As variation within species, such features have been considered not to reflect relationships above the species level, hence were termed "traits" by Nixon and Wheeler (1990). Examples of traits would be color variation over the geographic range of a taxon or protein polymorphisms within populations. Nixon and Wheeler argued that true characters must be invariant within a species, but variable across species. One of the key issues with this distinction is the definition of and distinction between species. Where is the line drawn to separate traits and characters (Vrana and Wheeler, 1992)?

Features that vary across taxa are referred to as phylogenetic features and are by consensus available for systematic analysis. The fraction of total variation this constitutes varies.

2.1.3 Characters and Character States

Traditionally, there are characters, which represent comparable features among organisms, and character states, which are the set of variations in aspect or expression of a character. The distinction between the two concepts has always been fluid (are crustacean biramous antennae absent/present or a state of appendages?) (Eldredge and Cracraft, 1980), and in dynamic homology (*sensu* Wheeler, 2001b), largely meaningless.

---

[1] Also used to describe the parent–offspring relationship, irrespective of reproductive mode.

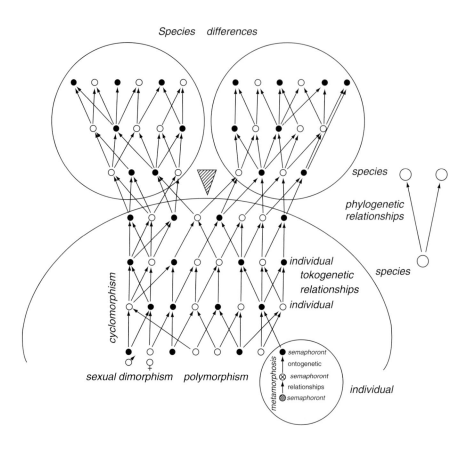

Figure 2.3: Tokogenetic and phylogenetic relationships of Hennig (1966).

As discussed by Platnick (1989), the distinction between character and character state (at least for anatomical features) can be entirely linguistic.

> In practice, objections like these sometimes amount only to linguistic quibbles. Abdominal spinnerets are not found in organisms other than spiders, and they are a valid synapomorphy of the order Araneae regardless of whether they are coded as spinnerets absent versus present, or as 'distoventral abdominal cuticle smoothly rounded' versus 'distoventral abdominal cuticle distended into spigot-bearing projections.' The more important point of these objections is that spinneret structure varies among different groups of spiders. One might treat each identifiable variant in the same way, resulting in a large number of binary characters, with each 'presence' representing a different, and additional, modification. If the sequence of modifications is detected correctly, an unobjectionable additive binary coding of the variable could be achieved. But as the

number of variants under consideration grows, the likelihood that some of the relationships among variants will be misconstrued also grows. If all the variants are coded in binary form, such misconstruals can produce erroneous cladograms, as Pimentel and Riggins [Pimentel and Riggins, 1987] demonstrated.

Platnick also raised the issue of alternate coding schemes. Traditional characters, those where character correspondences are at least thought to be known, are treated as additive (or ordered) (Farris, 1970), non-additive (or unordered) (Fitch, 1971), or matrix/Sankoff characters (Sankoff and Rousseau, 1975). In short (character types are discussed in detail in Chapter 10), these three multi-state character types (number of states > 2; there are no distinctions for binary characters) specify the relative costs of transformation between states.

For additive characters, the states are linearly ordered, with each successive state denoting a more restricted homology statement—requiring an additional transformation, or step. Therefore, a transformation between states 0 and 2 would require two steps—one step to account for the transformation from 0 to 1, and a second from 1 to 2. With additive characters, each state contains all of the homology information present in each of the preceding states. Consider the case of antennae in arthropods. The nearest extant relative of Arthropoda is the Onychophora (velvet worms; Giribet et al., 2001; Dunn et al., 2008), which do not possess antennae. Hexapods (including insects), myriapods (centipedes, millipedes, and kin), and crustaceans exhibit sensory antennae. Those of Crustacea may be biramous (two branches as opposed to one; Fig. 2.4). This feature

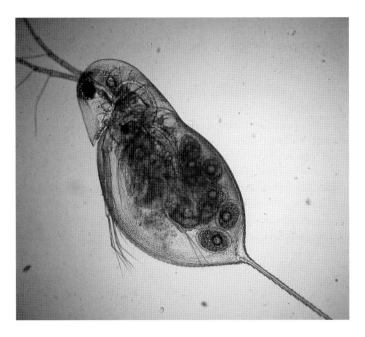

Figure 2.4: Biramous crustacean antennae of *Daphnia*.

could be coded in an additive fashion with state 0 = absence of antennae; 1 = presence of antennae; and 2 = presence of biramous antennae. Since biramous antennae are treated as a special case of the more general (uniramous) antennae and logically cannot occur independently of antennae, the character can be linearly ordered and transformations treated as additive.

In the above example, we might not feel (or be unwilling to assume) that biramous antennae logically require the underlying state of uniramous antennae. If this were the case, a transformation from state 0 to 2 would not require the intermediate state 1, and be no different from any other transformation. All transformations would cost the same—one step. This type of coding is referred to as non-additive or unordered.

A third possibility is that there are arbitrary transformation costs (perhaps limited by metricity; see Chapter 5). We might have an oracle that has revealed to us the derivation of biramous antennae *de novo* is three times as difficult (whatever that might mean) as the generation of uniramous antennae. Furthermore, it is revealed that to transform uniramous antennae into biramous is twice as costly as the origination of the uniramous antennae in the first place. These statements create a cost regime where transformations between 0 and 1 cost one step, between 0 and 2 cost three steps, and between 2 and 3 a single step. This more general cost regime character is referred to as a matrix, general, or Sankoff character. Additive and non-additive characters are special cases of this more flexible type.

In order to avoid the complexity and cost assumptions embedded in the three above types, Pleijel (1995) advocated coding features traditionally regarded as states as characters, thereby expanding the number of characters (each with two states). In the example above, the single character with three states would become two characters each with two states (presence/absence of uniramous antennae; presence/absence of biramous antennae). As Platnick (1989) pointed out (above), this really just moves the problem to a different level (that of characters) and the conflation of state information can easily cause errors (*e.g.* absence of all states).

The discussion so far assumed that we know *a priori* which states in different taxa correspond to each other (they are part of the same character); these are traditional "static" homology characters (*sensu* Wheeler, 2001b). There are situations in which these *a priori* correspondences are unknown or ambiguous. In such "dynamic" homology regimes (such as molecular sequence data), there are no input characters *per se*; feature (or state) correspondences are tree-specific, hence the result of an analysis, not an input (properties of this type of homology scheme are discussed in Chapters 7, 10, 11, and 12).

2.2 Taxa

Taxa are collections of organisms. *De minimus*, a taxon must consist of at least two creatures. Some, however, would limit the minimal taxon to be a "species" however defined, not allowing subspecific distinctions. Vrana and Wheeler (1992) advocated using individual specimens as terminals on trees. The term "terminal entities" was used specifically to avoid referring to a single creature as a taxon. Any other group, however, could be. Since all groups (size > 1) are arbitrary in their level (although "real" in the sense of monophyly), "higher taxa" are arbitrary. There are arguments surrounding the level of species, but complete consensus that the levels of all taxa above the species are arbitrary[2]. Hence, unless the taxa are sister-groups (below), they are incomparable. These groups can be "real" in the sense of reflecting hierarchical variation in nature, but are not equal in any way. An implication of this non-comparability is the senselessness of taxon enumeration (*e.g.* how many "families" went extinct, or the relative species richness of non-sister taxa).

There is a myriad of modifiers for taxa. Most refer to taxa on trees in different positions or levels of inclusion. These are discussed in more detail below. Two, however, are worth defining here since they are used a great deal in systematic literature. These are Operational Taxonomic Units (OTUs) and Hypothetical Taxonomic Units (HTUs) (Fig. 2.5). These terms were developed to be agnostic

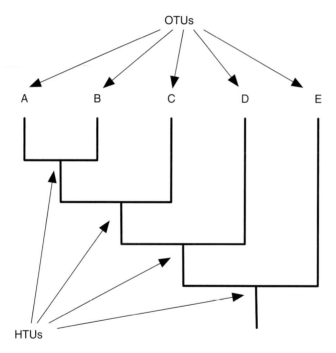

Figure 2.5: OTUs and HTUs on a tree.

---

[2]Hennig (1950) advocated a means of non-arbitrary definition of higher taxa. In this system, the rank of a taxon (Phylum, Class, *etc.*) would be determined by its age (first occurrence in fossil record)—the older, the higher. Although logical and consistent, the system never caught on largely for two reasons. First, the multiplication of higher taxa containing single or only a few taxa was regarded as ungainly (*e.g. Latimeria* as a monotypic Class); and second, the establishment of these dates is far from precise and subject to frequent revision based on the systematic relationships of other (*i.e.* sister) taxa and their own uncertain dating.

to discussions of species or other entities (mainly higher taxa) that appear on trees. An OTU is any object that occurs at the tips of a tree (also called terminal taxon or leaf). OTUs might be parts of individual creatures (*e.g.* molecular paralogs), individual organisms, species, collections of species, or higher taxa (*e.g.* Mammalia). These are normally the inputs to systematic analysis.

HTUs, on the other hand, are not observable, real things. They are abstractions created at the nodes of trees. They are "hypothetical" since they cannot be observed in nature and only exist as interpretive objects on a tree. At times, these HTUs can stand in, or be mistaken, for ancestral taxa. They are, however, constructed mathematical entities, and any path of transformation that may have occurred in nature need not have passed through the collection of attributes assigned to an HTU. These terms have origins in the early quantitative literature. OTU was coined for phenetic use (Sokal and Sneath, 1963), while Farris (1970) first used HTU in defining the Wagner tree-building procedure.

2.3   Graphs, Trees, and Networks

Trees are the central objects of systematic analysis. Taxa are ordered, characters explained, and hypotheses tested on trees. Since systematics informs and draws on other areas of science, there is a diversity of terminology for trees and their components. This section will lay out the basic definitions and descriptions that are commonly used.

Nelson (manuscript cited in Eldredge and Cracraft, 1980) made the distinction among cladograms, trees, and scenarios. This scheme differentiated the "what," "how," and "why" in phylogenetic diagrams (Fig. 2.6). *Cladograms* (following Hennig) were a representation of nested sister-group relationships. Each bifurcation on the tree signifies that the two descendants are sister taxa. These sister taxa are each other's closest genealogical (= phylogenetic) relative. Cladograms make no statements about character change, ancestors, or the evolutionary process—they are simply nested sets. Cladograms might seem to make relatively weak statements, but have embedded in them strong conclusions; among them, that sister-groups are coeval and comparable. For example, it makes little sense to compare aspects of diversity between the plant bug family Miridae and the carnivorous Reduviidae (Fig. 2.7). The only reasonable comparison would be with the sister-taxon lace bugs (Tingidae).

Trees are one step up in information content and inference. Trees, in Nelson's definition, are a series of ancestor–descendant statements. By this, Nelson meant that the points at bifurcations not only signified sister-groups, but also the ancestral condition of that larger group. With this, the changes between ancestors and descendants can be plotted on the branches that connect them. A tree remains a statement of pattern in that transformations are specified, and localized between ancestor–descendant pairs, but no motivation or biological explanation is offered.

A *scenario* is endowed with explanation in terms of evolution, ecology, or other biological or geological factors of the changes that are postulated to have

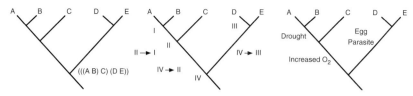

Figure 2.6: Nelson's distinction among cladogram, tree, and scenario. The clado-gram, left, conveys information of sister-groups alone. The tree, center, includes ancestor—(I–IV)—descendant statements and character transformations in ad-dition to genealogy. The scenario, right, incorporates biological motivation for evolutionary changes on the tree.

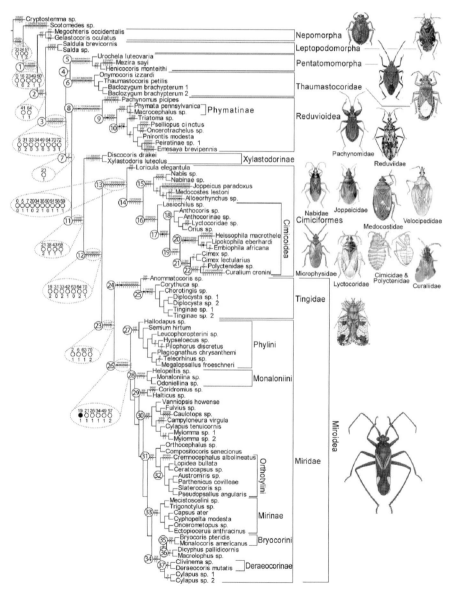

Figure 2.7: Cladogram of cimicomorpha relationships (Schuh et al., 2009).

occurred on the branches. In general, classifications deal with cladograms, systematics with trees, and evolutionary biology with scenarios.

2.3.1   Graphs and Trees

The mathematics and computer science literature defines trees in a manner nearly identical to that of Nelson, though in different terms. This field defines a tree as a connected, acyclic graph.

Sometimes called a free tree.

Graphs

Graphs are general mathematical objects consisting of a pair of sets (V, E) of *vertices* (points, V) and *edges* (lines between vertices, E). The edges can contain loops, and two vertices may be connected by multiple edges (Fig. 2.8), or by none. Simple graphs forbid parallel edges and loops. These are the sorts of graphs we will deal with here. By convention, edges are referred to by their incident vertices $[(u, v)\,\forall e_{u,v} \in E]$. The *degree* of a vertex is the number of edges incident upon it. In Fig. 2.8, vertex v_1 has degree 2. Directed graphs can specify in-degree and out-degree edges, their sum being the degree. Vertex v_3 (Fig. 2.8 left) has in-degree 1 and out-degree 2. It is possible to have a vertex that is unconnected to any other vertex, hence with degree 0. A graph is said to be *connected* if there are no vertices with degree 0, that is, if all vertices can be be visited by a path over edges. A directed graph can be created from an undirected graph by specifying an orientation to the edges. In systematics, the root (the only vertex with degree 2) directs all edges.

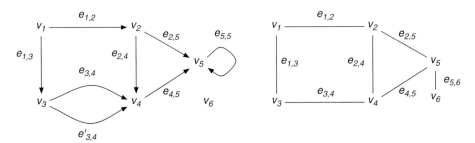

Figure 2.8: Graphs on vertex set $V = \{v_1, \ldots, v_6\}$ and edge set E. On the left, is a directed (digraph) including parallel edges between v_3 and v_4 $(e_{3,4}, e'_{3,4})$, a self loop edge on v_5 $(e_{5,5})$, and v_6 is unconnected. On the right, is an undirected, simple graph.

Trees

A tree $T = (V, E)$ is a connected graph without cycles. A *cycle* is a path over edges connecting two vertices via one or more additional vertices where each intermediate edge is visited exactly once. In Fig. 2.8, the set of vertices v_1, v_2, v_3, v_4

forms a cycle in the undirected graph (right) but not in the directed graph (left) since its edges can only be traversed in one direction. There are two types of vertices in V, *leaves* (often L) and internal ($V \setminus L$). The leaf vertices connect via a single (pendant) edge to another vertex (degree 1), whereas the internal vertices connect via at least three edges to other vertices (degree ≥ 3). In this scheme, OTUs and terminal taxa are L and HTUs, $V \setminus L$. Edges are equivalent to branches in biological terminology. Any edge may contain the root, thereby directing (or rooting) the entire tree. Hennig's concept of a "stem species" corresponds to a non-pendant edge. The tree is said to be *binary* if each non-leaf vertex has degree 3. A binary tree with $|L|$ leaves, contains $|L| - 2$ internal vertices and $2 \cdot |L| - 3$ edges ($|E| = |V| - 1$). If a tree is rooted, there is an additional internal node and an additional edge (since the root has degree 2). There is no requirement that the edges of a tree be directed. If there is a root, however, all edges become directed. The root will have in-degree 0 and out-degree 2, leaves in-degree 1 and out-degree 0, and all other (internal) vertices in-degree 1 and out-degree 2.

A *forest* is a set of trees (unconnected graph) over a set of vertices (Fig. 2.9).

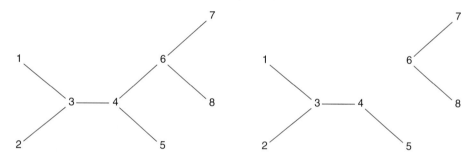

Figure 2.9: A tree, left, and forest, right, created by removing edge ($e_{4,6}$). The leaf vertices are 1, 2, 5, 7, and 8; internal vertices 3, 4, and 6.

2.3.2 Enumeration

There are many possible trees for any given set of leaves. It is straightforward to calculate the number of binary trees starting with three leaves and three edges connecting them. This is the only unrooted tree for three taxa. A fourth taxon may be added to each of the three edges (for n leaves, $2n - 3$ edges), yielding three trees. As taxa are added, the options multiply yielding Equation 2.1 (Schröder, 1870).

$$\text{for } n \geq 3 : \frac{(2n - 4)!}{(n - 2)! 2^{n-2}} \tag{2.1}$$

The number of rooted trees can be calculated by multiplying by the number of edges ($2n - 3$) or incrementing n by 1. This quantity becomes large very quickly (*e.g.* 3 trees for 4 taxa, 2027025 for 10, 2.84×10^{74} for 50) and for real data sets

n	unrooted	rooted
3	1	3
4	3	15
5	15	105
10	2,027,025	34,459,425
20	8.20×10^{21}	3.03×10^{23}
50	2.84×10^{74}	2.75×10^{76}
100	1.70×10^{182}	3.35×10^{184}

Table 2.1: Number of binary trees for n taxa.

in the hundreds or thousands of taxa analogy fails (Table 2.1). The number of forests can be generated from Equation 2.1 by observing that removing a set of k edges from a tree will generate a forest (*e.g.* 32 forests for a tree of only 4 leaves). Hence from each tree of n leaves can be generated

$$\sum_{k=0}^{2n-3} \binom{2n-3}{k} \tag{2.2}$$

forests. When k is 0, the forest contains only the single original tree. This, and other aspects of the mathematics of trees, is explored in Semple and Steel (2003).

Spanning Trees

A *spanning tree* of a graph G is a tree where all vertices in G are connected. A forest, since it does not connect all vertices (there is no sequence—path—of edges between vertices in different component trees), is not a spanning tree. As with trees in general, there can be no cycles or loops. For a complete graph with n vertices, there are n^{n-2} spanning trees.

If the edges have weights ($w(e)$), then the graph is a *weighted graph* and its weight, $w(G)$, is the sum of the weights of all its edges. The *minimum spanning tree* for a graph is the spanning tree whose weight is minimal. Kruskal (1956) described an algorithm to construct the minimum spanning tree in $O(m \log m)$ for m edges and Prim (1957) $O(n^2)$ for n vertices[3] (notation covered in Chapter 5).

The Steiner problem expands on minimal spanning trees. A *Steiner tree* allows for the addition of extra vertices and associated weighted edges[4]. This may further reduce the overall weight of the graph, but the problem becomes exponential in complexity (see Chapter 5). Gilbert and Pollak (1968) suggested (later proved) that the ratio between the weight of the best Steiner tree and that of the minimum spanning tree $\rho \geq \frac{\sqrt{3}}{2}$ (for Euclidean spaces, otherwise $\rho \geq \frac{1}{2}$).

---

[3]We can find the *maximum spanning tree* by replacing each $w(e)$ with $-w(e)$ and use Kruskal's or Prim's algorithms.

[4]For $n = 3$, the additional vertex is known as the Fermat point.

Since extant taxa cannot be ancestors of other extant taxa, and we do not have knowledge of actual ancestors, the tree reconstruction problem of systematics is a Steiner-type problem.

2.3.3 Networks

The word *network* has been used in a cavalier and confusing fashion in systematics literature. An early and persistent use describes an unrooted tree as a network (*e.g.* Bininda-Emonds et al., 2005). Current use is more precise, specifying a directed (rooted) tree with at least one node with in-degree 2 and out-degree 1 (Fig. 2.10).

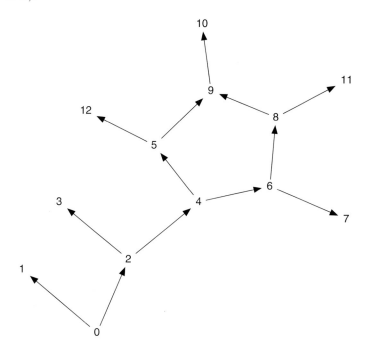

Figure 2.10: A network with directed edges, signified by arrows. This network has no cycles due to the direction of the edges. Note the presence of an in-degree 2 vertex (v_9).

Networks are most frequently used to describe reticulation events derived from either hybridization or horizontal gene transfer (HGT). In these cases, a network is created by adding edges to an existing tree (Fig. 2.11). Networks and their analysis are treated in later sections (Chapters 10, 11, and 12).

2.3.4 Mono-, Para-, and Polyphyly

When trees are directed (rooted) there are three types of groups (sets of taxa, subtrees) that can be delimited: *monophyletic*, *paraphyletic*, and *polyphyletic*.

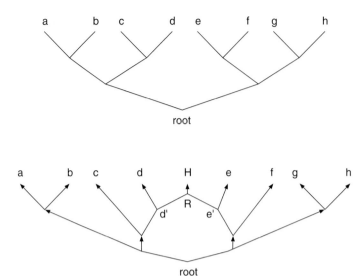

Figure 2.11: A tree, top, and network, below, created by adding a reticulate node R (and new pendant vertex H) and edges $(e_{d',R}, e_{e',R})$. The network edges are directed, signified by arrows. The tree edges have an implied direction due to the location of the root.

This scheme comes from Hennig (1950), but the term "monophyletic" has its origins with Haeckel (1868) (see Chapter 1). Hennig's monophyly was expressed in terms of the "stem" or ancestral species.

> A monophyletic group is a group of taxa in which each taxon is more closely related to every other taxon in the group than to any taxon that is classified outside the group.

> A taxon x is more closely related to another taxon y than it is to a third taxon z if, and only if, it has at least one stem species in common with y that is not also a stem species of z. (Hennig, 1966)

This definition of monophyly appears to be set theoretic, but is built on Hennig's stem species definition. Hennig's monophyletic group was the continuation of a stem species, containing all its attributes. Due to this requirement, a monophyletic group must contain *all* descendants of a common ancestor (including other descendent stem species). This is in opposition to Mayr (1942), who defined monophyly in a more catholic fashion, allowing monophyletic groups to have some, but not necessarily all, descendants of a common ancestor. By this definition, there may be taxa outside the group that are more closely related to some of the members inside the group (Fig. 2.12). The classic example of this is "Reptilia," not including Aves, but including its sister-taxon Crocodilia.

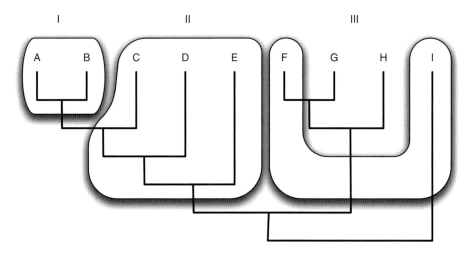

Figure 2.12: Groups. I. Monophyletic, II. Paraphyletic, and III. Polyphyletic.

Monophyletic groups are recognized, and defined, by synapomorphy (shared derived features—see below) and are the only "natural" (*i.e.* phylogenetic = directed subtree) groups.

Hennig defined two other sorts of groups: paraphyletic and polyphyletic. These were defined, in an alternate fashion, by the type of evidentiary error used to construct them. Paraphyletic groups are those based on symplesiomorphy (shared primitive features), and polyphyletic groups on convergence. In essence, paraphyletic groups are grades based on primitive features (often absence-based such as "Aptera" for basal hexapods without wings (Fig. 2.13), "Agnatha" for vertebrates without jaws (Fig. 2.14), and "Reptilia" for scale-covered tetrapods without feathers). The error is one of polarity; primitive features are mistakenly thought to be derived or mistakenly used as evidence of grouping. Paraphyletic groups of extinct lineages adjacent to extant monophyletic groups are often referred to informally as "stem" groups. Polyphyletic groups are based on errors in basic homology. Functional, convergent similarity (*e.g.* wings in insects and birds, herbivory) is mistaken for similarity based on descent.

The definitions of Hennig, with monophyly denoting naturalness based on shared descent and synapomorphy, paraphyly as a grade sharing primitive features, and polyphyly as bald error, are intuitive but informal. Their definitions are based on inconsistent terms and their application in specific cases (usually paraphyly and polyphyly) can be unclear. Farris (1974) offered the precise, formal definitions we use today. Farris used the concept of "group characters" and an algorithmic process to establish his definitions. In short, each member of a group is assigned a "1" and those not, a "0." The root of the tree is also assigned a "0." This binary character is then parsimoniously optimized on the tree (Fig. 2.15). If the group character has a single origin (transformation from 0 to 1), and no reversal (1 to 0), it is monophyletic (Group I of Fig. 2.15). A paraphlyetic group

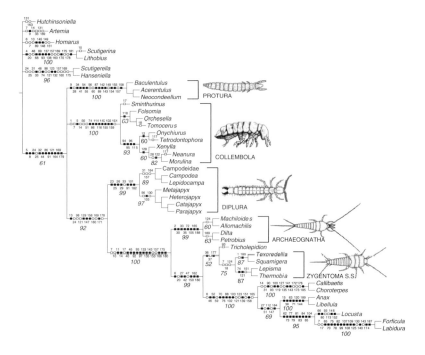

Figure 2.13: Basal paraphyly of "Aptera" (Giribet et al., 2004). Apterous hexapods are illustrated.

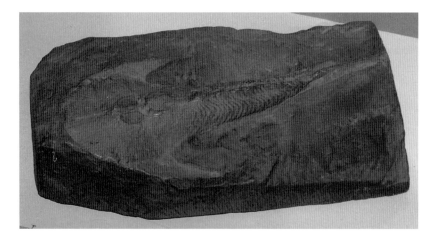

Figure 2.14: Devonian agnath *Cephalaspis*.

will have a single origin and at least one reversal (Group II). If there is more than one origin (0 to 1), the group is polyphyletic (Group III). Given the assignment of group states (0 or 1) to internal vertices, hypothetical ancestors are members of the tested groups if they are optimized to the derived state (1). Formalized, Farris's procedure is a series of simple operations (Alg. 2.1).

Algorithm 2.1: Farris1974GroupDetermination

Data: A tree, $T = (V, E)$, with leaf set, $L \subset V$ and a group of leaves $G \subset V$

Result: The group type (mono-, para-, or polyphyletic) of leaves labeled with 1.

Initialization;

forall *elements of L* **do**

    if $L_i \in G$ **then**

        | $V_i \leftarrow 1$;

    else

        | $V_i \leftarrow 0$;

    end

end

Down-pass;

while *there are unlabeled vertices* **do**

    if V_i *is unlabeled and its descendent vertices, V^L and V^R, are labeled*

    then

        if $V^L \cap V^R \neq \emptyset$ **then**

            | $V_i \leftarrow V^L \cap V^R$;

        else

            | $V_i \leftarrow V^L \cup V^R$;

        end

    end

end

Up-pass;

$V_{root} \leftarrow 0$;

while *there are vertices whose ancestors have not been set* **do**

    if *the ancestor of V_i, V_i^A has been set* **then**

        if $V_i = 0$ *or* 1 **then**

            | $V_i \leftarrow V_i$;

        else

            | $V_i \leftarrow V_i^A$;

        end

    end

end

Count origins and losses;

forall $e_{i,j} \in E$ **do**

    if $V_i = 0$ *and* $V_j = 1$ **then**

        | $origin = origin + 1$;

    else if $V_i = 1$ *and* $V_j = 0$ **then**

        | $loss = loss + 1$;

    end

end

Return result;

if $origin = 1$ *and* $loss = 0$ **then**

    | **return** *monophyletic*

else if $origin = 1$ *and* $loss > 0$ **then**

    | **return** *paraphyletic*

else

    | **return** *polyphyletic*

end

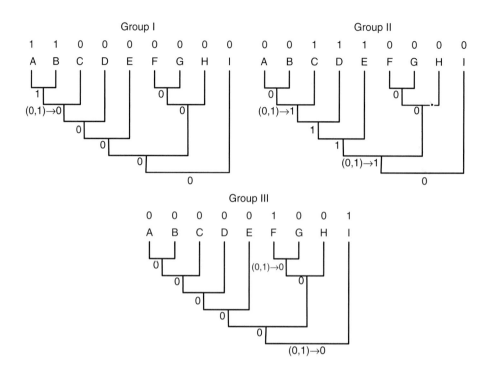

Figure 2.15: Groups of Figure 2.12 determined as in Algorithm 2.1 (Farris, 1974). Where there are arrows "→", the left value is from the first pass and the right from the second. Group I is monophyletic, Group II is paraphyletic, and Group III is polyphyletic.

2.3.5 Splits and Convexity

Efforts have been made to identify groups based on trees in the undirected state. Estabrook (1978) advocated "convex" groups of taxa. These were basically mono- and paraphyletic groups defined as taxa sharing a character state which was optimized to each internal vertex on the path between the taxa (Fig. 2.16). Other than giving more precision to the evolutionary taxonomists view of monophyly, this notion had little impact. This definition of convexity is different from that of Semple and Steel (2003) who defined a convex group as a group that can be created by a single split in a tree—in essence, a group that could be monophyletic given a rooting on one or the other side of the split.

Splits are divisions of trees (unrooted) into subtrees, frequently used in the mathematical literature (Buneman, 1971; Bandelt and Dress, 1986). A split is the division of a tree into two trees by the removal of an edge (Fig. 2.17). Biologists most frequently encounter splits in the Robinson and Foulds (1981) tree similarity metric and consensus techniques (Chapter 16), since these operations deal with shared and unique subtrees irrespective of root position. Splits are a very handy way to describe all manner of tree manipulations.

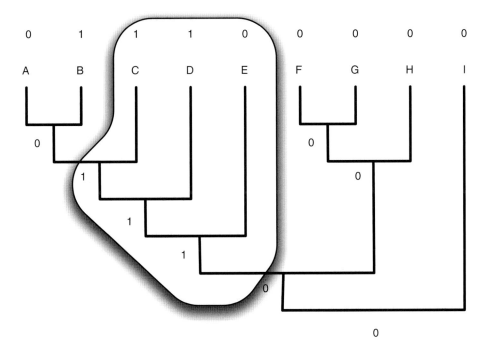

Figure 2.16: Convex group after Estabrook (1978).

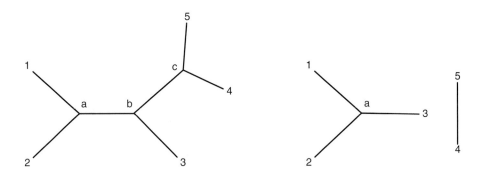

Figure 2.17: A tree (left) split on edge $e_{b,c}$ (right) (Buneman, 1971).

2.3.6 Apomorphy, Plesiomorphy, and Homoplasy

As mentioned above, the concepts of group status and the character evidence used to identify them, were intertwined in Hennig's discussions. There are two aspects of any character (or character state) that are used to describe the feature. The first is whether it is derived (apomorphic) or primitive (plesiomorphic), the second, whether the feature is shared with other creatures (syn- or

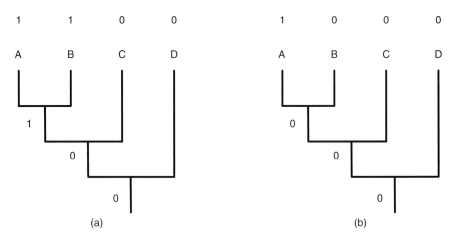

Figure 2.18: Primitive and derived character distributions. (a) State "1" is derived and a synapomorphy, while "0" is a symplesiomorphy. (b) State "1" is an autapomorphy.

sym-) or unique (aut-). Not all combinations are present, and we are left with synapomorphy, autapomorphy, and symplesiomorphy (Fig. 2.18)[5].

Since plesiomorphies reflect a feature of a larger group including the examined creatures and other more distantly related taxa, they are not evidence of relationship. The fact that humans and chimpanzees both have four legs and hair is based on their membership in Tetrapoda and Mammalia and has nothing to say about their relationship with respect to other primates. Similarly, a feature unique to a taxon (such as the collum in millipedes; Fig. 2.19) says nothing about

Figure 2.19: A diplopod with collum visible posterior to the head capsule.

---

[5]Although we might reserve *autplesiomorphy* as a pejorative.

kinship with other creatures. Only synapomorphy can be evidence of monophyly, demonstrating shared specific similarity among taxa. Clearly, the notions of synapomorphy, autapomorphy, and symplesiomorphy are level and question specific. An autapomorphy at one level is the synapomorphy of analysis at a lower level (collum as synapomorphy for Diplopoda), or symplesiomorphy when speaking about an even more restricted question (such as among millipede species).

Homoplasy is a term initially used to describe errors in character analysis (Lankester, 1870a); mistakes in homology—analogous features thought to be homologous. More precisely, homoplasy is any non-minimal (= parsimonious) change on a given tree irrespective of cause. On a particular tree, homoplasies may be described as parallelisms or convergences. Whereas on another tree, these same features may be optimized with minimal change, hence are no longer homoplastic (Fig. 2.20). Discussions of homoplasy can get bogged down in arguments over "good" homoplasy (*e.g.* adaptation) versus "bad" homoplasy (*e.g.* noise, error). The term itself refers only to non-parsimonious transformation.

2.3.7 Gene Trees and Species Trees

Although not a type of tree *per se*, the terms *gene tree* and *species tree* have entered the literature. The terms refer to the observation that trees reconstructed (by whatever means) from different genetic loci often do not agree, hence cannot all reflect the "true" historical genealogy ("species tree," Fig. 2.21; Fitch, 1970). This can occur for two reasons, one trivial, one less so. In the former case, differences among gene-based analyses reflect only the fact that data sets are finite and any combination of random character selection processes may result in non-identical results. This can be seen in any large real data set. Random partitions are unlikely to agree in every detail. In this sense, "gene trees" may reflect nothing more than homoplasy at the locus level. The latter case could be due to hybridization or *horizontal gene transfer* (HGT) between lineages as opposed to the expected vertical transfer between ancestors and descendants. HGT is no doubt a real effect, particularly among viruses and bacterial lineages, but ascribing all incongruence to HGT is cavalier and *ad hoc*. Certainly,

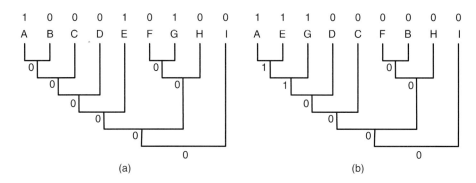

Figure 2.20: State "1" is homoplastic on tree (a) (three steps), and synapomorphic on tree (b) (one step).

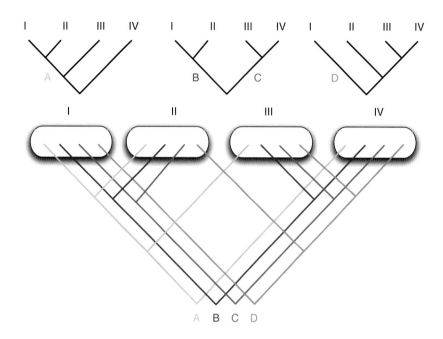

Figure 2.21: Four loci with three histories in four taxa.

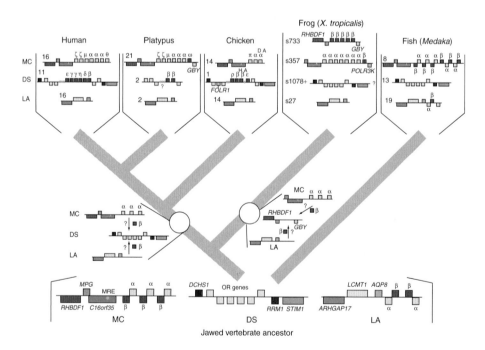

Figure 2.22: Globin duplication and diversification (Hardison, 2008). See Plate 2.22 for the color figure.

similar disagreement among anatomical character sets such as between larval and imaginal, or anterior and posterior features would not demand "horizontal" transfers of anatomical or developmental attributes across lineages.

Another use of "gene trees" concerns phylogenetic patterns of multigene families (*e.g.* vertebrate globins). This sort of tree is meant to reflect gene duplication and subsequent diversification (paralogy; Fitch, 1970). Patterns can become entangled and confusing when multigene families are analyzed for a variety of taxa. Patterns of "gene" and "species" trees can be difficult to distinguish (Fig. 2.22).

2.4 Polarity and Rooting

Systematic data and the trees constructed from them are inherently undirected[6]. That is, as graph theoretic objects, there are no directions to edges. Cladograms and Nelson's trees, however, are directed. The process of orienting the undirected trees is termed rooting (à la Haeckel). Rooting creates a special vertex (degree 2) on an edge—the root. This is based on establishing the polarity of individual characters—determining apomorphy from plesiomorphy. Three main methods have been employed: stratigraphic age, ontogenetic origin, and outgroup comparison[7].

2.4.1 Stratigraphy

Stratigraphic rooting was as simple a notion as it was incorrect. The idea was that taxa, and their features, found in lower (*i.e.* older) strata were likely to be more plesiomorphic (Hennig, 1966). This was advocated by many (especially, but not exclusively, paleontologists) up through the 1970s (*e.g.* Gingerich and Schoeninger, 1977; Harper, 1976; Szalay, 1977; Fig. 2.23). There was an immediate problem with this approach: it assumed a level of perfection in the fossil record (Schaeffer et al., 1972; Eldredge and Cracraft, 1980; Nelson and Platnick, 1981). Examples of this situation abound. If one were to use *Archaeopteryx* and its sister dromeosaurs to establish the polarity of feathers (Fig. 2.24), they would appear primitive since the derived *Archaeopteryx* is found in the Jurassic and the more basal dromeosaurs are known only from the Cretaceous (Norell and Makovicky, 2004).

2.4.2 Ontogeny

Since Haeckel (1866) and the "Biogenetic Law" (Fig. 2.25), ontogeny has been thought to be a window onto plesiomorphy. Earlier ontogenetic forms were

---

[6]The methods of Three-Taxon Analysis (Nelson and Platnick, 1991) are inherently rooted at the matrix stage. The matrix, however, is created with reference to an outgroup.

[7]Several other methods have been proposed and rejected including "common equals primitive"—repeatedly falsified (*e.g.* wings in hexapods), "complex equals derived"—falsified by empirical observations of reduction and loss, and "chorological progression"—basing polarity on geographic distribution and assumptions of the observability of advanced features.

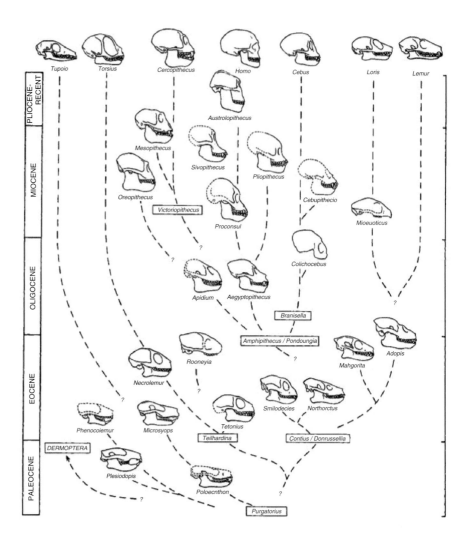

Figure 2.23: Primate stratigraphic polarity (Gingerich, 1984).

thought to be more primitive, or at least more general. Nelson (1978) reformed this idea into what he referred to as a "direct" method of inferring primitive character states. This had great influence, especially in vertebrate work, not only in determining polarity, but homology itself. This notion was less influential in invertebrate, specifically entomological, circles due to the radical changes in development in holometabolous insects that, in this view, would render genitalia, eyes, and wings non-homologous among insects. Ontogeny is much less relied upon for polarity determination currently, largely due to the fact that ontogentic

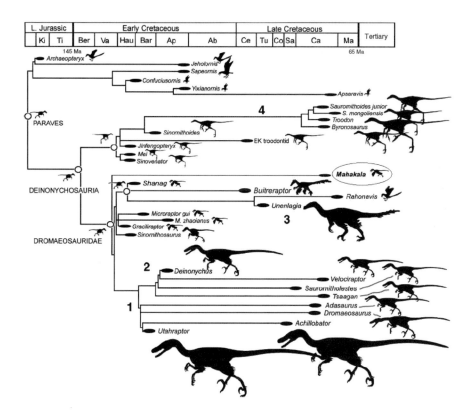

Figure 2.24: Dromeosaur phylogeny (Norell and Makovicky, 2004).

Quentin Wheeler

series are not in total agreement (hence they cannot all be correct) and that there is variation in development, which is itself informative. Developmental sequences are a fruitful source of systematic information (*e.g.* Velhagen, 1997; Schulmeister and Wheeler, 2004), but they no longer hold a special place in directing trees.

2.4.3 Outgroups

Today, the dominant technique for rooting trees is the outgroup method. The notion of character generality as indicating plesiomorphy (*e.g.* undifferentiated legs more general than wings) goes back to Hennig and before. As codified by Watrous and Wheeler (1983), outgroup comparison embraces the *ad hoc* nature of polarity determination and relies on the idea that features common to some members of the ingroup and its sister-group (ideally) or more distantly related

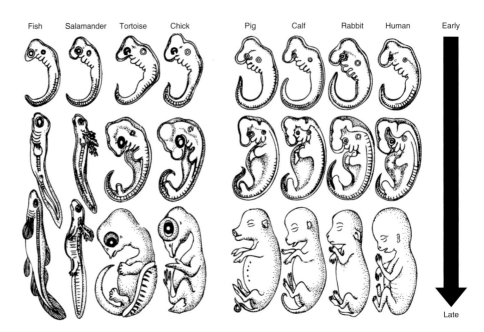

Figure 2.25: Haeckel's illustrations (later criticized) of the "Biogenetic Law" (Haeckel, 1866).

taxa are more likely to be plesiomorphic. After a "best" tree has been identified by whatever means[8], that edge which leads to the outgroup is said to contain the root node, and the tree is directed (Fig. 2.26).

Empirical comparisons between ontogenetic and outgroup criteria (*e.g.* Wheeler, 1990; Meier, 1997) have generally favored outgroup polarity establishment. Furthermore, the increased use of genomic characters, without the possibility of ontogeny, has solidified outgroup comparison as the dominant rooting technique (Nixon and Carpenter, 1993).

All-zero Outgroups

At times, analyses contain a hypothetical "all-zero" taxon and use this as an outgroup (*e.g.* Grimaldi and Engel, 2008) (Figs. 2.27 and 2.28).

Such pseudo-outgroups are simply the codification of the authors' notions of character polarity unmoored from the empirical restrictions of real taxa and,

---

[8]Usually the outgroup is treated as any other terminal or leaf taxon, but Lundberg (1972) suggested leaving the outgroup taxon outside of the primary analysis and only employing it for use in identifying the root edge.

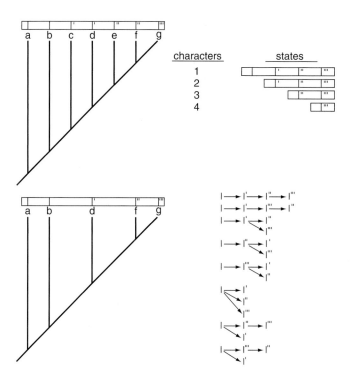

Figure 2.26: Progressively more derived character states as nodes move away from the outgroup (Watrous and Wheeler, 1983).

TABLE 1
Data Matrix for Phylogenetic Analysis of Piesmatidae
(see text for description of characters)

taxa	characters
	11111111112222
	12345678901234567890123
Piesma	
(*Piesma*)	10111110$11111111011111
(*Parapiesma*)	11111110211111111011111
(*Afropiesma*)	10111110011111111011111
Miespa	10110000201101?11011011
Mcateella	10110000201111011211011
Heissiana	10010000101111?11111?011
Eopiesma	10010001?01???0011?0111
Cretopiesma	00100110010000010000011
Outgroup*	00000000000000000000000

*The outgroup was generated from plesiomorphic states
inferred by comparison across basal pentatomorphs.
$ subset polymorphism (= 0/1)

Figure 2.27: The data matrix of Grimaldi and Engel (2008).

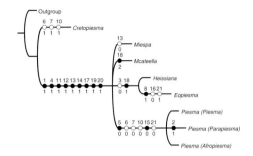

Figure 2.28: Tree of Grimaldi and Engel (2008).

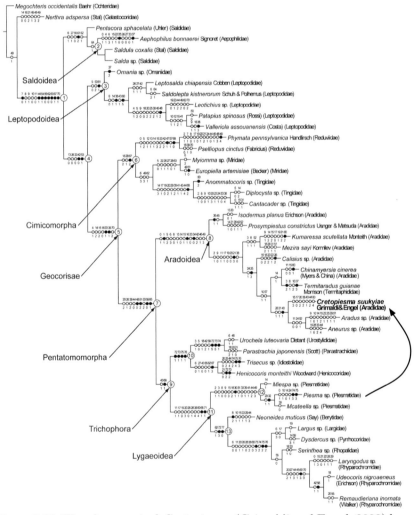

Figure 2.29: The placement of *Cretopiesma* (Grimaldi and Engel, 2008) based on an all-zero outgroup was revised into an entirely different family (Piesmatidae to Aradidae) based on character analysis of actual outgroup taxa (Cassis and Schuh, 2009).

usually, simply confirm their preconceived ideas. Often when character analysis of actual creatures is performed, conclusions based on such artificial outgroups are refuted (Cassis and Schuh, 2009) (Fig. 2.29).

2.5   Optimality

Much more will be said later about the specific calculus of optimality (Chapters 9, 10, 11, and 12). Here, the discussion centers around definitions and trees as hypotheses (Chapter 4). For trees to participate in hypothesis testing, we must be able to evaluate them and determine their relative quality. In order to do this, we require a comparable index of merit. We can argue about what that index should be, but we should all be in agreement that it must be some objective function based on data and tree $[c = f(D, T)]$. In other words, trees must have a cost and we are obliged to use this cost to determine if tree A is "better" than tree B and, transitively, identify the "best" tree or trees (Fig. 2.30(a)). Without such a cost, these objects are mere pictures—"tree-shaped-objects" (Wheeler et al., 2006a) of no use to science (Fig. 2.30(b)).

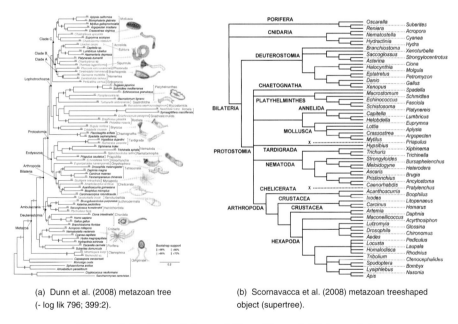

(a) Dunn et al. (2008) metazoan tree (- log lik 796; 399:2).

(b) Scornavacca et al. (2008) metazoan treeshaped object (supertree).

Figure 2.30: Tree, left, and Tree-Shaped-Object, right.

2.6   Homology

As with optimality, much more will be said later about homology, its definition, types, and identification criteria (Chapter 7). The definition of homology used here is that of Wheeler et al. (2006a).

Norman Platnick

Features are homologous when their origins can be traced to a unique transformation on the branch of a cladogram leading to their most recent common ancestor.

This definition grows out of that of Nelson and Platnick (1981), so far as homology is a synapomorphy of some group, and that transformation defines its origin. The difference here is the explicit reference to a specific edge of a specific tree. Tree, transformation, and synapomorphy are all components of homology. There can be no notion of homology without reference to a cladogram (albeit implicitly) and no choice among cladograms without statements of homology.

2.7   Exercises

1. How many unrooted trees are there for five taxa? How many rooted? How many forests (unrooted) per tree?

2. How many taxa will generate a mole of trees?

3. How many simple, undirected, unweighted graphs are there for n vertices?

4. Can a monophyletic group exist without apomorphies (Helen Keller paradox)?

5. Are polarity statements inferences or observations?

6. Which are more grievous: errors in polarity or homology?

7. Give an example of a group that is convex *sensu* Estabrook (1978) but not *sensu* Semple and Steel (2003).

8. Can synapomorphies be identified on the tree of all life?

9. What is the degree of each vertex in Figure 2.31?

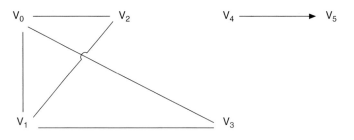

Figure 2.31

10. Is the graph in Figure 2.32 a tree? If not, what is it?

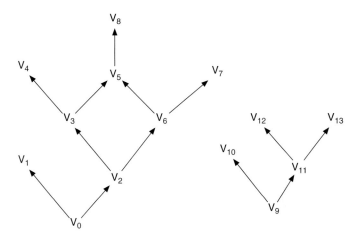

Figure 2.32

11. Show all directed trees derivable from the undirected tree in Figure 2.33.

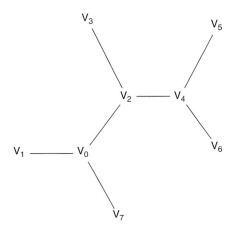

Figure 2.33

12. How many splits are possible in a tree of n leaves?

13. Using Farris's (1974) procedure, what is the status of groups A, B, and C in Figure 2.34?

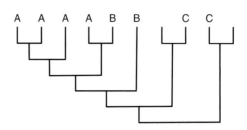

Figure 2.34

14. Contrast the statement "common equals primitive" with "broadly distributed equals primitive." Is either/both correct?

Chapter 3

Species Concepts, Definitions, and Issues

It is hard to imagine that a field as developed as systematics would have at its core such an unsettled and disputed concept as that of the species. There is an enormous literature and literally dozens of published concepts: biological, ecological, phylogenetic, monophyletic, evolutionary, Hennigian, internodal, taxonomic, and so on (Mayden, 1997; Lherminera and Solignac, 2000; Wilkins' list http://scienceblogs.com/evolvingthoughts/2006/10/a_list_of_26_species_concepts.php), many of which overlap in particulars, designed around specific or general categories, patterns, processes, and observational inference[1]. In addition to definitional differences, the point has even been made that perhaps species are not "real"[2] at all. The effects of these differences are not restricted to systematics alone. The definition, reality, and philosophical status of species have an impact on all areas in the study of biological diversity. How can "species" distribution, origin and extinction, historical gain and loss of diversity, even simple enumeration be accomplished without a consensual and concrete definition?

Without going into every nuance, modern species concepts can be differentiated along lines of emphasis of pattern or process and the primacy of monophyly. In pre-evolutionary discussions, a typological (static, descriptive) view of species was most often employed. The New Synthesis of the mid-20th century emphasized evolutionary coherence through interbreeding, yielding the Biological Species Concept. More recently, with the rise of cladistic thinking, explicit phylogenetic species concepts have been articulated, differing among themselves on the importance of monophyly and sex, and this has even led to a modern form

[1]To these we might add the *political* (designated by national boundaries) and *financial* (if there's money in it) species concepts.

[2]*Real* in the sense of being a component of nature separate from human perception.

Systematics: A Course of Lectures, First Edition. Ward C. Wheeler.
© 2012 Ward C. Wheeler. Published 2012 by Blackwell Publishing Ltd.

of species nihilism or nominalism and the debate between universal monoist and pluralist species concepts.

3.1   Typological or Taxonomic Species Concept

Plato, Aristotle, and the other ancients recognized that biological variation was not continuous (Chapter 1). There were observable natural kinds whose forms were similar and thought to be constant over time and space—"resemblance of the shapes of their parts or their whole body" (Aristotle, 350 BCE). The observation of external morphological similarity within and discontinuity between types (reflecting their "essence") underpinned the concept (and recognition) of what came to be called species[3]. This view held through to Linnaeus and later. Species were characterized by a suite of anatomical features that were viewed as universal and unchanging evidence of the plan of Providence. After Darwin, this view of species morphed into what has been termed the typological species concept—one grounded in anatomy and construing species as the basic taxonomic unit. Species in this view were recognized through similarity while defined and characterized by a specific suite of features.

As Darwinian ideas of natural selection and interbreeding entered species thinking, this typological definition hit its operational apex with Regan (1926): "A species is a community, or a number of related communities, whose distinctive morphological characters are, in the opinion of a competent systematist, sufficiently definite to entitle it, or them, to a specific name."

Charles Tate Regan
(1878–1943)

The typological species concept was (and is) absent from the two main foci of modern species debate: first, species as the smallest unit in the phylogenetic hierarchy of life; and second, species as the fundamental unit of evolution.

3.2   Biological Species Concept

Claimed to be the most generally influential species concept, the Biological Species Concept (BSC[4]) was a product of the New Synthesis (Sect. 1.14) of the mid 20th century. The concept was one of individuals in populations linked by sexual reproduction and gene flow developed by Dobzhansky (1937), Wright (1940), and Mayr (1940) among others. The most familiar version of the BSC is from Mayr (1942): "Species are groups of actually or potentially interbreeding natural populations, which are reproductively isolated from other such groups." In its most recent form, the BSC emphasized the actuality of reproductive isolation: "A species is a group of interbreeding natural populations that are reproductively isolated from other such groups" (Mayr and Ashlock, 1991). As compared to later, phylogenetic concepts, the BSC is a *process*-based definition as it relies on the reproductive habits of sexual organisms.

---

[3]The word "species" came into use in the 16th century via Middle English based on the Latin *specere*, to look.

[4]These naming and abbreviation (BSC, MSC, PSC, *etc.*) conventions may not be identical in other works.

The importance of interbreeding was recognized much earlier by Ray (1686): "no surer criterion for determining species has occurred to me than the distinguishing features that perpetuate themselves in propagation from seed." Dobzhansky (1937) emphasized this in his definition of speciation based on the restriction of gene flow between populations through barriers to interbreeding. Patterson (1980, 1985) emphasized the complement by placing importance on species mate recognition as opposed to avoiding interbreeding (Dobzhansky's reinforcement). The BSC also jibed well with the main allopatric mode of New Synthesis speciation and population thinking that Mayr and others thought crucial to understanding biological diversity.

John Ray
(1627–1705)

It is an important point, Mayr's assertions aside, that the BSC was never widely employed outside of Metazoa. In plants, for instance, the BSC was largely irrelevant (Levin, 1979; Donoghue, 1985; Gornall, 1997), and for non-Eukaryotes (as mentioned above) inapplicable.

3.2.1   Criticisms of the BSC

A multitude of criticisms (reviewed in Wheeler and Meier, 2000) has been leveled at the BSC focusing, appropriately, on the emphasis on sexual reproduction.

Agamotaxa (asexual creatures) by definition have no exchange of genetic information and each individual, for all intents and purposes, represents an independent lineage. Although less common (though far from unknown) in the metazoan taxa for which the BSC was formulated, an enormous (and perhaps majority) of taxa are asexual. These creatures lie outside the BSC. Furthermore, as Hull (1997) points out, during perhaps the first half of the history of life on Earth there was no sexual reproduction. Does this mean there were no species?

Operational difficulties immediately arose where creatures were not sympatric. How could one determine if allopatric individuals and populations could interbreed? Other information regarding morphology, ecology, behavior and so forth would have to be used as a proxy for the interbreeding criterion (Mayr, 1963). Similarly, allochronic taxa would face the same difficulty. It is impossible to test reproductive compatibility in creatures that do not reproduce contemporaneously.

Hybridization of taxa that are largely allopatric, but show some level of hybridization in sympatry, present an acknowledged problem to species delineation (Cracraft, 1989). Are these creatures one or two species? Since there are no precise conditions for species recognition, it will come down to the judgement of Regan's "competent taxonomist" as to whether the degree of hybridization is significant enough to assign the creatures to a single species.

Emphasis on reproduction as the sole delimiter of species implied that the origin of reproductive isolation was *the* mechanism of speciation and the generation of biological diversity. The most trenchant, perhaps even fatal, observation was made by Rosen (1979) in showing that reproductive compatibility could, and perhaps often was, plesiomorphic (Fig. 3.1). This would lead to species with members that were more phylogenetically related to members of other species than to their own. This type of species paraphyly is due to the reliance on a single character, reproductive compatibility, over the totality of information.

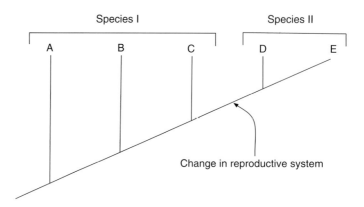

Figure 3.1: Two biological species (I and II) with individuals or populations A–E. A, B, and C can interbreed with each other, but not with D and E (which can interbreed with each other). C is genealogically more closely related to D and E than A or B, yet is a member of a different biological species. (after Rosen, 1979).

In sum, these factors yield a situation in which the BSC offers no precise way to group creatures or assign the species rank.

3.3   Phylogenetic Species Concept(s)

Starting with Donn Rosen in the late 1970s, cladistic ideas were applied to the species problem. These resulted in a series of "phylogenetic" species definitions that differed in their emphasis on monophyly and the tokogenetic relationships (Sect. 2.1.2) between parents and offspring. Unlike the BSC, reproductive compatibility has no special place in these species concepts, but each emphasizes aspects of historical or parental patterns of relationship.

3.3.1   Autapomorphic/Monophyletic Species Concept

Donn Rosen
(1929–1986)

As mentioned above, Rosen (1979) pointed out that sole reliance on reproductive compatibility could impede the understanding of biological diversity. Rosen (1978) discarded the BSC in favor of a more restrictive definition based on geography and apomorphy, "a geographically constrained group of individuals with some unique apomorphous characters, is the evolutionary unit of significance." Rosen argued for monophyly (via apomorphy) to establish the group, and biogeography the rank.

Mishler and Donoghue (1982), Mishler (1985), and Donoghue (1985) followed Rosen's criticisms of the BSC as leading to non-monophyletic species and erected a strictly monophyletic notion of species recognized by apomorphy.

A species is the least inclusive taxon in a formal phylogenetic classification. As with all hierarchical levels of taxa in such a classification, organisms are grouped into species because of evidence of monophyly. Taxa are ranked as species rather than at some higher level because they are the smallest monophyletic groups deemed worthy of formal recognition, because of the amount of support for their monophyly and/or because of their importance in biological processes operating on the lineage in question. (Mishler and Theriot, 2000)

Brent Mishler

This definition of a Monophyletic Species Concept (MSC), although precise in grouping individuals, leaves the assignment of the species rank as Reganeskly arbitrary (a point elaborated by Vrana and Wheeler, 1992). Apomorphy defines the group, but whether or not it is a species depends on the systematist's opinions on the level of support and the importance of the clade. If such an assignment is arbitrary, are species (as a level, rather than as monophyletic groups) real?

Two main criticisms are leveled at monophyletic species concepts. The first, discussed more below, is that the entire notion of monophyly may have no application below the level of species due to the reticulate nature of bisexual taxa. The importance of tokogenetic relationships is emphasized in the Phylogenetic Species Concept below. The second point is due to the fact that the ancestral or stem species (in Hennig's terminology) cannot have any apomorphies not shared by its descendants (although it could be diagnosably distinct; Fig. 3.2). Consequently, this lineage cannot be recognized by apomorphy[5].

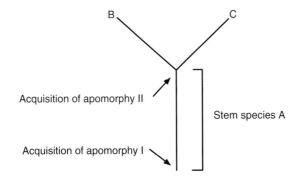

Figure 3.2: The stem species "A" will share apomorphy I with descendent species "B" and "C", but not apomorphy II. Hence, "A" can have no unique apomorphies (Hennig, 1966).

---

[5]Of course, at a time horizon during the existence of the stem species, it would have recognizable apomorphies and would be monophyletic, since its descendants would not yet exist.

Joel Cracraft

3.3.2 Diagnostic/Phylogenetic Species Concept

The term Phylogenetic Species Concept (PSC) was coined by Cracraft (1983) to differentiate his definition as different from those of Rosen and Mayr. One of his key points was that monophyly did not apply at the level of species. This idea was promulgated earlier with Nelson and Platnick (1981) stating that species were "the smallest detected samples of self-perpetuating organisms that have unique sets of characters." This definition was thought to be compatible with cladistic phylogenetic analysis, but not dependent on it, in that these clusters of organisms need not share apomorphies.

Cracraft (1983) defined species as "the smallest diagnosable cluster of individual organisms within which there is a parental pattern of ancestry and descent," the "parental pattern" being the ancestor–descendant relationship between parent and offspring used by Eldredge and Cracraft (1980) in their definition. This concept was general to sexual or asexual taxa, but the parental component proscribed hybrids (unless they were to create a new group that continued to reproduce on its own).

A central feature of this definition is the absence of apomorphy, which is replaced by diagnosability. This was further amplified by Nixon and Wheeler (1990), "the smallest aggregation of populations (sexual) or lineages (asexual) diagnosable by a unique combination of character states in comparable individuals (semaphoronts)." This definition also shifted the definition from one referring to individuals to one based on aggregations of populations or lineages.

These instantiations of the PSC allow, on the face of it, both monophyletic and paraphyletic groups to be recognized as species (but would forbid polyphyletic species). The authors of these definitions, however, argue strenuously that the concept of monophyly does not apply to species, hence species-paraphyly is no sin[6]. This idea can be traced back to Hennig (1966) in his distinction between phylogenetic relationships (those between species) and tokogenetic (those between individuals within a species) (Fig. 3.3). Nixon and Wheeler refer to features that vary within a species and reflect tokogeny as "traits" while those between, reflecting phylogeny, were recognized as "characters," which are fixed for all individuals. Species are diagnosed on the basis of unique combinations of traits (as opposed to the apomorphies of the MSC).

As Nixon and Wheeler (1992) point out, this distinction allows for phyletic speciation (without cladogenesis) as the diagnosis of a lineage changes through time and new features originate and become fixed in populations. A single lineage could undergo multiple "speciations" without any cladogenesis. This is directly contrary to speciation modes envisioned by the MSC and other concepts (below) that require splitting events for new species and render the ancestral species extinct. Furthermore, in the PSC, species may survive a splitting event if their diagnosable suite of features is unchanged (as opposed to the new sister species) (Fig. 3.4). The potential for creating species based on plesiomorphy, the possibility of phyletic speciation, and the potential perpetuation of species

---

[6]Hennig himself viewed species in a similar manner to the BSC as "reproductive communities" (Hennig, 1966) that were bounded in origin and extinction by splitting (cladogenic) events (see below).

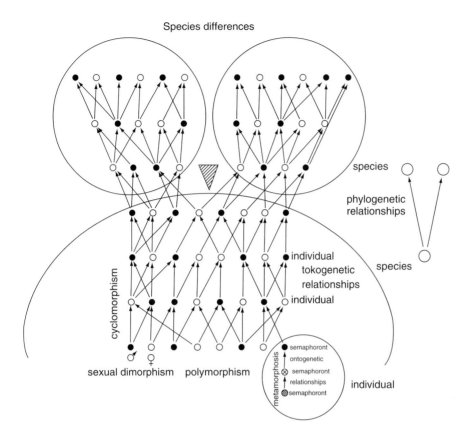

Figure 3.3: Diagram showing the phylogenetic relationships between species (upper) and tokogenetic relationships between individuals within a sexually reproducing species (Hennig, 1966).

through cladogenesis are the three implications of the PSC that attract the most criticism.

3.4 Lineage Species Concepts

These species concepts define species not as groups, but as lineages through time. Both Hennig and Simpson employed lineage concepts, but with very different results.

3.4.1 Hennigian Species

The Hennigian lineage concept (Hennig, 1950, 1966; Meier and Willmann, 2000) was of a collection of individuals linked by tokogenetic relationships with its origin at one splitting event and its demise at a second. Such a species could only

Rudolf Meier

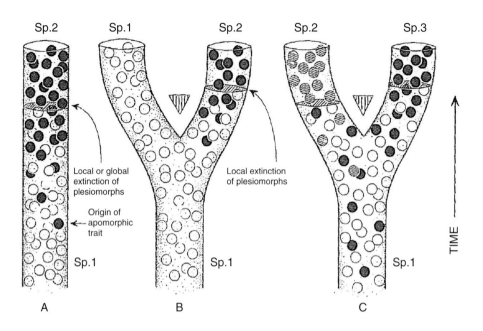

Figure 3.4: Three modes of speciation consistent with the PSC: phyletic A, ancestral persistence B, and ancestral extinction C (Nixon and Wheeler, 1992).

persist in the absence of splitting because cladogenesis disrupts the tokogenetic relationships between the daughter groups, now independent lineages (Fig. 3.3). Unlike the BSC or PSC, there can be neither phyletic speciation nor persistence of species through splitting. In the former case, tokogeny is maintained throughout the time between splits, and in the latter, tokogeny is necessarily disrupted. These notions were formalized by Kornet (1993) as "internodal" species. The Hennigian species concept emphasizes the reproductive continuity of the lineage and isolation of post-split sister taxa.

3.4.2   Evolutionary Species

Simpson (1961) was dissatisfied with the atemporal aspect of the BSC and proposed the Evolutionary Species Concept (ESC) to remedy this shortcoming. He defined a species as "a lineage (an ancestor–descendant sequence of populations) evolving separately from others and with its own unitary role and tendencies." Many of these terms are, to put it mildly, vague. Wiley (1978) and Wiley and Mayden (2000) sought to make the concept more precise with:

> An evolutionary species is an entity composed of organisms that
> maintains its identity from other such entities through time and

over space and that has its own independent evolutionary fate and historical identities.

Wiley and Mayden (2000) defined the maintenance of identity as did Hennig via tokogeny among sexually reproducing organisms. Their idea of independent evolutionary fate and identity comes from the separation of this lineage from all others, its independence. By these definitions, the ESC is little different from the Hennigian concept above (as Wiley and Mayden say)[7]. The main difference lies in the lack of specific beginning and end points. In the Hennigian concept, splitting defines both the birth and death of a species lineage. The ESC has no such restriction. As with the PSC, species may persist through splitting for an undetermined term. A crucial problem with this concept lies in how one would operationally identify such a species and differentiate it from other lineages either antecedent or descendent?

E. O. Wiley

3.4.3 Criticisms of Lineage-Based Species

Although the temporal component of lineage concepts is attractive, it also raises an operational problem. How are such species to be recognized? Lineages that exist between splitting events (Hennigian species) will not be observable, even in principle because 1) they cannot have unique apomorphies and 2) any newly observed taxon creates a new splitting event, hence new lineage species. Even the existence of such species in the first place is the result of a cladogram constructed from observed terminals. The splitting points (hypothetical ancestors) are operational, mathematical constructs (graph vertices) as are the lineages themselves (graph edges). Like ancestors, they must have been there, but cannot be identified (Fig. 3.5).

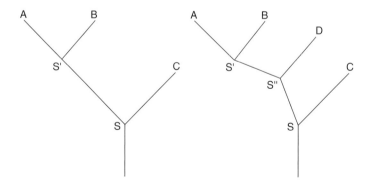

Figure 3.5: The Hennigian lineage species defined by splitting event S and S' no longer exists when terminal taxon D is observed. $S \to S'$ is replaced by $S \to S''$ and $S'' \to S'$.

---

[7]Wiley and Lieberman (2011) now say that the ESC is identical to the Hennigian concept.

3.5   Species as Individuals or Classes

One of the prominent issues in the discussion of species as ontological objects is whether species are *individuals* or *classes*. This may seem an arcane topic, but it has far-reaching ramifications in studies of historical diversity, extinction, and conservation.

An ontological class is a universal, eternal collection of similar things. A biological example might be herbivores, or flying animals that are members of a set due to the properties they possess. Classes are defined in this way intentionally, by their specific properties as necessary and sufficient, such as eating plants or having functional wings. Such a class has no beginning or end and no restriction as to how an element of such a set got there. A class such as the element Gold (in Hull's example) contains all atoms with 79 protons. It does not matter if those atoms were formed by fusions of smaller atoms or fission of larger, or by alchemy for that matter. Furthermore, the class of Gold exists without there being any members of the class. Any new atoms with atomic number 79 would be just as surely Gold as any other. One of the important aspects of classes is that scientific laws operate on them as spatio-temporally unrestricted generalizations (Hull, 1978). Laws in science require classes.

Michael Ghiselin

Individuals on the other hand, have a specific beginning and end, and are not members of any set (other than the trivial sets of individuals). Species, however defined, are considered to have a specific origin at speciation and a specific end at subsequent speciation or extinction (or at least will). As such, they are spatio-temporally restricted entities whose properties can change over time yet remain the same thing (as we all age through time, but remain the same person). A particular species (like a higher taxon) is not an instance of a type of object; each is a unique instance of its own kind.

Much of the thinking in terms of law-like evolutionary theory at least implicitly relies on the class nature of species. Only with classes can general statements be made about speciation, diversity, and extinction. Ghiselin (1966, 1969, 1974) argued that species were individuals and, as such, their names were proper names referring to specific historical objects, not general classes of things. As supported by Hull (1976, 1978) and others, this ontology has far-reaching implications. This view of species renders many comparative statements devoid of content. While it might be reasonable to ask why a process generated one gram of Gold while another one kilogram, the question "why are there so many species of beetles and so few of aardvarks?" has no meaning at all if each species is an individual. General laws of "speciation" become impossible, and temporally or geographically based enumerations of species meaningless.

David Hull
(1935–2010)

Although the case for species as individuals has wide acceptance currently (but see Stamos, 2003), biologists often operate as if species were classes. As an example, species descriptions are based on a series of features and those creatures that exhibit them are members of that species. This implies that species are an intensionally defined set and would exist irrespective of whether there were any creatures in it or not.

3.6   Monoism and Pluralism

Given the great diversity of species concepts and definitions, some have argued that different concepts should be used for different situations, offering a solution for the differences between sexual, asexual and hybrid origin taxa. This reasoning has been extended to allow for a diversity of concepts to match the diversity of reproductive systems, life history strategies, and patterns of variation (Cracraft, 2000).

A less extreme version of pluralism, advocated by Mishler and Brandon (1987) and Mishler and Theriot (2000), would have a single rule for grouping—monophyly—but allow a pluralistic view of what level in a tree hierarchy is to be assigned the rank species, and what the rationale would be. Agreeing with these authors on monophyly, but unwilling to accept an arbitrarily defined entity in systematics, Vrana and Wheeler (1992) urged the separation of pattern statements (monophyly) from process (species), removing the entire issue from concern. Wheeler and Platnick (2000) criticized this view, saying that due to the reticulating nature of the historical patterns of sexual creatures, there is no assurance that different sources of data would yield the same set of relationships. This argument loses force, however, with the realization that this can, and usually does, occur in all data sets, at all levels, with sexual or asexual creatures.

The monoist perspective is that there must be a single definition of species, universally applied to all scenarios presented by living things. Even though this does not currently exist (at least consensually), systematists have been working to identify one in the thought that it is attainable and desirable, and the present "a particularly unfortunate time to give up" (Wheeler and Platnick, 2000). Furthermore, of what use is a word, like a pluralistically defined species, that means different things to different people in different situations? If species are a component of nature, and scientists are to make general statements about them, the word must have a single meaning.

If, however, species are defined as in the MSC and they are ontological individuals, a non-arbitrary designation of a precisely comparable level in the hierarchy of individuals will be difficult to identify. For this reason, Mishler and Brandon (1987) have embraced their convenience-based pluralism.

3.7   Pattern and Process

One of the difficulties in constructing a single species definition relates to the sometimes conflicting goals of incorporating pattern and process information. The BSC is the most notable—based as it is on gene-flow through interbreeding—but not the only (e.g. ecological) concept based on an explicit notion of biological process. In essence, the concepts state that evolution occurs this way, so species should be defined in this manner.

Other definitions emphasize pattern phenomena, specifically monophyly. In such a concept, the patterns that drive diversification are irrelevant to whether a group is monophyletic or not. Such ideas began with Rosen (1978, 1979) and continued through the MSC of Donoghue (1985) and Mishler (1985) and the species nihilism of Vrana and Wheeler (1992). In each of these systems, monophyly is the only criterion for grouping taxa. Mishler and Brandon (1987) then proposed a secondary, potentially process-based assignment of the species rank.

The PSC and ESC both emphasize the tokogenetic relationships among taxa as a fundamental aspect of conspecifics. Tokogeny, due to the process of interbreeding, links the members of a species. As Wheeler and Platnick (2000) state, "a concept fully compatible with phylogenetic theory but not dependent on prior cladistic analysis."

No definition can simultaneously optimize the description of both pattern and process (not that they will always disagree, but they certainly can). Perhaps we should only use the word monophyly for statements of pattern and reserve "species" solely for statements of evolutionary units and process.

3.8   Species Nominalism

Some authors have argued whether species as things actually exist in nature, or are merely a notion imposed by the human mind to aid us in organizing information. This argument of species nominalism usually does not progress to the level where species are thought to be completely arbitrary collections of organisms as constellations are of stars. The focus of most arguments are on the level of species. Are they arbitrary in the manner of higher taxa (*e.g.* genera, families), or are they *real* components of nature?

Darwin (1859b) has been thought to be such a nominalist: "I look at the terms species as arbitrarily given, for the sake of convenience, to a set of individuals closely resembling each other." At the very least, for species to be real in any sense they must first be non-arbitrary. Each of the concepts above, certainly from the BSC through MSC and PSC to lineage definitions claim to be precise in their identification of the groups of creatures (or populations) to be gathered into a species. The BSC, PSC, and lineage definitions are also quite specific about assignment of these groups to the species rank. Those ideas based on strict monophyly, however clear they are in defining the groups, are inherently arbitrary in the choice of which level to assign the rank of species. Rosen (1979) used geographic restriction as a criterion; Mishler and Brandon (1987) and Mishler and Theriot (2000) allow for multiple criteria to be used (quoted above).

Vrana and Wheeler (1992) took this argument to its extreme, arguing that the species level was entirely arbitrary and in no way different from higher taxa. These authors argued for phylogenetic analysis of the pattern of diversification based solely on monophyly. The notion of species as phylogenetic units would be abandoned in favor of using individual organisms as terminals (leaves). The designation of species would then be free to be used on any convenient basis for the investigation of evolutionary process.

These last two views disagree on extent, but are in agreement that the assignment of species rank will always be arbitrary, hence cannot be *real* in the ontological sense of being natural components of the biological world.

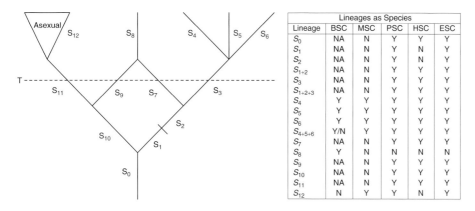

Lineage	BSC	MSC	PSC	HSC	ESC
S_0	NA	N	Y	Y	Y
S_1	NA	N	Y	N	Y
S_2	NA	N	Y	N	Y
S_{1+2}	NA	N	Y	Y	Y
S_3	NA	N	Y	Y	Y
S_{1+2+3}	NA	N	Y	Y	Y
S_4	Y	Y	Y	Y	Y
S_5	Y	Y	Y	Y	Y
S_6	Y	Y	Y	Y	Y
S_{4+5+6}	Y/N	Y	Y	Y	Y
S_7	NA	N	Y	Y	Y
S_8	Y	N	N	N	N
S_9	NA	N	Y	Y	Y
S_{10}	NA	N	Y	Y	Y
S_{11}	NA	N	Y	Y	Y
S_{12}	N	Y	Y	N	Y

Figure 3.6: A comparison of species concepts given a cladogram. The solid lines represent tokogenetically related groups of individual creatures. The triangle marked "Asexual" contains multiple independent asexual lineages. The abbreviations for the various concepts BSC, MSC, PSC, HSC, and ESC are contained in the text. The cell is marked "Y" if the lineage is a species in that concept, "N" if not, "NA" if inapplicable (*e.g.* no temporal component). S_{4+5+6} is marked "Y/N" due to the fact that it would be "Y" if the lineages interbreed and "N" if not. If the diagram were to be cut at time "T," the MSC would have "Y" for S_3, S_7, S_9, and S_{11}.

3.9   Do Species Concepts Matter?

As mentioned above, it seems amazing that such a fundamental idea as species seems to provoke so much disagreement and apparent chaos among systematists (Fig. 3.6). Is it worth it? Some would say no and embrace the idea that species are unique, potentially arbitrary, and in no way comparable. However, there are those who feel that making statements and testing hypotheses involving the comparison and enumeration of objects called "species" as natural evolutionary units is possible and desirable.

Studies in biodiversity, historical patterns of diversification and extinction, "speciation" theory, and adaptation all require that species be real components of nature, that there is a unique and precise method to identify a species, and that this method applies to all creatures and all time. If such a species concept exists, not everyone agrees we have found it.

3.10   Exercises

1. Is it important to have a single species definition, or are multiple concepts allowable or desirable?

2. How do the practical issues involved in identification of taxa interact with the species concepts described in the chapter?

3. Are asexual taxa species? How might horizontal gene exchange affect this?

4. Suppose that each organism in a study group is unique at the whole genome level (and you have these data). What would be the issues involved in applying the various species concepts to identifying species level taxa?

5. Give examples of studies where species are treated as a class and as individuals. If all species were individuals how would these studies have been effected? If classes?

Chapter 4

Hypothesis Testing and the Philosophy of Science

This chapter briefly presents several philosophical topics relevant to systematic analysis. Although there is a large universe of philosophical ideas discussed in the systematics literature, I have limited the discussion here to a number of areas that have direct impact on how analyses are done—how decisions are made as to choices among analytical options and the process of reasoning from data acquisition to choice of final tree hypothesis. The discussion is far from complete, but should serve as an entrée into the underlying epistemology of systematics.

Plato
(*c.*427–*c.*347 BCE)

4.1 Forms of Scientific Reasoning

4.1.1 The Ancients

Although scientific reasoning has existed as long as science itself, the hypothesis-driven approach is more recent. Plato argued the cause of natural laws in his ideas of universals (Chapter 1). Universals were ideal forms with inherent reality. The observed world, however, consisted of shadows, imperfect realizations of absolute and unchanging forms. One could learn about ideals from their observed representations and hence study the natural laws that governed Plato's universals, but this was never embodied in any hypothesis-testing framework.

Aristotle rejected the ideal forms of Plato, placing the real world and observations of it at the center of scientific inquiry. In this, Aristotle was the founder of *Ontology*, the metaphysical science of being itself. Aristotle rejected platonic ideals and universal forms. He argued that observable objects around us were the reality of nature, ideals the mental abstractions. Aristotle was not entirely pure in his reasoning, however, and though we associate him with Ontology, his ideas were something of a mix.

Aristotle
(384–322 BCE)

Systematics: A Course of Lectures, First Edition. Ward C. Wheeler.

In biology, the classifications of Aristotle can be viewed as hypotheses of pattern in biological variation with *terata* as counter, if not falsifying, data. Aristotle did not weigh alternate scenarios based on specific observations, however, he produced narrative classifications to explain what he saw. Aristotle did have a notion of minimal necessary explanation (parsimony, see below) often thought to have originated with Ockham.

4.1.2 Ockham's Razor

William of Ockham
(*c.*1285–*c.*1347)

Ockham was an English Franciscan monk who wrote widely in the intellectual areas of medieval thought including logic, metaphysics, physics, and theology. He eventually ran afoul of Pope John XXII (branding Ockham a heretic) over apostolic poverty. For this, Ockham was excommunicated (though rehabilitated after his death) and fled to Germany, residing there the remainder of his life. His political writings urge an early form of separation of church and state, no doubt not a coincidence. Of most interest to systematics is Ockham's nominalism and ontological parsimony.

Ockham's nominalism (as with Aristotle) rejected ideal universal forms as having reality. He focused, instead, on the individuality of objects. Any generalizations were intensions of the human thought process. These generalizations might have had more weight than mere words, but remained human-generated abstractions, not components of the natural world. Such discussions in his *Summa Logicae* (Ockham, 1323) form part of the foundation of modern epistemology.

We look to Ockham most, however, for *Ockham's razor*. This is the idea that in explaining phenomena the simplest explanation is most favored.

> For nothing ought to be posited without a reason given, unless it is self-evident (literally, known through itself) or known by experience or proved by the authority of Sacred Scripture.

Ockham felt that human reasoning was incapable of complete understanding of not only nature but theological principles such as the soul, and these could only be understood through divine revelation.

Although Ockham did express the idea of simplicity and minimalism in logical inference, much more has been ascribed to him than he actually wrote and the idea of minimal inference goes back as far as Aristotle[1]. The most commonly encountered quotes from Ockham:

> Pluralitas non est ponenda sine neccesitate [Purality should not be posited without necessity]

and

> Frustra fit per plura quod potest fieri per pauciora [It is pointless to do with more what can be done with less]

---

[1]The often quoted "Entia non sunt multiplicanda sine necessitate." is nowhere to be found in Ockham's writings.

refer to simplicity in elements, parameters, or causations. Anything beyond the minimal is unsupported, hence unnecessary.

The parsimony criterion makes explicit reference to Ockham's razor in justifying its simplicity argument, but in reality, all optimality-based methods operate on this basis (lower cost, higher likelihood or probability). Likelihood analyses incorporating model selection explicitly seek to minimize parameters in balance with enhanced tree likelihoods. Parsimony methods, in using a minimal set of operational assumptions (*e.g.* lack of a statistical model), are most explicit in minimizing operational assumptions as well as explanatory events (see below).

The minimization principle of inference runs through much of formal scientific logic as an operational principle. This is not because it will necessarily lead to truth, but because it is precise and efficient[2].

4.1.3 Modes of Scientific Inference

Scientific inference is the process of generating explanations of data with hypotheses—here trees. Among the many forms of scientific inference, there are four of most direct interest to systematics: induction, deduction, abduction, and the synthetic hypothetico-deduction. Modern systematics employs these forms of logic in multiple guises and often in impure, combined forms.

4.1.4 Induction

Francis Bacon
(1561–1626)

Induction is a process whereby multiple observed instances of a phenomenon lead to a generalization (sometimes ascribed in origin to Francis Bacon). In essence, for a series of observations a^i each leading to result b: $a^0 \rightarrow b, a^1 \rightarrow b, \ldots, a^k \rightarrow b \therefore A \rightarrow b$. In a commonly used example, "Every swan I see is white, therefore all swans are white."

Hume (1748) cited two problems with this mode of inference. First, since certainty would require exhaustive observation of all data (A above), which is impossible for non-trivial problems, absolute proof is impossible. Second, induction assumes that all future observations will agree with those already gathered.

In systematics, we can see this play out in Bayesian analysis of clades (Chapter 12). Clade posterior probabilities are estimated through additional observations generalizing to a high probability (in terms of the data) result. Even though very high probabilities (often 1.0) are recovered, a very small fraction of trees are evaluated and the probabilities depend on specific data sources. Given that clade posterior probabilities can vary with data source (such as molecular sequence loci), the identical behavior of past and future data sources cannot be assured.

David Hume
(1711–1776)

4.1.5 Deduction

A deductive inference (Socrates) is constructed by building a series of logical statements into an argument or syllogism. If the premises are true, the result

---

[2]After all, which of the non-minimal solutions should be chosen—and why?

Socrates
(469–399 BCE)

must be true. A syllogism is based on a major and minor premise, which then produce a conclusion. There are four premise types: all are, all are not, some are, and some are not. From these elements, a proof system is built. Aristotle produced the canonical example of deductive syllogism:

> All men are mortal.
> Socrates is a man.
> Socrates is mortal.

The major premise "All men are mortal" and the minor "Socrates is a man" leads inexorably to the conclusion "Socrates is mortal." In the case of swans, one might construct the following:

> All swans are white.
> Your dinner is a swan.
> Your dinner is white.

Deduction is an absolute proof system, not directly applicable to empirical systematic questions. There are cases, however, where deductive logic can be useful. The case of ghost taxa (Norell, 1987) is an example:

> Sister taxa are coeval.
> These taxa are sister taxa.
> These taxa are coeval.

By this syllogism, if we have a fossil of one taxon, even if we do not have one from its sister-group, we know the sister-group is at least that old.

4.1.6　Abduction

Charles S. Peirce
(1839–1914)

Abductive inference chooses that hypothesis which is most consistent with empirical instances. The method was championed by Charles Peirce in the late 19th century (Houser et al., 1997); philosophy of science being among his widely varied interests in metaphysics, logic, semiotics, mathematics, and physics. Although without an academic position for most of his life (his appointment at Johns Hopkins being terminated due to his living with a woman who was not, at that time, his wife) his work was regarded as of the highest caliber. Unfortunately for Peirce, much of this was not known until after his death with Bertrand Russell referring to him as one of the most original thinkers of the 19th century. After his death, his papers were purchased by Harvard University, yielding over 1650 unpublished manuscripts totaling more than 100,000 pages.

The abductive process is one of identifying a set of potential hypotheses and, given a series of observations that could have been generated by the hypotheses, choosing that which is best (Eq. 4.1).

$$
\begin{aligned}
\text{Set of hypotheses} \quad & H \\
\text{Data } D \ &= \ f(H) \\
H' \subseteq H \text{ such that } D' \ &\subseteq \ f(H')
\end{aligned}
\tag{4.1}
$$

With respect to swans, one might say "All I see are white swans, it must be that all swans are white."

For science in general, alpha-level hypotheses are by their nature abductive. The intuiting of patterns from observation is a crude abductive practice. In systematics, maximum likelihood (ML) is the foremost example of abductive inference. In ML, that hypothesis that maximizes the probability of the data, *i.e.* observations, is chosen as the best hypothesis.

4.1.7 Hypothetico-Deduction

Hypothetico-deduction was proposed as a method of scientific inference by Whewell (1847)[3]. The method improves upon inductive reasoning and solves the flaws described by Hume by use of the *modus tollens* ("manner that denies by denying"), or, more simply, falsification.

The basic notion of *modus tollens* is expressed simply for statements P and Q and observation x as:

$$
\begin{aligned}
P &\subseteq Q \\
x &\notin Q \\
\therefore x &\notin P
\end{aligned}
$$

William Whewell
(1794–1866)

Hypothetico-deduction proceeds in several steps involving observation, hypothesis creation, prediction/retrodiction, and testing through falsification.

1. Make observations.

2. Erect a hypothesis to explain the observations.

3. Deduce a prediction or retrodiction from the hypothesis.

   If false (Falsification) goto step 2.

   Else if true (Corroboration) goto 3.

Repeated lack of falsification (corroboration) is a measure of the strength of the hypothesis. Empirical falsification can mean several things, however. The conclusion of the *modus tollens* could be correct (as above, $x \notin P$), or the entire original hypothesis could be incorrect ($P \not\subseteq Q$), or it could be that the hypothesis requires modification. If we refer to the swan example, we have the hypothesis "All swans are white" and we have a black bird in front of us. The *modus tollens* yields three alternatives: "This is not a swan," or "I'm wrong about swans," or "All swan are white, except this one."

Karl Popper

It is hard to overstate the influence of Popper on practicing scientists today, especially in systematics. Popper (1934, 1959, 1963, 1972, 1983) extended and formalized the operations of hypothetico-deductive inference and applied it to many

Karl Popper
(1902–1994)

---

[3]Whewell coined many terms, among them "scientist."

areas of science. For this work, the AMNH presented Popper with a gold medal in 1979. Popper made precise several concepts crucial to the hypothetico-deductive method and science in general. These ideas were discussed in terms of four basic concepts: background knowledge, probability, hypothesis, and evidence.

- Background knowledge b—This is the sum total of knowledge not subject to test or falsification. This may include axiomatic statements and previously tested hypotheses. Any hypotheses included in background knowledge must be highly corroborated since they are not subject to test at this stage. An example of this would be the inclusion of Aves within Dinosauria as a hypothesis in 1863 (Huxley, 1863), but background knowledge now after much corroboration.

- Probability p—Popper employed the "relative frequency" of events definition of probability, eschewing logical and subjective concepts (see Chapter 6).

- Hypothesis h—The causative scenario currently subject to test.

- Evidence e—Data gathered to test the hypothesis.

Popper then defined three measures of aspects of a test of a hypothesis given evidence and background knowledge.

- Support (Eq. 4.2)—quantifies the difference between the probability of the evidence given the hypothesis and background information and the probability of the evidence alone. The first term is the likelihood of the hypothesis (Popper, 1959) given other information such as a model (not subject to test, hence should be highly corroborated).

$$p(e|h, b) - p(e|b) \tag{4.2}$$

- Severity of Test (Eq. 4.3)—quantifies the relative level of support on $[0,1]$, normalized to the total probability of the evidence given the two alternates of with and without the hypothesis.

$$\frac{p(e|h, b) - p(e|b)}{p(e|h, b) + p(e|b)} \tag{4.3}$$

An alternate formulation (Eq. 4.4) is related to the likelihood ratio test discussed in Chapters 11 and 15.

$$\frac{p(e|h, b)}{p(e|b)} \tag{4.4}$$

- Corroboration (Eq. 4.5)—also normalized support on $[-1,1]$ with the probability of the evidence and the hypothesis given the evidence.

$$\frac{p(e|h, b) - p(e|b)}{p(e|h, b) - p(e, h|b) + p(e|b)} \tag{4.5}$$

This measure can be negative if an alternate hypothesis is favored over that which is tested.

All three of these measures share the same numerator, and hence are positively correlated. Additionally, all involve the likelihood of the hypothesis (Grant and Kluge, 2007, 2008a notwithstanding). Much discussion about the relative merits of hypothesis testing and optimality criterion choice are based on these concepts. The idea of Total Evidence Analysis, for instance (Kluge, 1989), is grounded in severity of test. The greater the amount and diversity of evidence brought to bear on a question, the greater the opportunity for falsification and the more severely the hypothesis is tested.

Thomas Kuhn
(1922–1996)

Popper, though highly respected in systematics, is not without his critics, some severe within the philosophy of science community. Feyerabend (1975, 1987) wrote explicitly that he thought Popper's ideas were without justification. Kuhn (1962) argued that theories were not rejected based on their falsification, but on social factors among scientists. Bartley (1976) analogized Popper's ideas to narrative stories.

> Sir Karl Popper is not really a participant in the contemporary professional philosophical dialogue; quite the contrary, he has ruined that dialogue. If he is on the right track, then the majority of professional philosophers the world over have wasted or are wasting their intellectual careers. The gulf between Popper's way of doing philosophy and that of the bulk of contemporary professional philosophers is as great as that between astronomy and astrology. (Bartley, 1976)

Champion (1985) judged Popper to have failed due to (he felt) the quixotic nature of his goals.

> Popper's ideas have failed to convince the majority of professional philosophers because his theory of conjectural knowledge does not even pretend to provide positively justified foundations of belief. Nobody else does better, but they keep trying, like chemists still in search of the Philosopher's Stone or physicists trying to build perpetual motion machines. (Champion, 1985)

Falsification in Systematics

In systematics, trees (or statements conditioned on them) are the main form of hypothesis, which are tested with characters. In the context of parsimony, characters with non-minimal change on a tree (homoplasy) act as falsifiers of that tree hypothesis. However, there are no non-trivial homoplasy free data sets (at least that I know of). A strict falsificationist perspective would require that any tree with homoplasy is falsified. In fact, this would result in *all* trees begin falsified. All hypotheses of relationship would then be rejected. Clearly, this is an absurd position. This situation has been referred to as *naïve falsification*. Naïve in the sense that it is absolutely (deductively) correct, yet leads to an improper (or empty) result in non-trivial empirical cases.

As mentioned above, a single non-conforming observation need not sink an entire theory. The theory can be modified to account for this disagreeing observation. Such an exception is referred to as an *ad hoc* statement in that it is

Arnold Kluge

specific to a particular observation and has no (or at least limited) generality. The discovery of a black swan, for instance, might be accommodated in the theory that all swans are white, by the *ad hoc* statement "except for this one here" (as opposed to eating the swan quickly and invisibly). This process can be extended without limit rendering falsification meaningless (as W. V. Quine argued). This would seem fatal, until it is realized that not all hypotheses require the same degree of *ad hoc* rescue.

The realization and use of the observation that all observations support (and reject) all hypotheses, but not to equal extent, leads to *sophisticated* or *methodological* falsification. This is the form of falsification pursued by Farris (1983) and Kluge (*e.g.* Kluge, 2009). In this framework, the *least* falsified (in terms of homoplasy) hypothesis (= tree) is most favored. Farris (1983) emphasized the minimization of "extra-steps" as *ad hoc* hypotheses of character convergence or parallelism. This concept is slightly recast as the minimization of *total* cost in dynamic homology since the calculation of minimal cost for a character (sequence character for instance) is tree specific (Wheeler, 2001b, 2011). According to Farris (1983), sophisticated falsification equates to parsimony and Popper and justifies its use. The appeal to Popperian logic underpins much of the argumentation in favor of parsimony.

Synthesis of Inference

In light of the above discussion, the model of hypothetico-deduction in systematics can be recast as a combination of methods. Steps 1 and 2 (making observations and erecting a hypothesis) are abductive. This is followed (step 3) by the deduction of a predictive statement. This statement is falsified in terms of an objective function of tree optimality. The prediction is that the current tree (hypothesis) optimizes an objective function more effectively than an alternate tree. The falsification comes in comparing the value of the objective function between the two. If the current tree is "better," it is corroborated (induction) and the method returns to the previous step (3), identifying another competing hypothesis to be tested against the current best (deductive step). If the current tree is less optimal than an alternate, it is falsified and the challenger becomes the current best hypothesis. This process is repeated until all alternate hypotheses are tested and as long as the objective function is transitive [if $f(A) \leq f(B)$ and $f(B) \leq f(C)$ then $f(A) \leq f(C)$] the set of optimal (least falsified) hypotheses will be identified.

1. Make observations.

2. Erect a hypothesis (tree T) to explain the observations (initially the optimal tree $T^o \leftarrow T$) and compete with T^o.

3. Deduce the prediction that T^o will optimize $f()$ better ($f(T^o) \leq f(T)$).

4. Compare $f(T)$ and $f(T^o)$

   If $f(T^o) \nleq f(T)$(Falsification) then $T^o \leftarrow T$

   Else true (Corroboration).

5. If there are trees remaining to be tested, goto step 2, else return T^o.

The objective function could in principle take many (transitive) forms. The above formalism allows model-based methods such as likelihood and topology-based posterior probability (Bayesian *Maximum A Posteriori*, MAP) as optimality criteria ($f()$). The choice of criterion must then be defended on other grounds, but the hypothetico-deductive process would remain unchanged.

4.2 Other Philosophical Issues

There are many issues where philosophical considerations affect how we think about problems and what we do to address them. Several are discussed in other sections, including the arguments over whether species are individuals or classes and species nominalism (Chapter 3), the combination ("Total Evidence") or partition of data (Sect. 16.2.7), and optimality criteria (Chapter 13).

One topic that is an element of many discussions is the operational meaning of minimization and weighting.

4.2.1 Minimization, Transformation, and Weighting

Gonzalo Giribet

When we speak of minimizing the cost of a cladogram, especially in a Popperian context, what are the components to be minimized? Are we to minimize transformations, or overall cost—allowing for weighted transformations? At what level do we apply the parsimony criterion? There is no current consensus on these issues and they definitely affect systematic results.

Kluge and Grant (*e.g.* Grant and Kluge, 2005; among many others) have argued that the entity to be minimized when evaluating competing tree hypotheses is transformations. This form of minimization requires that all events be weighted equally since only this scheme will minimize both overall cost and absolute numbers of transformations. Wheeler (1995) and Giribet (2003) argued that alternate transformation weighting schemes should be considered and in some cases favored over homogeneous weighting. They advocated sensitivity analysis (Section 10.11) to evaluate and potentially choose hypotheses based on alternate weighting schemes. There is no theoretical, in the sense of mathematical, limitation (other than perhaps metricity) on these weights. Are both these scenarios—minimizing "steps" and minimizing "cost"—parsimony? What of larger-scale genomic events such as locus insertion–deletion and rearrangement? If these transformations were to be weighted equal to nucleotide substitutions, insertions and deletions, grossly non-metric costs would be implied and trivial (all large indel) results produced (Fig. 4.1).

The issue is unsettled[4].

Ockham's razor appears even within the context of the minimization of transformations. Consider the nucleic acid sequences AAC and TT. If we allow substitution, insertion, and deletion events of one nucleotide, there are a minimum of three events (transformations) required to edit one sequence into the other. If, however, we choose a simpler model, one with only insertions and deletions, a minimum of five events are required (there are multiple five transformation

[4]This same argument applies to "Implied Weights" (Goloboff, 1993b).

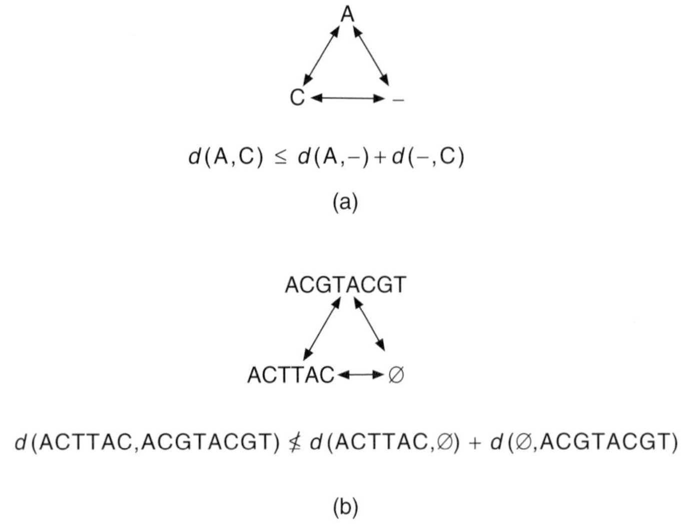

$$d(A,C) \leq d(A,-)+d(-,C)$$

(a)

$$d(\text{ACTTAC},\text{ACGTACGT}) \nleq d(\text{ACTTAC},\varnothing) + d(\varnothing,\text{ACGTACGT})$$

(b)

Figure 4.1: Giribet and Wheeler (2007) scenario for non-metric large scale transformation costs implied by equal transformation weighting (Grant and Kluge, 2005).

scenarios). The former model is more complex, yet yields a lower number of transformations; the latter, lower model complexity, yet a greater number of events. At which point is "plurality" to be avoided? Both solutions are more and less parsimonious, but in different components of their arguments.

4.3 Quotidian Importance

Although the nomenclature of inferential methods may not seem of particular relevance to an empirical scientist, scientific hypothesis testing is based on specific rules to ensure the validity of results. Tree searching is a direct application of hypothetico-deductive reasoning. Systematic analysis relies on hypothesis testing in many guises, hence it is crucial that these tests conform to the rigors of logical systems, without which valid inference is impossible.

4.4 Exercises

1. Give examples of inductive, deductive, and abductive inferences in systematics.

2. Can single observations falsify a systematic hypothesis? Would Patterson's angels qualify?

3. Is weighted parsimony parsimony?

4. Who's cuter, Gonzalo or Taran?

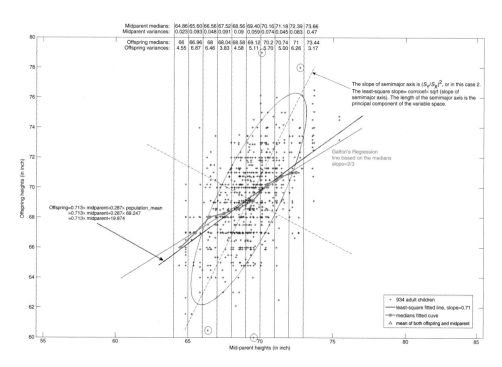

Plate 2.1: Heritability *sensu* Galton (1889).

Systematics: A Course of Lectures, First Edition. Ward C. Wheeler.
© 2012 Ward C. Wheeler. Published 2012 by Blackwell Publishing Ltd.

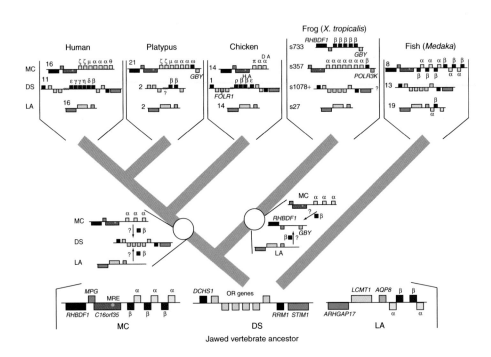

Plate 2.22: Globin duplication and diversification (Hardison, 2008).

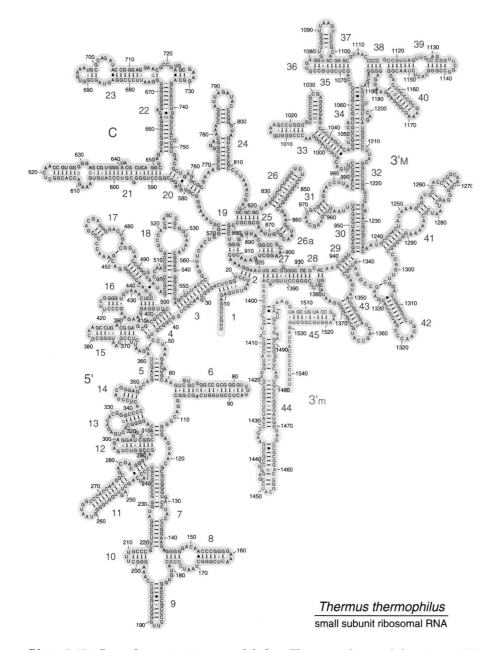

Thermus thermophilus
small subunit ribosomal RNA

Plate 8.12: Secondary structure model for *Thermus thermophilus*. http://rna.ucsc.edu/rnacenter/ribosome_images.html.

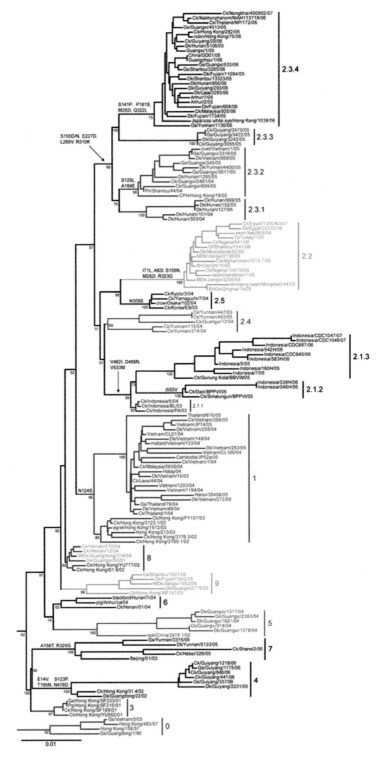

Plate 9.21: Neighbor-Joining tree of H5N1 "Avian" flu virus of WHO/OIE/ FAO H5N1 Evolution Working Group http://www.cdc.gov/eid/content/14/ 7/e1-G2.htm.

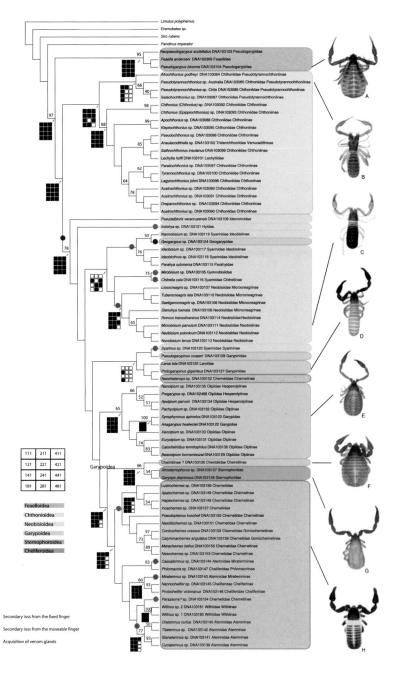

Plate 10.20: Pseudoscorpion analysis of Murienne et al. (2008). The base tree is that which minimized incongruence among multiple molecular loci. The "Navajo rugs" show the presence or absence of each vertex in parameter space.

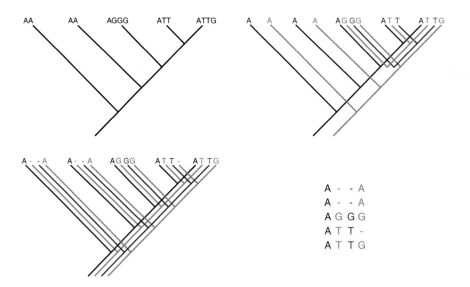

Plate 10.21: Implied alignment (Wheeler, 2003a) of five sequences: AA, AA, AGGG, ATT, and ATTG. The original optimized tree is shown on the upper left; the implied traces upper right; implied traces with traces extended and gap characters filled in lower left; and the final implied alignment in the lower right.

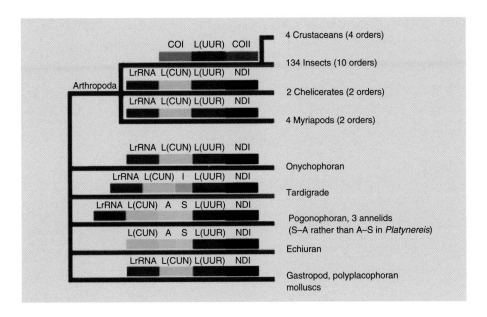

Plate 10.22: Mitochondrial gene order variation in protostome taxa (Boore et al., 1998).

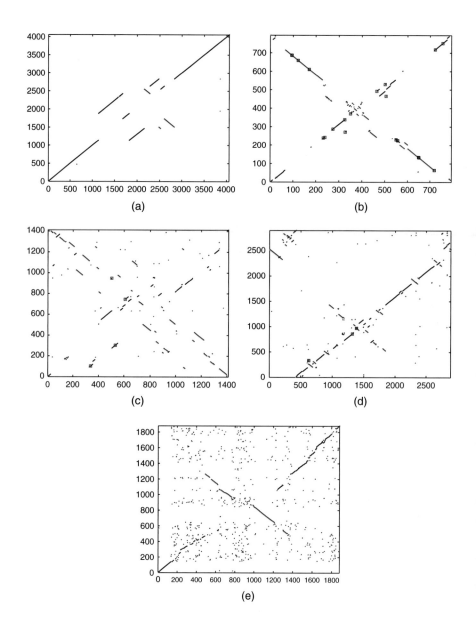

Plate 11.14: Genomic rearrangement locus dot-plot scenarios of Dalevi and Eriksen (2008): (a) = "Whirl," (b) = "X-model," (c) = "Fat X-model," (d) = "Zipper," and (e) = "Cloud."

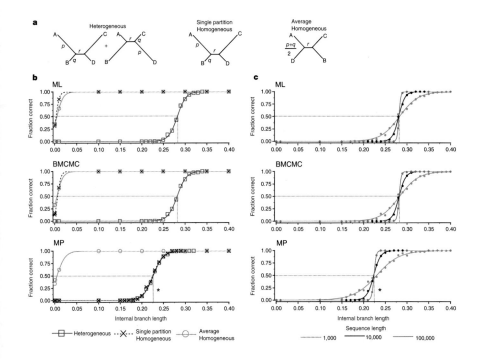

Plate 13.3: Results of Kolaczkowski and Thornton (2004) showing the superior performance of parsimony over likelihood and Bayesian methods under a condition of heterotachy. ML = maximum likelihood, BMCMC = Bayesian MCMC, and MP = parsimony.

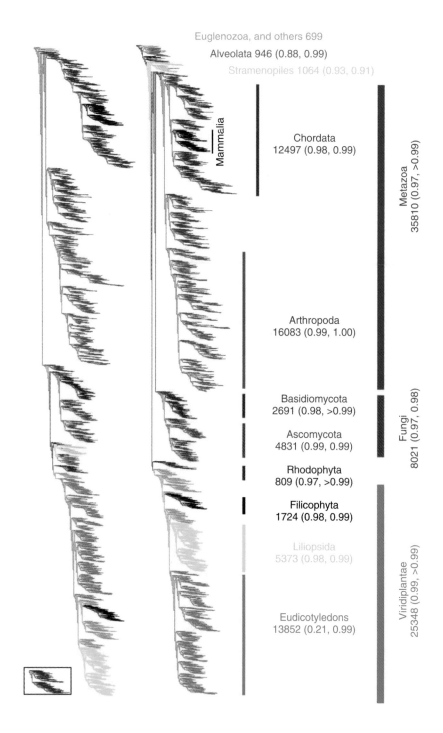

Euglenozoa, and others 699
Alveolata 946 (0.88, 0.99)
Stramenopiles 1064 (0.93, 0.91)

Mammalia

Chordata
12497 (0.98, 0.99)

Metazoa
35810 (0.97, >0.99)

Arthropoda
16083 (0.99, 1.00)

Basidiomycota
2691 (0.98, >0.99)

Fungi
8021 (0.97, 0.98)

Ascomycota
4831 (0.99, 0.99)

Rhodophyta
809 (0.97, >0.99)

Filicophyta
1724 (0.98, 0.99)

Liliopsida
5373 (0.98, 0.99)

Viridiplantae
25348 (0.99, >0.99)

Eudicotyledons
13852 (0.21, 0.99)

Plate 14.19: Combined eukaryote analysis tree at 730,435 steps of Goloboff et al. (2009). The values beneath the taxon names are the number of taxa included and % placement in agreement with GenBank taxonomy.

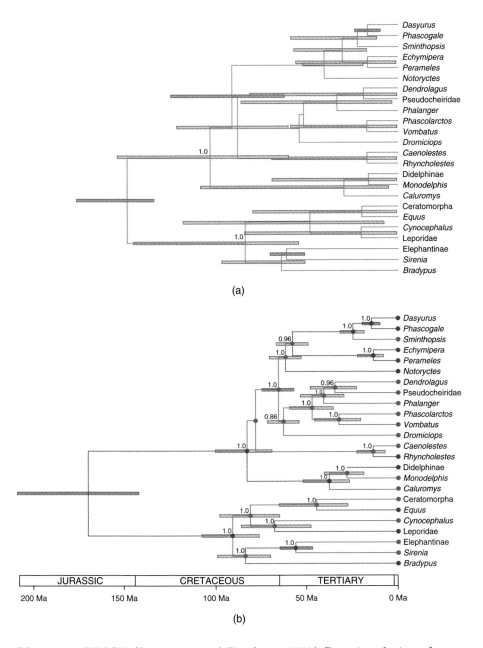

(a)

JURASSIC | CRETACEOUS | TERTIARY

200 Ma 150 Ma 100 Ma 50 Ma 0 Ma

(b)

Plate 17.6: BEAST (Drummon and Rambaut, 2007) Bayesian dating of trees with priors on dates (above) and posterior probabilities (below).

Chapter 5

Computational Concepts

Systematics is a synthetic science, and several fundamental problems fall among the most difficult faced by computer science. In order to better understand the issues involved and the techniques developed to deal with them, an introduction to computer science basics is reviewed.

5.1 Problems, Algorithms, and Complexity

5.1.1 Computer Science Basics

The theory of computation begins with (Gödel, 1931), Kleene (1936), Post (1936), Church (1936a,b), and Turing (1936, 1937). These works answered the *Entscheidungsproblem* ("Decision Problem") proposed by Hilbert in 1900[1] of whether a method could always be devised to output correctly whether a statement were true or false. In other words, were there unsolvable problems? Gödel proved that there were statements that were true, but could not be proven so. Simply put, for any self-consistent system sufficient to describe the arithmetic of natural numbers, there are true statements that cannot be proven. Church ("undecidable" problems) and Turing[2] ("halting" problem) followed this result and independently proved that there were problems that could not be computed (Church's Theorem), answering the *Entscheidungsproblem* in the negative. These results, and the mechanisms used to arrive at them, are the heart of computation theory.

David Hilbert
(1862–1943)

Church developed a system, the Lambda Calculus, to study computation that, with his student Turing's machine-based approach, form the basis of modern computer science. Both these systems were "universal" in that they could

Kurt Gödel
(1906–1978)

---

[1]This was one of Hilbert's 23 unsolved problems that drove much of mathematics in the 20th century.

[2]Church's Ph.D. student.

Systematics: A Course of Lectures, First Edition. Ward C. Wheeler.

Alonzo Church
(1903–1995)

compute anything that was computable and eventually lead to "functional" (such as ML) and "imperative" (such as C) schools of computer programming languages. Church, Kleene, Rosser, and Turing showed that these alternate approaches were in fact interconvertible (Church and Rosser, 1936). The Turing Machine was the first physical conception of a universal computing device and its development and use are at the basis of algorithmic complexity analysis relevant to systematics.

Turing Machines

In his computing model, Turing imagined a person with a pencil, paper, and a brain. The person (computer) could move about the paper (left and right); could read and write on the paper; could store an internal state, a series of actions based on that state, and what was read from the paper (Fig. 5.1)—an elegant and very human model. More specifically, there was:

- A *Tape* one cell wide and potentially infinitely long onto which symbols could be written and from which they can be read.

- A *Head* that reads the state of the cell under it and can write to it. The head can move to the left or right.

- A *Table* (transition function) containing the action of the head given the symbol on the tape cell and the state of the Register.

- A *Register* that contains an internal state (finitely many including an initial, start state).

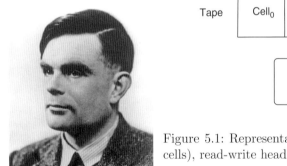
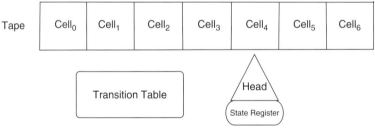

Figure 5.1: Representation of a Single Tape Turing Machine with tape (seven cells), read-write head, transition table, and head state register.

Alan Turing
(1912–1954)

One of the most amazing results of Turing's work was that this simple construct was proven to be able to compute anything that was computable (receive input, operate on that input, and halt); it was "universal" (Church–Turing Thesis)[3].

---

[3]Turing's construct was not intended to be fabricated, but provided the basis for mathematical analysis of problems. There are many variations of Turing Machines that have been created for convenience or to model specific problems.

5.1.2   Algorithms

In order to discuss the algorithms important to systematic analysis, it is important to define the term first. There are a variety of definitions of algorithm in use today from the intuitive and informal ("a series of steps taken by a computer to solve a problem") to precise, mathematical statements. One definition useful to our discussions is derived from the Church–Turing thesis: *an algorithm is a Turing machine that always halts.* By this we mean, a set of instructions (transition function) that will yield a definite result for any input. All problems that are algorithmically solvable are solvable by Turing machines; those that are not, are not.

Computability

The converse of the universality of the Turing machine was the definition of computability. If no Turing Machine could be defined that was guaranteed to halt (algorithm), the problem was uncomputable. This was the algorithmic incompleteness of Gödel, and Church and Turing's solution to the *Entscheidungsproblem.* It may seem cruelly arcane to discuss non-computable problems, but they do occur in every-day situations. An example is data compression. A general algorithm for the guaranteed maximum compression of a (non-trivial) string cannot exist (Kolmogorov complexity). It is uncomputable (Solomonoff, 1964).

5.1.3   Asymptotic Notation

Before discussing the complexity of algorithms (below), we require a language to describe the relative size of functions that tend to infinity (since all problems will grow without bound with unbounded input). This is referred to as O-notation (Bachman, 1894) and is extremely useful in the discussion and analysis of computer algorithms.

O-notation

One of the main strengths of the O-notation is that it hides less important details and focuses on how functions change. In essence, the O-notation concentrates on the component of a function that grows most quickly. Hence, a function like:

$$f(x) = ax^2 + bx + c \tag{5.1}$$

will be dominated by the x^2 term as x grows. If we define a function $g(x) = x^2$, we then say that $f(x)$ is of the order of $g(x)$, written $f(x)$ is $O(g(x))$ or in this case $O(x^2)$. The fact that f and g differ by the constant factor a, or that f has other, lower order terms is irrelevant. They will grow at the same rate and differ only by a constant factor (asymptotically). A cubic function will have $O(x^3)$ and so forth. By definition, f is of order at most g if for positive real c and real $x > x_0$ (Fig. 5.2):

$$|f(x)| \leq c \cdot |g(x)| \tag{5.2}$$

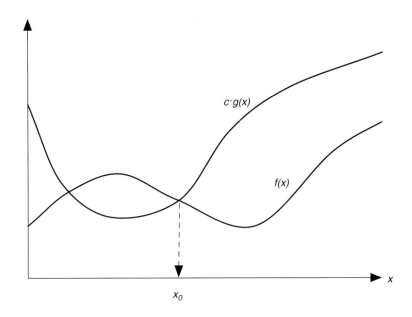

Figure 5.2: The function $f(x)$ is of at most order g.

Algorithms are characterized as $O(log\ n)$, $O(n)$, $O(n \cdot log\ n)$, *etc.* depending on their growth in execution time with input size[4]. Additional precision was given to this form of analysis by Knuth (1973), defining $\Omega()$ and $\Theta()$ (for $a, b \in \mathbb{R}^+$):

$$f(x) \text{ is } \Omega(g(x)) \quad [f \text{ is of order } \textit{at least } g] \quad a \cdot |g(x)| \le |f(x)|$$

$$f(x) \text{ is } O(g(x)) \quad [f \text{ is of order } \textit{at most } g] \quad |f(x)| \le b \cdot |g(x)|$$

$$f(x) \text{ is } \Theta(g(x)) \quad \quad [f \text{ is of order } g] \quad a \cdot |g(x)| \le |f(x)| \le b \cdot |g(x)|$$

In systematics, we primarily use $O()$, "big-O."[5].

5.1.4   Complexity

One useful way to describe the performance of algorithms is the time or storage space they require. These are called *time complexity* and *space complexity* and are expressed in the *O*-notation (above). We focus here on time complexity, but space complexity can be a significant issue in some forms of computation. Certain biomolecules (*e.g.* DNA) theoretically can be used to compute difficult problems in low complexity time, but may require exponentially large amounts of space—significantly curtailing their potential utility.

---

[4] The function *log* is assumed to be log_2 unless otherwise noted.

[5] There are other, more rarely used, complexity measures *e.g.* ω, o.

Time Complexity

The time complexity of an algorithm is the growth of the execution time of an algorithm as the problem size (the input size n) grows. A procedure that takes four times as long to complete when the problem size is doubled (*e.g.* calculating all pairwise distances between n points on a map) is said to have time complexity $O(n^2)$. Often, it is impossible to calculate the exact time complexity of an algorithm. For this reason, "best case," "worst case," and "average case" time complexities can be calculated. It can be extremely difficult (or practically impossible) to calculate best case complexity and, in most cases, we are satisfied with worst and average case time complexity calculations.

Donald Knuth

It may also be important to understand the growth in memory requirements of an algorithm, and its space complexity captures this (of course a computer program can only access as much memory as it has time, limited by the time complexity, so time and space complexity are closely linked). Algorithm choice may require trade-offs between time and space complexity. This type of analysis was pioneered by Knuth (1973) and others in the 1950s and 1960s.

An Example: Loops

Consider the following algorithm fragments. Algorithm 5.1 would have a complexity $O(n)$ since the number of operations would grow linearly with n. Consider this slightly more elaborate fragment (Alg. 5.2). This procedure would have a complexity $O(nm)$ since the number of operations would grow with the product of n and m. If m were $O(n)$, the algorithm would be $O(n^2)$.

Algorithm 5.1: SingleLoop

---

for $i = 1$ **to** n **do**
  | $sum \leftarrow sum + i$;
end
return sum;

Algorithm 5.2: NestedLoops

---

for $i = 1$ **to** n **do**
  for $j = 1$ **to** m **do**
    | $sum \leftarrow sum + i + j$;
  end
end
return sum;

Time Complexity and Systematics

When algorithms are presented in later sections, their time complexity will be discussed (*e.g.* trajectory searches using SPR—$O(n^2)$, or TBR—$O(n^3)$). Knowledge of the time complexity of the algorithms used in phylogenetic analysis is

Time Factors				
Complexity	Problem Size			
	2	4	8	16
$O(1)$	1	1	1	1
$O(\log n)$	1	2	3	4
$O(n)$	2	4	8	16
$O(n^2)$	4	16	64	256
$O(2^n)$	4	16	256	65536
$O(n!)$	2	24	40320	2×10^{13}

Table 5.1: Time complexity and problem size.

important for both planning and evaluation of systematic results. If a tree search algorithm is $O(n^3)$ in number of leaves, then the investigator knows that if they were to double the number of taxa in their analysis, they should expect to spend a factor of eight more time (Table 5.1). A more effective (in terms of optimality) procedure of higher time complexity might be favored over one of lower complexity for small data sets, but not for large. Only through understanding the underlying complexity inherent in systematic algorithms can an investigator evaluate results and make best use of available time and resources.

5.1.5 Non-Deterministic Complexity

The above description was for the calculation of *deterministic complexity*—informally, the time required on a Turing Machine where at each step, the machine performs a single operation, each in turn until it halts. Consider a series of decisions, such as searching a house for a lost item. At each step, a different (but unvisited) room is chosen and examined. Assuming that you would like to make as short a search as possible (in terms of moving around the house), what path should be taken? If there were n rooms, there would be $n!$ potential routes. In order to evaluate these routes, a naïve deterministic algorithm might evaluate each order of room visits, hence an $O(n!)$ process. If, however, instead of evaluating each path in turn, all room choices at each point were evaluated simultaneously, the time complexity would grow only as $O(n)$—a vast improvement. This is non-deterministic time complexity.

Since a non-deterministic machine is able to perform many (even infinitely, but countably many) operations simultaneously, complexity is determined assuming the algorithm makes the "best" choice at each point (since it evaluates all choices), and that of minimum complexity. Hence, a non-deterministic machine can be viewed as a (potentially exponentially large) collection of deterministic machines. The complexity of a non-deterministic machine, then, is that of the minimum complexity of all the deterministic machines implied in its operation.

5.1.6 Complexity Classes: P and NP

A desire for more fine-grained classification of problems than computable or not has yielded partitions of computable problems into classes depending on their complexity. Some of these problems are easily solvable or at least *tractable*, and others, as far as we know, are not. We would like to know where our particular problems lie. For our purposes, we can limit discussion to the classes *P*, *NP*, and *NP–complete*.

An intuitive definition of tractable[6] problems is "those that have polynomial-time algorithms." By these we mean (Eq. 5.3) that for some problem size n and constant k, there exists an algorithm A (deterministic Turing machine), whose time complexity is of the order n^k.

$$\text{Time}_A n \in O(n^k) \qquad (5.3)$$

The time to solve the problem grows as an exponent of the problem size. This exponent may be large, but it is still polynomial. These problems constitute the class *P*.

There are two reasons why this definition is used. First, it might seem that if k were very large, say 100 or 1000, n^{1000} would still be unfeasible for non-trivial n. However, once a polynomial-time algorithm has been identified, even with large k, experience shows that improvements soon follow, reducing the exponent to more manageable levels. Second, this definition preserves the invariance with respect to computing models and languages. All universal computing models can be converted into each other with some complexity (polynomial-time reducible[7]), hence the definition will work for all languages. Conversely, if there is no polynomial-time algorithm for any computing model, none exists for any of them.

The second class we are concerned with is *NP*. These are those problems that cannot be solved in polynomial-time on a deterministic machine, but are solvable in polynomial-time by a non-deterministic machine (with unlimited space), hence *N*on-deterministic *P*olynomial. The deterministic time complexity of these problems is at least exponential ($O(2^n)$)—intractable for non-trivial inputs. One of the most important outstanding problems in theoretical computer science and mathematics in general (since it has other implications) is whether *P* and *NP* are equal or *P* is a subset of *NP*.

Whether $P = NP$ is equivalent to whether it is easier to verify a mathematical proof than create one for a given theorem.

$$P = NP ? \; P \subsetneq NP \qquad (5.4)$$

A prize of US\$1,000,000 is available at http://www.claymath.org if you feel you have insights. A proof was claimed in 2008, but is widely thought to be flawed.

If these classes are equal, then there must exist polynomial-time deterministic algorithms for all problems. This is strongly suspected to not be the case since there are many problems for which there are no known polynomial-time algorithms after considerable effort, but it has not been proven to be true.

Among those problems in $NP \setminus P$ (assuming this is non-empty), are those that are most difficult. Each one of them is at least as hard (in terms of polynomial-time reducibility; Karp, 1972) as all others in *NP*. These problems are called *NP–complete* (Fig. 5.3). If a polynomial time solution can be found for any one of these problems, it will be applicable to all. Unfortunately, many of the problems encountered in systematics are NP–complete. These include tree search and tree alignment, implying that we are unlikely ever to find polynomial-time algorithms for their solution (Day, 1987; Wang and Jiang, 1994).

Richard Karp

---

[6]More precise definitions of algorithmic tractability are used in the computer science literature (*e.g.* Hromkovic, 2004).

[7]Usually $\leq O(n^3)$

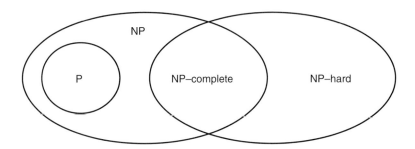

Figure 5.3: Assuming $P \neq NP$, the relationship among computable classes of problems.

Technically, the term NP–complete refers to a decision problem (*i.e.* Turing true or false). The term *NP–hard* is used to refer to a decision, optimization, or search problem which is as hard (or harder) as any NP problem.

5.2 An Example: The Traveling Salesman Problem

Julia Robinson (1919–1985) published one of the first papers on TSP (Robinson, 1949)

A classical NP–complete problem is the *Traveling Salesman Problem* (TSP). Consider a salesman wishing to visit a collection of cities, yet minimize his or her travel time. The number of possible tours is very large—exponentially in fact. If there are n cities to visit, there are n choices for the first city to visit, $n - 1$ for the second, $n - 2$ for the third and so on (Fig. 5.4) yielding $n!$ total tours ($(n - 1)!$ if you start at one of the cities). The problem has received a great deal of attention, with problem sizes increasing dramatically over the last 50 years (Table 5.2). The largest problem solved to date is for 24,978 Swedish cities (a course of 72,500 km) completed in 2004 (Fig. 5.5). Since many commonly encountered NP–hard optimizations can be reduced to the Traveling Salesman Problem, TSP heuristics are generally useful (Graham et al., 1985).

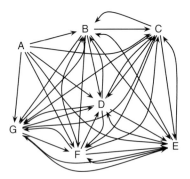
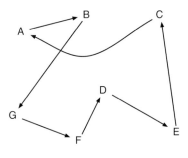

Figure 5.4: Traveling Salesman Problem. Tours after two cities have been visited left, a complete tour, right.

Year	Research Team	Problem Size
1954	G. Dantzig et al.	49 cities
1971	M. Held and R.M. Karp	64 cities 64 random points
1975	P.M. Camerini et al.	67 cities 67 random points
1977	M. Grötschel	120 cities
1980	H. Crowder and M.W. Padberg	318 cities
1987	M. Padberg and G. Rinaldi	532 cities
1987	M. Grötschel and O. Holland	666 cities
1987	M. Padberg and G. Rinaldi	2,392 cities
1994	D. Applegate et al.	7,397 cities
1998	D. Applegate et al.	13,509 cities
2001	D. Applegate et al.	15,112 cities
2004	D. Applegate et al.	24,978 cities

Table 5.2: Progress in Solving Traveling Salesman Problems. `http://www.tsp.gatech.edu//history/milestone.html`.

5.3 Heuristic Solutions

As mentioned above, exact solutions to NP–complete problems can require exponential time. Therefore, we are extremely unlikely (unless $P = NP$) to identify exact solutions to these problems. Since such problems are so frequently encountered (especially in systematics), techniques have been developed that yield useful, if inexact solutions. These heuristic techniques come in several flavors.

- Approximation—Algorithms that will yield results not exactly optimal, but within an acceptable distance (hopefully guaranteed) from the optimal solution.

- Local Search—A technique to refine and improve an initial solution by varying elements in its local space (defined for the problem), accepting better solutions repeatedly until a stable, if local, result is found. Also referred to as "Hill Climbing."

- Simulated Annealing—By mimicking the process of annealing metals, locally optimal solutions (often identified via Local Search) are improved by escaping to more global optima by transitioning through less optimal intermediates in a probabilistic fashion (Metropolis et al., 1953).

- Genetical Algorithm—This procedure mimics the evolutionary generation of variation with genetic recombination and selection to improve a pool of local solutions (Fraser and Burnell, 1970).

- Randomization—Potential solutions are generated and improved by various Monte Carlo-type (Metropolis and Ulam, 1949) random processes.

The application of these techniques to systematic problems will be discussed in greater depth in later sections (Chapters 8 and 14).

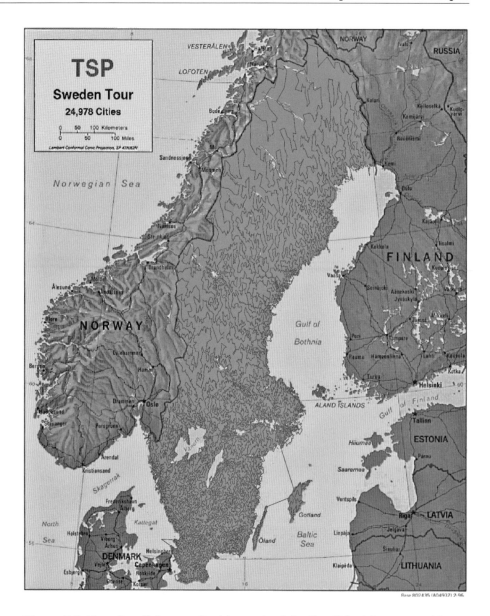

Figure 5.5: Traveling Salesman Problem solved for Swedish cities. http://www.
tsp.gatech.edu//sweden/index.html.

5.4   Metricity, and Untrametricity

A component of many NP–hard optimizations is a distance function and the
objective of the problem is often to maximize or minimize this function summed
over elements or choices. Examples are the segments of a tour, as in the

Traveling Salesman Problem, or edges in other graph problems such as phylo-genetic trees.

In order for approximation algorithms and heuristic procedures to have bounded behavior (guaranteed to be within some factor of the optimal solu-tion), these distances or costs used by optimization problems usually must be metric[8]. The limitations of metricity on these distances is natural and has far-reaching influence on phylogenetic problems (*e.g.* Wheeler, 1993). There are four conditions for metricity (Eq. 5.5).

$$\begin{aligned}
\forall x \quad d(x,x) &= 0 \\
\forall x,y; x \neq y \quad d(x,y) &> 0 \\
\forall x,y \quad d(x,y) &= d(y,x) \\
\forall x,y,z \quad d(x,y) &\leq d(x,z) + d(z,y)
\end{aligned} \tag{5.5}$$

Put simply, the distance between any element and itself must be zero, all other distances must be greater than zero, all distances must be symmetrical, and the most direct distance between two elements must be lower cost than any route through a third element (triangle inequality). Non-metric distances can have unforeseen and sometimes bizarre effects. An example of this would be to imagine a TSP with a city hovering somewhere—in an alternate universe with zero distance to all other cities (since it resides in another dimension, such things are possible). All cities could be reached by traveling first to this bizzaro-city and then to any other at zero cost. Analogous pathological situations can occur with sequence data when indels (gaps) are treated as missing data (Sect. 8.3.3; Wheeler, 1993).

There are further constraints that can be placed on distances such as that of Equation 5.6, resulting in an *ultrametric distance.*

$$\forall x,y,z \quad d(x,y) \leq max(d(x,z), d(z,y)) \tag{5.6}$$

These topics are pursued in the context of distance-based tree reconstruction techniques (Chapter 9) and molecular clocks (Chapter 17).

5.5 NP–Complete Problems in Systematics

The core problem of phylogeny reconstruction, the Phylogeny Problem, is *NP–complete* (Foulds and Graham, 1982). In this problem, we seek to find a tree ($T = (V, E)$), with cost C_T (metric cost d summed over edges in E), that minimizes C_T over all trees. Given the exponential number of trees (Eq. 2.1), it is not surprising that this problem is beyond polynomial solutions (except under certain unlikely conditions; Chapter 9). This problem is familiar to most systematists, but there are many other *NP–hard* optimizations including the Tree Alignment Problem (Sankoff, 1975) shown to be NP–complete by Wang and Jiang (1994), ML tree reconstruction (Roch, 2006), and many problems in genomic analysis (such as chromosomal inversion; Caprara, 1997).

---

[8]In fact, most provable results for these problems require metric distances.

Systematists must understand the theoretical basis and implications of the complexity of the problems they desire to solve. Only with a solid grounding in these computational concepts, can systematists evaluate the tools and techniques required to attack the myriad of hard optimizations presented by phylogeny reconstruction.

5.6   Exercises

1. If a rooted tree has n leaves, what is the time complexity of locating a specific edge?

2. What is the time complexity of writing down all possible DNA sequences of length n? Up to length n?

3. If the length of a tree could be determined by examining a single edge, what would be the time complexity for evaluating m trees with n taxa?

4. If the evaluation of a genome (size m loci) on a vertex of a tree (n leaves) is $O(m^3)$, the evaluation of the entire tree requires visiting each vertex, and a search in tree space requires $O(n^2 \log n)$, what is the time complexity of the entire operation?

5. If given the optimal phylogenetic tree by an oracle, how could this be verified (decided)?

6. Give examples of metric and non-metric distances for DNA sequence data with indels.

7. Show that $P \subsetneq NP$, or barring that, $P = NP$[9].

---

[9]This will guarantee a passing grade in the class.

Chapter 6

Statistical and Mathematical Basics

Many techniques in systematics involve statistical approaches and employ basic mathematical tools to evaluate phylogenetic trees[1]. Minimum evolution, likelihood, and Bayesian approaches are built on these concepts and this section touches on several of the core ideas required to understand and evaluate these methods.

6.1 Theory of Statistics

The discussion here is a precis of some of the background concepts used in statistical methods in systematics. This is not meant to provide complete or in-depth coverage of these topics, but to introduce the basic ideas encountered in likelihood and Bayesian phylogenetic methods. There are numerous excellent treatments of this subject (*e.g.* DeGroot and Schervish, 2006—upon which this discussion is based) available to the curious.

Pierre de Fermat
(1601–1665)

6.1.1 Probability

The concept of probability is usually thought to originate with Pascal and Fermat in the 17th century in studies of dice games, although some calculations were performed over a century earlier. Intuitive definitions of probability abound from the highly personal ("It will probably rain tomorrow") to the general ("These planes are nearly always late"). Fortunately, the mathematics of probability are unaffected by this variation.

Blaise Pascal
(1623–1662)

---

[1]Notation reference in Appendix A.

Systematics: A Course of Lectures, First Edition. Ward C. Wheeler.
© 2012 Ward C. Wheeler. Published 2012 by Blackwell Publishing Ltd.

There are three general flavors of interpretations of probability in common use:

1. *Frequency*—The relative frequency of events is commonly used to describe the probability of an event. For example, how many times will heads come up in a coin toss, given the previous tosses? This interpretation has the problems of being imprecise (how many events are required to determine probability) and unclear as to how much variation is allowed around a probability value. Another obvious shortcoming of this notion is that it implies a large number of trials or events, hence is difficult to apply in situations where events are few, rare, or in the future (probability of a meteor falling).

2. *Classical*—This concept comes from the notion of *equally likely outcomes*, such as those in a coin toss. There are two possible outcomes (head or tails) that seem equally likely, hence have the same probability. Since these probabilities must sum to 1, heads and tails are each assigned a probability of $\frac{1}{2}$. In general, if there are n possible outcomes, their probabilities will be $\frac{1}{n}$. Problems with this notion include that it is basically circular—the likelihood of outcomes is their probability—and that it is unclear how to proceed when events are not equally likely.

3. *Subjective*—This interpretation comes from an individual's own ideas or intuition as to the probabilities of events (do I think a meteor will fall). Although this notion can be expressed numerically, there is no way to assure accuracy, precision, that any individual will be consistent in assigning probabilities, or that any two individuals would assign the same probability to the same event.

Set Theory and Probability

The basis of probability theory resides in sets. The set of all possible events is the *sample space*, S. A specific event s then must reside in S, $s \in S$. If a coin is tossed once, there are two possible events, heads or tails ($s_0 = H, s_1 = T$), in the sample universe ($S = \{H, T\}$). If a coin were tossed n times there would be 2^n events (sequences of tosses) in the sample space.

Events may overlap, complement, or be disjoint (Fig. 6.1) and the union, intersection, and complement operators describe these relationships. Consider the sets of hexapods (A) and herbivores (B). Those creatures that are both hexapods and herbivores would be defined by their intersection ($A \cap B$), those that are either hexapods or herbivores their union ($A \cup B$), and those that are neither their complement ($(A \cup B)^c = S \setminus (A \cup B)$).

Axioms

No matter to which interpretation of probability one cleaves, mathematically, a probability of an event x, $\Pr(x)$, must comply with three axioms (S is the

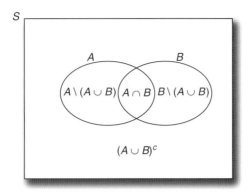

Figure 6.1: The relationship between events A and B in sample space S.

sample space or set of all events):

$$\text{For any event } A, \Pr(A) \geq 0 \tag{6.1}$$

$$\Pr(S) = 1 \tag{6.2}$$

For every infinite sequence of disjoint events $A_0, A_1, \ldots$

$$\Pr\left(\bigcup_{i=1}^{\infty} A_i\right) = \sum_{i=1}^{\infty} \Pr(A_i) \tag{6.3}$$

Where disjoint events are mutually exclusive ($A \cap B = \emptyset$). From these axioms, a *probability distribution* or probability on sample space S, is defined as a specification of $\Pr(A)$ that satisfies Equations 6.1, 6.2, and 6.3.

6.1.2 Conditional Probability

The probability of A given that B has occurred is referred to as the *conditional probability* of A given B ($\Pr(A|B)$). This may occur in a situation where the outcome of a coin toss experiment seems bizarre (assuming a fair coin) but given that the coin was biased, not so strange after all. This would be the conditional probability of an outcome given that the coin was unfair.

Conditional probability is defined by Equation 6.4 for $\Pr(B) > 0$. The conditional probability is undefined when $\Pr(B) = 0$.

$$\Pr(A|B) = \frac{\Pr(A \cap B)}{\Pr(B)} \tag{6.4}$$

Conditional probability is most prominently applied in systematics via Bayes, Theorem (see below) in which the probability of a parameter (such as a tree) is conditioned upon a set of observations.

6.1.3   Distributions

Random Variables

The fundamental entity of probability theory is the *random variable*. Consider a sample space S. A random variable X is a function that assigns a positive real number $X(s)$ to each outcome $s \in S$. As an example, consider a coin toss experiment. If a coin were tossed 100 times, there would be 2^{100} possible outcomes. X could be the the the number of heads (0 to 100) found in the 100 trials.

A probability distribution is the set of probabilities of all events in the sample space. Distributions of random variables may be *discrete* or *continuous*. Discrete distributions can take a finite number of different values, or an infinite (but countable) sequence of values. Continuous distributions, on the other hand, may achieve all values on an interval.

For a discrete distribution, we can define a probability function f of a random variable X such that,

$$f(x) = \Pr(X = x) \tag{6.5}$$

The probability of any x that is not possible is 0, and the sum of the probabilities of all possible events is 1 ($\sum_{x \in S} f(x) = 1$). For the 100 coin-toss experiment above, f would assign a probability to each of the 2^{100} possible outcomes. Distributions can be discrete or continuous, in either case the summed or integrated probabilities over S must be 1.

The definition for a continuous random variable is somewhat different, since the probability of any individual value is zero. Probabilities are more properly assigned to intervals $(a, b]$ where the probability function f (where $\forall x, f(x) \geq 0$) is integrated over the interval (Eq. 6.6).

$$\Pr(a < X \leq b) = \int_a^b f(x)dx \tag{6.6}$$

The interval containing all values of x has probability 1 ($\int_{-\infty}^{\infty} f(x)dx = 1$). We encounter both discrete and continuous distributions in statistical phylogenetic methods.

Mean and Variance

Two properties of distributions of particular interest are the mean and variance. These values give information about the general behavior of a distribution compactly. For a random variable X, the mean or *expectation* ($E(X)$) of a distribution is defined as Equation 6.7 or 6.8:

$$E(X) = \sum_{x \in S} x \cdot f(x) \quad \text{for discrete distributions} \tag{6.7}$$

$$E(X) = \int_{-\infty}^{\infty} x \cdot f(x)dx \quad \text{for continuous distributions} \tag{6.8}$$

The variance ($Var(x)$) of a distribution is defined as Equation 6.9.

$$Var(x) = E\left[(X - E(x))^2\right] \tag{6.9}$$

Probability Distributions

There are several probability distributions commonly used in systematics. These include the Uniform, Normal or Gaussian, Binomial, Poisson, Exponential, Gamma, and Dirichlet.

Uniform—This distribution describes events with equal probability over the sample space (Fig. 6.2). This distribution is used when values are drawn "at random" from some set of possibilities (k of them from a to b) for the discrete form (Eq. 6.10) or an interval ($[a, b]$) in the continuous (Eq. 6.11).

$$f(x) = \begin{cases} \frac{1}{k} & \text{for } x = a, a+1, a+2, \ldots, a+k-1 = b \\ 0 & \text{otherwise} \end{cases} \tag{6.10}$$

$$f(x) = \begin{cases} \frac{1}{b-a} & \text{for } a \le x \le b \\ 0 & \text{otherwise} \end{cases} \tag{6.11}$$

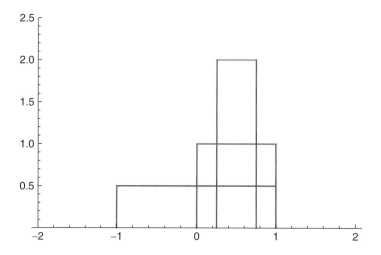

Figure 6.2: Discrete uniform distributions with $[a, b] = [-1, 1], [0, 1]$, and $[0.25, 0.75]$.

The mean and variance are calculated as:

$$E(X) = (a+b)/2 \tag{6.12}$$

$$Var(X) = (b-a)^2/12 \tag{6.13}$$

Gaussian—The Gaussian or Normal distribution (Eq. 6.14) is a continuous distribution that describes observations tending to cluster around a mean value (Fig. 6.3). One reason for the utility of this distribution is the fit it has to many real-world situations. This is due, in part, to the central limit theorem. This result states that for a random sample of size n taken from *any* distribution with mean μ and variance σ^2, the sample mean $\overline{X}_n$ will have a distribution

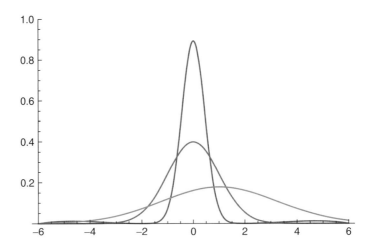

Figure 6.3: Gaussian distributions with $(\mu, \sigma) = (0, 0.45), (0, 1)$, and $(1, 2.24)$.

that is approximately normal with mean μ and variance σ^2/n. As a result, the Gaussian distribution can be used to describe a broad variety of phenomena.

$$f(x|\mu, \sigma^2) = \frac{1}{(2\pi)^{\frac{1}{2}}\sigma} e^{-\frac{1}{2}\left(\frac{x-\mu}{\sigma}\right)^2} \text{ for } -\infty < x < \infty \tag{6.14}$$

With mean and variance:

$$E(X) = \mu \tag{6.15}$$

$$Var(X) = \sigma^2 \tag{6.16}$$

Binomial—When events can have two outcomes, and the probability of the occurrence of one outcome is p, and the other (or non-occurrence) is $(1 - p)$ or q, the probability of exactly x occurrences in n trials is described by Eq. 6.17. This discrete distribution (Fig. 6.4) is used to calculate such familiar scenarios as coin tossing.

$$f(x) = \begin{cases} \binom{n}{x} p^x q^{n-x} & \text{for } x = 0, 1, 2, \ldots, n \\ 0 & \text{otherwise} \end{cases} \tag{6.17}$$

The mean and variance are calculated as:

$$E(X) = np \tag{6.18}$$

$$Var(X) = npq \tag{6.19}$$

Poisson—The occurrence of random arrival events that occur during fixed time (or space) intervals at an average rate are Poisson distributed (Eq. 6.20, Fig. 6.5). A Poisson process is one in which the number of events occurring in a fixed interval t with mean λt and the number of events in disjoint time intervals are independent. Poisson processes are used in likelihood analyses to describe

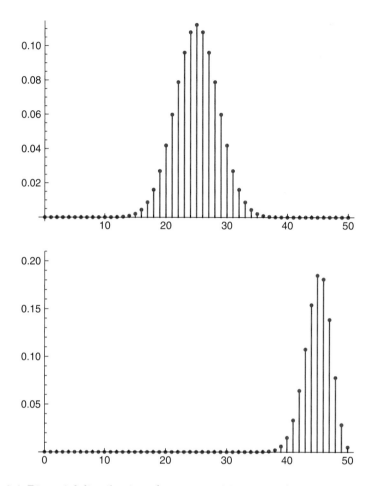

Figure 6.4: Binomial distributions (upper $p = 0.5, n = 50$, lower $p = 0.9, n = 50$).

the distribution of transformation events in time (numbers of changes) and the location of changes in a gene or genome. The Poisson distribution is a discrete distribution and has the interesting property that its mean and variance are the same.

$$f(x|\lambda) = \begin{cases} \frac{e^{-\lambda}\lambda^x}{x!} & \text{for } x = 0, 1, 2, \ldots \\ 0 & \text{otherwise} \end{cases} \tag{6.20}$$

The mean and variance are calculated as:

$$E(X) = \lambda \tag{6.21}$$

$$Var(X) = \lambda \tag{6.22}$$

Exponential—The exponential distribution (Fig. 6.6, Eq. 6.23) is a continuous distribution that describes the time intervals between a series of independent

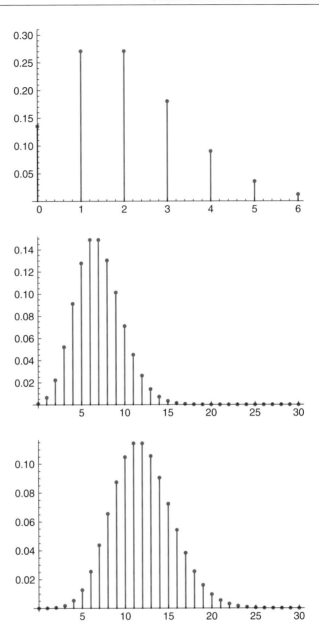

Figure 6.5: Poisson distributions (upper $\lambda = 2$, middle $\lambda = 7$, lower $\lambda = 12$).

events that follow a Poisson process. This distribution is often used to describe events such as the failure of light bulbs or edge weights in Bayesian analysis.

$$f(x|\beta) = \begin{cases} \beta e^{-\beta x} & \text{for } x > 0 \\ 0 & \text{otherwise} \end{cases} \tag{6.23}$$

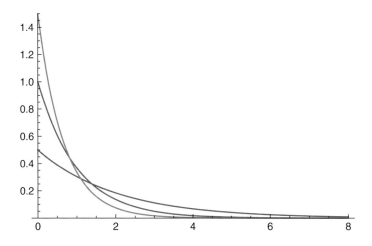

Figure 6.6: Exponential distributions with $\beta = 0.5, 1.0$, and 1.5.

The mean and variance are calculated as:

$$E(X) \quad = \quad \frac{1}{\beta} \tag{6.24}$$

$$Var(X) \quad = \quad \frac{1}{\beta^2} \tag{6.25}$$

Gamma—The continuous gamma distribution (Fig. 6.7, Eq. 6.26) describes the sum of (in the integer form) α exponentially distributed variables, each with a mean of β^{-1}. The gamma distribution is used to describe rate variation among classes of characters in likelihood calculations (in the discrete form with $\alpha = \beta$).

$$f(x|\alpha, \beta) = \begin{cases} \frac{\beta^\alpha}{\Gamma(\alpha)} x^{\alpha-1} e^{-\beta x} & \text{for } x > 0 \\ 0 & \text{otherwise} \end{cases} \tag{6.26}$$

Where the normalization term $\Gamma(\alpha)$ insures integration to 1:

$$\Gamma(\alpha) = \int_0^\infty x^{\alpha-1} e^{-x} dx \tag{6.27}$$

The mean and variance are calculated as:

$$E(X) \quad = \quad \frac{\alpha}{\beta} \tag{6.28}$$

$$Var(X) \quad = \quad \frac{\alpha}{\beta^2} \tag{6.29}$$

The exponential distribution is the same as the gamma distribution with $\alpha = 1$.

Dirichlet—The Dirichlet distribution (Eq. 6.30) is used most frequently in systematics as a prior for parameter values in General-Time-Reversible (GTR)

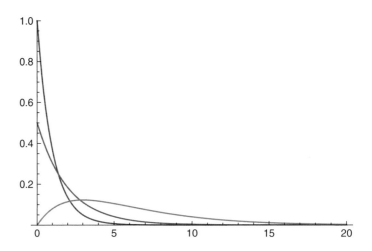

Figure 6.7: Gamma distributions with $(\alpha, \beta) = (1, 1), (1, 2)$, and $(2, 3)$.

and other character change models (Chapter 12). This continuous distribution is used to describe the scenario where the numerous parameter combinations are evenly distributed.

$$f(x_1, \ldots, x_{k-1} | \alpha_1, \alpha_2, \ldots, \alpha_k) = \begin{cases} \frac{1}{B(\alpha)} \prod_{i=1}^{k} x_i^{a_i - 1} & \text{for } x > 0 \\ 0 & \text{otherwise} \end{cases} \quad (6.30)$$

Where :

$$B(\alpha) = \frac{\prod_{i-1}^{k} \Gamma(a_i)}{\Gamma(\sum_{i=1}^{k}) a_i} \text{ for } \alpha = (\alpha_1, \ldots, \alpha_k) \quad (6.31)$$

The mean and variance are calculated as:

$$E(X) = \frac{\alpha_i}{\alpha_0} \quad (6.32)$$

$$Var(X) = \frac{\alpha_i(\alpha_0 - \alpha_i)}{\alpha^2(\alpha_0 + 1)} \quad (6.33)$$

and

$$\alpha_0 = \sum_{i=1}^{k} \alpha_i \quad (6.34)$$

6.1.4 Statistical Inference

Statistical inference is the process of estimating characteristics of an unknown probability distribution from a set of observations. The inference may concern the type of distribution (*e.g.* Gaussian) or its parameters (*e.g.* μ and σ). The basic problem is that the data are finite and absolute certainty is impossible. The goal is to choose the "correct" parameter or distribution with high probability.

6.1.5 Prior and Posterior Distributions

Priors and Problems

A typical inference problem would be to infer the specific value of a parameter θ in the parameter space Ω. Any observations (x) gathered would have been generated from the distribution function f and parameter θ. Observations have been drawn from $f(x|\theta)$.

Before any observations are gathered from $f(x|\theta)$, an investigator may have knowledge about where in Ω the parameter θ may lie; that θ is more likely to be found in one area than another. As an example, the height of an adult human female is less likely to be less than one meter or greater than three than between one and three meters. This knowledge may be based on previous experimental data, experiences, or even subjective opinion and can be expressed as a distribution of θ on Ω, $\xi(\theta)$. That distribution is known as the *prior distribution of θ*. "Prior" is used because the distribution is specified before any observations are drawn from $f(x|\theta)$.

Disagreements over the validity, and even existence, of prior distributions are extreme within the statistical community. Bayesian statisticians adhere to the proposition that prior distributions can be defined in all circumstances and these probabilities are as valid as any other in statistics. This parallels the subjective interpretation of probability itself. If all probabilities are subjective, they are all equally valid no matter their source.

Others disagree, stating that a parameter θ is a fixed, if unknown, value and is not drawn from a distribution. The only way, then, to establish priors is through extensive previous observation (such as the historical performance of a machine). Both factions would agree that in the presence of good prior information, it should be used.

Posterior Distribution

The conditional probability of a set of observations given a specific value of θ multiplied by the probability (prior, $\xi(\theta)$) that θ has that value, $f(x|\theta)\xi(\theta)$, is their joint distribution. The total probability over all possible values of θ in the parameter space Ω for a given set of observations x is then:

$$g(x) = \int_{\theta \in \Omega} f(x|\theta)\xi(\theta)d\theta \tag{6.35}$$

The conditional probability of θ given the observations x is then:

$$\xi(\theta|x) = \frac{f(x|\theta)\xi(\theta)}{g(x)} \tag{6.36}$$

which is known as the *posterior probability* of θ given x. In its discrete form, this is known as Bayes' (1763) Theorem (Eq. 6.37).

$$p(\theta = \theta_i|x) = \frac{p(x|\theta_i)p(\theta = \theta_i)}{\sum_{j \in \Omega} p(x|\theta_j)p(\theta = \theta_j)} \tag{6.37}$$

Thomas Bayes
(1702–1761)

Since $g(x)$ is always a constant (depending on x not θ):

$$\xi(\theta|x) \propto f(x|\theta)\xi(\theta) \tag{6.38}$$

When the probability of the observations x are treated as a function of θ, $f(x|\theta)$ is referred to as the *likelihood function*.

6.1.6 Bayes Estimators

Let us suppose that there is a universe of possible sets of observations X from which we will draw (observe) a specific set x. An *estimator* of the parameter θ is a real values function $\delta(X)$ that specifies the *estimate* of θ for each x. In general, we are interested in estimates that are close to parameter values. More specifically, we desire an estimator δ such that $\delta(X) - \theta$ will be near zero with high probability. We can define a "loss" function $L(\theta, a)$ that measures the cost of an estimate a of θ. Most likely, as the difference between a and θ grows, so does L. For a particular estimate a, the expected loss is:

$$E[L(\theta, a)] = \int_\Omega L(\theta, a)\xi(\theta)d\theta \tag{6.39}$$

Conditioned on a set of observations x:

$$E[L(\theta, a)|x] = \int_\Omega L(\theta, a)\xi(\theta|x)d\theta \tag{6.40}$$

If an estimator (δ^*) is chosen that minimizes the expected loss in Equation 6.40, it is the *Bayes estimator* of θ (Eq. 6.41).

$$E[L(\theta, \delta^*(x))|x] = \min_{a\in\Omega} E[L(\theta, a)|x] \tag{6.41}$$

In systematics, the use of Bayes estimators would be in identifying the specific value of a parameter such as the tree. If the loss function is uniform over all "incorrect" parameter values (all bad choices are equally bad), we can choose the parameter value that maximizes the posterior probability. Such an estimator is referred to as the *Maximum A Posteriori* (MAP) estimate. Since the denominator of Bayes' Theorem (Eq. 6.37) is a constant, the MAP estimator can be found by maximizing the numerator of Equation 6.37 over the parameter space (Eq. 6.42).

$$\theta^{MAP} = \underset{\theta\in\Omega}{\operatorname{argmax}} \left[p(x|\theta) \cdot p(\theta)\right] \tag{6.42}$$

As an example, consider a coin toss. Let us suppose the coin comes from a mint that, in the past, has produced coins with two tails or two heads $\frac{1}{20}$ of the time each, and the remainder are fair coins with heads on one side and tails on the other. A coin is tossed three times, producing heads each time. What is the MAP estimate of the type of the coin (two headed—$\theta = 1$, two tailed—$\theta = 0$, or fair—$\theta = 0.5$)? Since there are only three possibilities, we can calculate them easily. Using Equation 6.42, the *a posteriori* probabilities would be:

$$
\begin{aligned}
\theta = 0 : \quad & 0 \cdot 0.05 && = 0 \\
\theta = 0.5 : \quad & 0.125 \cdot 0.9 && = 0.1125 \\
\theta = 1 : \quad & 1.0 \cdot 0.05 && = 0.05
\end{aligned}
$$

Hence, the MAP estimate would be that the coin was fair ($\theta^{MAP} = 0.5$). It would take a run of five heads to change the MAP result to 1, showing the initially great, but waning influence of the prior.

6.1.7   Maximum Likelihood Estimators

Bayes estimators form a complete and precise system for estimation of parameters. However, they require two things that may be difficult or impossible to acquire—a specific loss function and the prior distribution of the parameter. In avoiding these issues, other systems usually have serious defects and limitations. One simple and broadly popular method to construct estimators without loss functions and priors is maximum likelihood (ML).

Fisher (1912) employed the likelihood function $f(x|\theta)$ from Equation 6.38, the idea being that the value of θ that maximized the probability of the observed data (x) should be a good estimate of θ. For each observed sequence $x \in X$, we can define $\delta(x)$ to signify a value $\theta \in \Omega$ such that $f(x|\theta)$ is maximal. The estimator defined this way $(\widehat{\theta} = \delta(x))$ is the *maximum likelihood estimator* (Eq. 6.43).

$$
\theta^{ML} = \widehat{\theta} = \operatorname*{argmax}_{\theta \in \Omega} p(x|\theta) \tag{6.43}
$$

If we reconsider the coin toss example above, the likelihood values would be:

Ronald A. Fisher
(1890–1962)

$$
\begin{aligned}
\theta = 0 : \quad & 0 \\
\theta = 0.5 : \quad & 0.125 \\
\theta = 1 : \quad & 1
\end{aligned}
$$

Leading to the result that $\widehat{\theta} = 1$.

Given that $\widehat{\theta}$ is a maximum point of a function, there may be multiple such points or none (yielding estimation problems). These situations are not encountered frequently, but can lead to non-identifiability of parameters.

6.1.8   Properties of Estimators

When evaluating estimators, there are three properties that are commonly discussed: consistency, efficiency, and bias.

Consistency—A comforting property of an estimator would be that it converges on the parameter value as sample size grows ($|\widehat{\theta} - \theta| < \epsilon$ [*i,e.* arbitrarily small] as $n \to \infty$ [2]). When this is the case, the estimator is said to be consistent. Bayes and ML methods are generally, but not exclusively consistent. In systematics, proofs of consistency rely on specific models. If the conditions of these models are violated by the data, the proofs do not hold (Chapter 13).

---

[2]The estimator is said to be *strongly* consistent if $\lim_{n \to \infty} |\widehat{\theta} - \theta| = 0$.

Efficiency—Informally, an estimator with a low variance is said to be efficient.

Bias—An estimator with symmetrical error distribution (*i.e.* equally probable to be erroneously high or erroneously low). Unbiased estimators are relatively inefficient. In fact, there is always a biased estimator for a parameter with greater efficiency than an unbiased.

6.2   Matrix Algebra, Differential Equations, and Markov Models

Here, a brief introduction to the solution of simultaneous linear differential equations is outlined as it relates to systematic analysis. The topics covered here are restricted to those that touch on distance, likelihood, and Bayesian methods. An excellent general text on the subject is Strang (2006).

6.2.1   Basics

Recall that systems of linear equations and unknowns (Eq. 6.44)

$$a_{00}x_0 + a_{01}x_1 + a_{02}x_2 = y_0 \tag{6.44}$$
$$a_{10}x_0 + a_{11}x_1 + a_{12}x_2 = y_1$$
$$a_{20}x_0 + a_{21}x_1 + a_{22}x_2 = y_2$$

can be represented in matrix form (Eq. 6.45).

$$\begin{bmatrix} a_{00} & a_{01} & a_{02} \\ a_{10} & a_{11} & a_{12} \\ a_{20} & a_{21} & a_{22} \end{bmatrix} \cdot \begin{bmatrix} x_0 \\ x_1 \\ x_2 \end{bmatrix} = \begin{bmatrix} y_0 \\ y_1 \\ y_2 \end{bmatrix} \tag{6.45}$$

6.2.2   Gaussian Elimination

Systems of n equations and n unknowns can be solved via manipulation of their matrix representation in a process called *Gaussian Elimination*. An example would be the determination of edge weights on an additive tree. Given the additive distance matrix of Table 6.1 and tree of Figure 6.8, we can solve for edge weights.

	A	B	C	D
A	0	3	8	9
B	3	0	9	10
C	8	9	0	9
D	9	10	9	0

Table 6.1: Additive distances for tree in Figure 6.8.

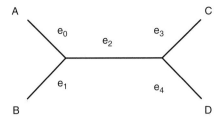

Figure 6.8: An additive tree with distances of Table 6.1.

The edge weights and observed distances are a series of unknowns and linear equations (Eq. 6.46).

$$
\begin{aligned}
dAB &= e_o + e_1 \\
dAC &= e_o + e_2 + e_3 \\
dAD &= e_o + e_2 + e_4 \\
dBC &= e_1 + e_2 + e_3 \\
dBD &= e_1 + e_2 + e_4 \\
dCD &= e_3 + e_4
\end{aligned}
\tag{6.46}
$$

These equations can be represented in matrix form (Eq. 6.47).

$$
\begin{bmatrix}
1 & 1 & 0 & 0 & 0 \\
1 & 0 & 1 & 1 & 0 \\
1 & 0 & 1 & 0 & 1 \\
0 & 1 & 1 & 1 & 0 \\
0 & 1 & 1 & 0 & 1 \\
0 & 0 & 0 & 1 & 1
\end{bmatrix}
\cdot
\begin{bmatrix}
e_0 \\
e_1 \\
e_2 \\
e_3 \\
e_4
\end{bmatrix}
=
\begin{bmatrix}
3 \\
8 \\
9 \\
9 \\
10 \\
9
\end{bmatrix}
\tag{6.47}
$$

The last row of the coefficients (left) and distances (right) can be removed since they contain redundant information (there are $\binom{n}{2}$ distances and only $2n - 3$ edges) leaving 5 equations and 5 unknowns (Eq. 6.48).

$$
\begin{bmatrix}
1 & 1 & 0 & 0 & 0 \\
1 & 0 & 1 & 1 & 0 \\
1 & 0 & 1 & 0 & 1 \\
0 & 1 & 1 & 1 & 0 \\
0 & 1 & 1 & 0 & 1
\end{bmatrix}
\cdot
\begin{bmatrix}
e_0 \\
e_1 \\
e_2 \\
e_3 \\
e_4
\end{bmatrix}
=
\begin{bmatrix}
3 \\
8 \\
9 \\
9 \\
10
\end{bmatrix}
\tag{6.48}
$$

By adding and subtracting complete rows, we can transform the expression into a diagonal series of 1's and 0's elsewhere with the result vector (left) containing the edge weights (Eq. 6.49).

$$\begin{bmatrix} 1 & 0 & 0 & 0 & 0 \\ 0 & 1 & 0 & 0 & 0 \\ 0 & 0 & 1 & 0 & 0 \\ 0 & 0 & 0 & 1 & 0 \\ 0 & 0 & 0 & 0 & 1 \end{bmatrix} \cdot \begin{bmatrix} e_0 \\ e_1 \\ e_2 \\ e_3 \\ e_4 \end{bmatrix} = \begin{bmatrix} 1 \\ 2 \\ 3 \\ 4 \\ 5 \end{bmatrix} \tag{6.49}$$

Given that most distance matrices are non-additive, and the systems overdetermined (more equations than unknowns), the least-squares approach of Section 9.5.2 is normally employed.

6.2.3   Differential Equations

Let us suppose we have a coupled pair of linear differential equations (Eq. 6.50) that we wish to solve by finding functions of v and w in terms of t only.

$$\frac{dv}{dt} = 5v - 6w \tag{6.50}$$

$$\frac{dw}{dt} = 3v - 4w$$

With initial $(t = 0)$ conditions $v = 5$ and $w = 6$. These can be represented in matrix form (Eq. 6.51):

$$u(t) = \begin{bmatrix} v(t) \\ w(t) \end{bmatrix}, \text{ with } A = \begin{bmatrix} 5 & -6 \\ 3 & -4 \end{bmatrix}, \text{ and } u(0) = \begin{bmatrix} 5 \\ 6 \end{bmatrix} \tag{6.51}$$

with

$$\frac{du}{dt} = Au \text{ with } u = u(0) \text{ at } t = 0 \tag{6.52}$$

In order to solve this linear differential equation in two unknowns, we can look to the simpler case of a single equation

$$\frac{du}{dt} = au \tag{6.53}$$

where there would be a simple solution:

$$u(t) = e^{at}u(0) \tag{6.54}$$

If we generalize to the vector case:

$$v(t) = e^{\lambda t}y \tag{6.55}$$

$$w(t) = e^{\lambda t}z$$

or as a vector:

$$u(t) = e^{\lambda t}x$$

We can then substitute back from the scalar case and have:

$$\lambda e^{\lambda t}y = 5e^{\lambda t}y - 6e^{\lambda t}z$$

$$\lambda e^{\lambda t}z = 3e^{\lambda t}y - 4e^{\lambda t}z$$

This can be represented as the *eigenvalue problem*:

$$5y - 6z = \lambda y \tag{6.56}$$

$$3y - 4z = \lambda z$$

yielding the *eigenvalue equation*

$$Ax = \lambda x \tag{6.57}$$

where λ is an eigenvalue and x is its associated eigenvector.

6.2.4 Determining Eigenvalues

λ is an eigenvalue of A if and only if $A - \lambda I$ is a "singular" matrix, meaning that the determinant must be equal to zero (Eq. 6.58).

$$\det(A - \lambda I) = 0 \tag{6.58}$$

The determinant is a value associated with a matrix, which can be quite laborious (naïvely $O(n^3)$, not so naïvely $O(n^{2.376})$) to calculate. In this simple case of a 2 x 2 matrix, the determinant can be calculated easily (Eq. 6.59).

$$B = \begin{bmatrix} a & b \\ c & d \end{bmatrix} \tag{6.59}$$

$$\det|B| = ad - bc$$

Hence, we seek the determinant of

$$B = \begin{bmatrix} a - \lambda & b \\ c & d - \lambda \end{bmatrix} \tag{6.60}$$

yielding:

$$\det|B - \lambda I| = (a - \lambda)(d - \lambda) - bc = 0$$

So for Equation 6.51:

$$\det|A - \lambda I| = (5 - \lambda)(-4 - \lambda) - (-18) = 0 \tag{6.61}$$

This rearranges to $\lambda^2 - \lambda - 2$, the characteristic polynomial of A with solutions $\lambda = (-1, 2)$. By solving back into the original matrix form, we determine the eigenvectors x_1 and x_2.

$$\lambda_1 = -1 \rightarrow (A - \lambda_1 I)x_1 = \begin{bmatrix} 6 & -6 \\ 3 & -3 \end{bmatrix} \begin{bmatrix} y \\ z \end{bmatrix} = \begin{bmatrix} 0 \\ 0 \end{bmatrix} \tag{6.62}$$

$$x_1 = \begin{bmatrix} 1 \\ 1 \end{bmatrix} \tag{6.63}$$

$$\lambda_2 = 2 \rightarrow (A - \lambda_1 I)x_2 = \begin{bmatrix} 3 & -6 \\ 3 & -6 \end{bmatrix} \begin{bmatrix} y \\ z \end{bmatrix} = \begin{bmatrix} 0 \\ 0 \end{bmatrix} \tag{6.64}$$

$$x_2 = \begin{bmatrix} 2 \\ 1 \end{bmatrix} \tag{6.65}$$

In general, there will be k eigenvalues and eigenvectors for k equations of k variables.

The complete solution has the form:

$$u(t) = \sum_{i=1}^{k} c_i e^{\lambda_i t} x_i \tag{6.66}$$

We can then solve back given each eigenvalue and generate our final solution (Eq. 6.67).

$$u(t) = c_1 e^{-t} \begin{bmatrix} 1 \\ 1 \end{bmatrix} + c_2 e^{2t} \begin{bmatrix} 2 \\ 1 \end{bmatrix} \tag{6.67}$$

For initial conditions $(t = 0)$:

$$c_1 x_1 + c_2 x_2 = u(0) \tag{6.68}$$

$$\begin{bmatrix} 1 & 2 \\ 1 & 1 \end{bmatrix} \begin{bmatrix} c_1 \\ c_2 \end{bmatrix} = \begin{bmatrix} 5 \\ 6 \end{bmatrix} \tag{6.69}$$

Resulting in:

$$u(t) = 7 e^{-t} \begin{bmatrix} 1 \\ 1 \end{bmatrix} - e^{2t} \begin{bmatrix} 2 \\ 1 \end{bmatrix} \tag{6.70}$$

or in the original non-matrix form Eq. 6.71.

$$v(t) = 7 e^{-t} - 2 e^{2t} \tag{6.71}$$

$$w(t) = 7 e^{-t} - e^{2t}$$

6.2.5 Markov Matrices

A stochastic process where the probability of a successor event is solely dependent on its current state (not past) is called a Markov process. Markov processes are employed in character transformation models due to this *memoryless* property. As an example from DNA sequences, the probability of an A substituting to a T has nothing to do with whether the A was previously a C, G, or anything else.

A special class of matrix differential equations that models this sort of process employs a Markov or stationary matrix. In these matrices, the sum of the solutions is fixed (nothing is gained or lost—only transitioned among states) so each column sums to 1, and all entries are positive (*e.g.* Eq. 6.72).

Andrei Markov
(1856–1922)

$$M = \begin{bmatrix} .7 & .1 & .1 & .1 \\ .1 & .7 & .1 & .1 \\ .1 & .1 & .7 & .1 \\ .1 & .1 & .1 & .7 \end{bmatrix} \tag{6.72}$$

In the discrete case, the maximum eigenvalue is always equal to 1 and the others ≤ 1. As a result, as time steps increase (in essence, multiplication by the

Markov transition matrix) the terms derived from the non-dominant (*i.e.* < 1) eigenvalues diminish and limit to zero, resulting in the steady-state condition.

Continuous time Markov processes are a related form where the net rate of change is zero. In this case, the transition probabilities represented in the columns of the matrix sum to zero. As an example, we might have a mutation matrix with equal forward and reverse rates between states 0 and 1 ($\mu_{0 \leftrightarrow 0} = \mu_{0 \leftrightarrow 1}, \mu = \mu_{0 \leftrightarrow 0} + \mu_{0 \leftrightarrow 1}$) (Eq. 6.73).

$$M = \begin{bmatrix} -\frac{\mu}{2} & \frac{\mu}{2} \\ \frac{\mu}{2} & -\frac{\mu}{2} \end{bmatrix} \tag{6.73}$$

The characteristic polynomial and eigenvalues of M are determined as above (Eq. 6.75).

$$\lambda^2 + \mu\lambda = 0 \tag{6.74}$$

$$\lambda = \{0, -\mu\}$$

For continuous Markov processes, $\lambda_{max} = 0$ in all cases.

This yields eigenvectors $\{\begin{bmatrix} 1 \\ 1 \end{bmatrix}, \begin{bmatrix} 1 \\ -1 \end{bmatrix}\}$. Solving for the general form with initial conditions $\begin{bmatrix} 1 \\ 0 \end{bmatrix}$, yields the familiar equation for a two state character.

$$P_{i=j}(t) = 0.5 + 0.5e^{-\mu t} \tag{6.75}$$

$$P_{i \neq j}(t) = 0.5 - 0.5e^{-\mu t}$$

6.3 Exercises

1. What is the probability of fair coin toss outcome H, H, T, T, T?

2. What are the mean and variance of the estimate of probability of heads in the coin toss experiment of the previous question?

3. What is the probability of tossing a fair coin five times with outcome 2H, 3T?

4. Consider the coin toss outcome H, T, H, T, T. There are seven possible values for the probability of heads ($\theta = 0.01, 0.2, 0.4, 0.5, 0.6, 0.8, 1.0$). Consider two prior distributions on θ, uniform and $\frac{1}{\theta}$. What are the values for θ^{MAP} and θ^{ML} for these observations and the two priors? What are the posterior probabilities for the MAP solutions? What is the effect of the uniform prior on θ^{MAP} compared to θ^{ML}?

5. Consider the tree of Figure 6.8. Using additive distance matrix, Table 6.2, determine the edge weights.

	A	B	C	D
A	0	5	11	12
B	5	0	12	13
C	11	12	0	11
D	12	13	11	0

Table 6.2: Additive distances for the tree in Figure 6.8.

6. Determine the complete time equations for the following matrix (initial conditions $v(0) = 4, w(0) = 6$:

$$v(t) = 8v + 2w \tag{6.76}$$
$$w(t) = 2v + 8w$$

7. Determine the complete time equations for the following matrix (initial conditions $a(0) = 1, b(0) = 0$:

$$a(t) = -0.1a + 0.1b \tag{6.77}$$
$$b(t) = 0.1a - 0.1b$$

Part II

Homology

Chapter 7

Homology

Homology is perhaps the central concept of comparative biology and systematics. First defined by Owen (1843, 1847), the idea has undergone redefinition and refinement in the succeeding century and a half from pre-evolutionary "same-ness" to the modern cladistic incarnation as an optimized character transformation.

7.1 Pre-Evolutionary Concepts

Before evolutionary ideas presented a generative explanation, the observation that organisms possessed similar parts in similar positions offered biologists a basis for classification. The patterns of similarity and difference were taken as evidence of a natural or super-natural plan and allowed comparative biologists to construct classifications and compete them on their ability to explain observed variation (Hull, 1988).

7.1.1 Aristotle

The use of anatomical features (Aristotle, 350BCE) to create his classification implies comparability between observed features of organisms (*e.g.* blooded, egg-laying). These similarities among features were organizational and empirical, and perhaps functional, but implied nothing about their origins. For Aristotle, the similarities and their classifications were statements of organization of nature. As a result, similar statements could be made about minerals or other non-living natural phenomena. Aristotle saw a progression of perfection, so to speak, in moving from "lower" to "higher" forms, but these held no notion of what we would call origins today.

7.1.2 Pierre Belon

Belon's comparison between the skeleton of a human and bird is striking (Sect. 1.3) not only in its general aspects but in its specific aspects as well.

Pierre Belon
(1517–1564)
Father of comparative biology, murdered in the Bois de Boulogne.

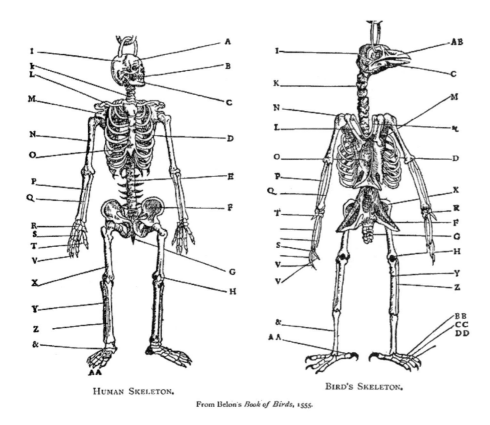

Figure 7.1: Belon's Big Bird (Belon, 1555).

Belon (1555) labeled corresponding structures in a way we recognize as homology today (Fig. 7.1). It is unclear whether for Belon this was the result of an analytical procedure, or simply an observational statement (Rieppel, 1988). Nonetheless, his Figure still impresses and is emulated in modern work.

7.1.3 Étienne Geoffroy Saint-Hilaire

One of the most important principles in the recognition of comparable anatomical components is their topological (*i.e*, positional) similarity. This idea is first found in Geoffroy Saint-Hilaire's "principe des connexions" (Fig. 7.2). This positional criterion for structures as variants of the same "type" is more or less equal to what we now call homology. Saint-Hilaire (1830) differed from Cuvier in the lack of requirement of a functional similarity, hence vestigial structures could be part of the same "unity of type" as functional ones and constitute evidence of the "uniformity of type." For Geoffroy, these were philosophical, intellectual constructs akin to archetypes. This is in opposition to the ideas of Cuvier who based his ideas on empirical observation of natural pattern. Cuvier, via careful analysis, falsified some of Geoffroy's ideas (*e.g.* structures in cephalopods and vertebrates).

Étienne Geoffroy Saint-Hilaire (1772–1844)

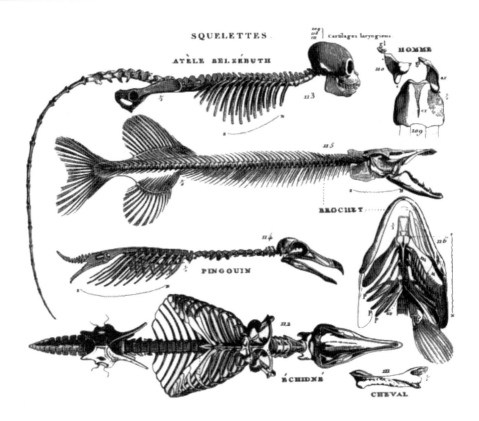

Figure 7.2: Illustration of Geoffroy Saint-Hilaire's idea of topological "connections" among anatomical features (Saint-Hilaire, 1818).

Darwin cited the "principe des connexions" in the *Origin* as evidence for common origin of groups and support for his evolutionary explanation of homology.

> What can be more curious than that the hand of a man, formed for grasping, that of a mole for digging, the leg of the horse, the paddle of the porpoise, and the wing of the bat, should all be constructed on the same pattern, and should include the same bones, in the same relative positions? Geoffroy Saint-Hilaire has insisted strongly on the high importance of relative connexion in homologous organs: the parts may change to almost any extent in form and size, and yet they always remain connected together in the same order. (Darwin, 1859b)

7.1.4 Richard Owen

Richard Owen
(1804–1892)

Though inimical to the idea of evolution (his views were less consistent later in life), the modern idea, and often definition, of homology used today is that of Owen. Owen coined the terms "archetype" (Fig. 7.3) (Owen, 1847, 1848, 1849) and "homology" (Owen, 1843). He did this through synthesizing the functional

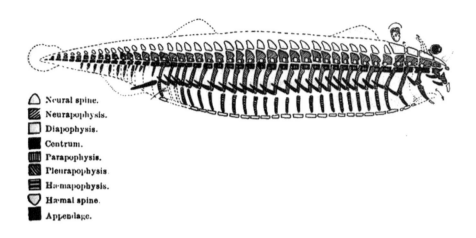

○ Neural spine.
▨ Neurapophysis.
▫ Diapophysis.
■ Centrum.
▥ Parapophysis.
▧ Pleurapophysis.
▤ Hæmapophysis.
▽ Hæmal spine.
■ Appendage.

Figure 7.3: Owen (1843) archetype.

ideas of Cuvier and the topological relations of Geoffroy Saint-Hilaire. Owen still discussed pattern in his archetype as an abstraction. This was not an ancestor in the evolutionary sense, but a general plan for a group such as classes of vertebrates.

The definitions of Owen (1843):

- Analogue—A part or organ in one animal which has the same function as another part or organ in a different animal.

- Homologue—The same organ in different animals under every variety of form and function.

- Analogy—Superficial or misleading similarity.

In order to recognize homologues, Owen used the criteria of connections and composition derived from Geoffroy Saint-Hilaire. Owen further recognized both special (between organisms) and general (within organisms) homology as subtypes. The latter is usually referred to as serial homology today.

7.2 Charles Darwin

Darwin (1859b) did not redirect or refine the concept of homology, but used homology as evidence for and an explanation of common descent. In proposing a natural explanation for the generation of homologues, Darwin transformed the discussion of archetypes to that of ancestors. This had little effect on *what* comparative biologists actually did in an operational sense, but it completely changed *why* they did it. The determination of archetypes was replaced by the search for ancestors.

Charles Darwin
(1809–1882)

> All the foregoing rules and aids and difficulties in classification are explained, if I do not greatly deceive myself, on the view that the natural

system is founded on descent with modification; that the characters which naturalists consider as showing true affinity between any two or more species, are those which have been inherited from a common parent, and, in so far, all true classification is genealogical; that community of descent is the hidden bond which naturalists have been unconsciously seeking, and not some unknown plan of creation, or the enunciation of general propositions, and the mere putting together and separating objects more or less alike. (Darwin, 1859b)

7.3   E. Ray Lankester

Ray Lankester
(1847–1929)

Lankester, in addition to defining homoplasy (Lankester, 1870a), felt that the "homology" defined by Owen required refinement, since it was non-evolutionary. Lankester defined "homogeny" as a structure in organisms that was similar due to shared ancestry (Lankester, 1870b). This evolutionary concept of homology has its direct descendant in Hennig's homology concept, and is the one we use today.

7.4   Adolf Remane

Adolf Remane
(1898–1976)

Remane (1952) described the criteria by which homologues were to be recognized. This work was very influential in the mid 20th-century New Synthesis for its summary and synthesis of the ideas of Geoffroy Saint-Hilaire, Cuvier, Owen, and Haeckel.

Remane (1952)(translation of Reidl, 1978) proposed six criteria—three principles:

1. Position—"Homology can be recognized by similar position in comparable systems of features."

2. Structure—"Similar structures can be homologized without reference to similar position, when they agree in numerous special features. Certainty increases with the degree of complication and of agreement in the structures compared."

3. Transition—"Even dissimilar structures of different position can be regarded as homologous if transitional forms between them can be proved so that in considering two neighboring forms, the conditions under (1) and (2) are fulfilled. The transitional forms can be taken from ontogeny of the structure or can be true systematically intermediate forms."

and three auxiliary:

1. General conjunction—"Even simple structures can be regarded as homologous when they occur in a great number of adjacent species."

2. Special conjunction—"The probability of the homology of simple structures increases with the presence of other similarities, with the same distribution among closely similar species."

3. Negative conjunction—"The probability of the homology of features decreases with the commonness of occurrence of this feature among species which are not certainly related."

Remane, however, presented no definition of homology—only a method to recognize homologues. This lack of a specific definition of homology has led to a great deal of fruitless argumentation due to the conflation of the definition and recognition criteria of homology.

7.5 Four Types of Homology

Wiley (1975), followed by Patterson (1982), divided definitions of homology into four[1] types: classical, evolutionary, phenetic, and cladistic. This classification of ideas tracks the pre- and post-evolutionary schools of classification.

7.5.1 Classical View

This view, exemplified by Haas and Simpson (1946) and Boyden (1973), is basically that of Owen: "The same organ in different animals under every variety of form and function." As such it suffers from the imprecise nature of Owen's idea of "same" and "essential." The concept has intuitive appeal, but provides no means to exclude or include alternate hypotheses of homology.

7.5.2 Evolutionary Taxonomy

Closely tracking the ideas of the evolutionary synthesis (Sect. 1.14) (Mayr, 1982), the Evolutionary Taxonomy school employed the Remane-type rationale of position, structure, and transformation. Similarity was the only test, yet not all similarities were accepted as homologies since "analogies" (due to convergence) were differentiated from homologues as evolutionary similarities (Bock, 1974, 1977). Parallelisms ("two or more character states derived from a common ancestral state," Hecht and Edwards, 1977), however, could be permitted as homologous. This would seem to exclude very few homology hypotheses, including those of states with independent origins (*e.g.* "wings" of bats and birds), since there was no restriction on the recency of ancestors.

Ernst Mayr
(1904–2005)

Evolutionary Taxonomy as a school was devoted to the study of process and such focus served as an impediment when it came to homology. As made clear by Brady (1985), such evolutionary taxonomists as Ernst Mayr confused the definition of homology with its causative explanation:

> One very important methodological aspect of science is frequently misunderstood and has been a major cause of controversy over such concepts as homology or classification. It is the relation between a definition and the evidence that the definition is met in a particular instance. This is best illustrated by an example: The term

---

[1]$n + 1$ where $n =$ Gaul.

"homologous" existed already prior to 1859, but it acquired its currently accepted meaning only when Darwin established the theory of common descent. Under this theory the biologically most meaningful definition of "homologous" is: "A feature in two or more taxa is homologous when it is derived from the same (or a corresponding) feature of their common ancestor." What is the nature of the evidence that can be used to demonstrate probable homology in a given case? There is a whole set of such criteria (like the position of a structure in relation to others), but it is completely misleading to include such evidence in the definition of "homologous," as has been done by some authors. (Mayr, 1982)

7.5.3 Phenetic Homology

In an effort to avoid the causal and potentially circular definitions of homology presented by Evolutionary Taxonomy, an "operational" homology definition was used by pheneticists Sokal and Sneath (1963); Jardin (1970) and Sneath and Sokal (1973). By avoiding causative, process statements, in principle all "similarities" could be treated as homologies, whether primitive or derived. In practice, however, "compositional and structural correspondence" were used to select those similarities useful for analysis. Basically, this was a return to Owen (1843).

7.5.4 Cladistic Homology

Willi Hennig (1913–1976) receiving AMNH Gold Medal from Director Thomas Nicholson

Donn Rosen (1929–1986), Gareth Nelson, and Norman Platnick.

Hennig (1950, 1966) made more specific distinctions between primitive (plesiomorphy) and derived homology (apomorphy) and non-homologous similarity (analogy, homoplasy) (see Sect. 2.3.6). For Hennig, these distinctions arose out of a transformational notion of homology, "Different characters that are to be regarded as transformation stages of the same original character are generally called homologous." Wiley (1975) extended this concept: "Two (or more) characters are said to be homologous if they are transformation stages of the same original character present in the ancestor of the taxa which display the characters." This differs from the classical ideas of similarity, since the features need not be similar in any particular way, only have common origins. This is also at variance with the evolutionary taxonomist view that two features can have origins in a third (if common) state in a common ancestor. The emphasis is on transformation from generally distributed states to more restricted, derived states. Homologies are apomorphies—synapomorphies in fact.

Homology = Synapomorphy

Regarded as the rebel yell of "pattern" cladistics, the equality of homology and synapomorphy (Nelson and Platnick, 1981) has an older (cf. Patterson, 1982) and more nuanced meaning.

The focus of this definition of homology is in features shared by monophyletic groups. Synapomorphies are, by definition, features (or their states) restricted to a group of taxa that contains all descendants of a common ancestor. As with

Wiley's definition above, an attribute found in the common ancestor and its descendants is homologous and a synapomorphy (*e.g.* wings in pterygote insects). A symplesiomorphy is simply a synapomorphy of a more inclusive group, homologous at that level (*e.g.* having six legs in Pterygota, plesiomorphic at that level, but a synapomorphy for the more including Hexapoda). Hence, a homology is always a synapomorphy, (if at a more general level) and a synapomorphy is always a homology. Furthermore, this definition immediately offers a test. Each homology statement, as a synapomorphy, is tested through congruence with other features. If the weight of evidence were to support an alternate scheme, where members of the group possessing a feature were not monophyletic, the hypothesis of homology would be falsified. Homologies can be proposed by multiple routes, but are always tested and potentially falsified by the same congruence-based process[2]. Brady (1985) traces this idea back to at least Darwin:

> The importance, for classification, of trifling characters, mainly depends on their being correlated with several other characters of more or less importance. The value indeed of an aggregate of characters is very evident in natural history. Hence, as has often been remarked, a species may depart from its allies in several characters, both of high physiological importance and of almost universal prevalence, and yet leave us in no doubt where it should be ranked. Hence, also, it has been found, that a classification founded on any single character, however important that may be, has always failed; for no part of the organization is universally constant. The importance of an aggregate of characters, even when none are important, alone explains, I think, that saying of Linnaeus, that the characters do not give the genus, but the genus gives the characters; for this saying seems founded on an appreciation of many trifling points of resemblance, too slight to be defined. (Darwin, 1859b)

7.5.5 Types of Homology

In his review of homology, Patterson (1982) enumerated several types and modes of homology identification and distinguished these from *tests*. Patterson (1988) further associated terms in use in molecular evolution with their corresponding relation in morphological comparisons. These relations are organized around three "tests" of homology: congruence (agreement with other characters), similarity (likeness in form), and conjunction (discussed above). Only homology (or the molecular, orthology) meets all three criteria (Table 7.1).

DePinna (1991) refined the ideas of Patterson, defining "primary" and "secondary" homology. A primary homology was a putative homology statement that had passed the tests of similarity and position, but was as yet untested on a cladogram. DePinna pointed out that the only real test was that of congruence.

---

[2]Patterson (1982) also proposed "conjunction," the presence of two features in a single organism (as in wings and forelimbs in "angels"), as a potential falsifier of homology. DePinna (1991) and others have pointed out that such angels can occur as forms of homology (*e.g. de novo* character origination), leaving congruence as the only absolute test of homology.

Relation	Congruence Test	Similarity Test	Conjunction Test	Molecular Term
Homology	Pass	Pass	Pass	Orthology
Homonomy	Pass	Pass	Fail	Paralogy
Complement	Pass	Fail	Pass	Complement
Two homologies	Pass	Fail	Fail	Two orthologies
Parallelism	Fail	Pass	Pass	Xenology
Homoeosis	Fail	Pass	Fail	Paraxenology
Convergence	Fail	Fail	Pass	Convergence
Endoparasitism	Fail	Fail	Fail	-

Table 7.1: Combinatorics of homology and non-homology relations after Patterson (1982, 1988).

A secondary homology had passed this test with a unique origin on a parsimonious cladogram.

7.6   Dynamic and Static Homology

Molecular sequence data presented an interesting challenge to homology determination. These sequences could vary not only in state (elements of the sequence) but in length as well. At first, when compared among multiple taxa, such variation was placed in the traditional context of putatively homologous (= comparable) columns via a formal or informal *alignment* process (Chapter 8). However, when these sequences were examined directly on trees, it was immediately obvious that the putative homology schemes could vary among tree topologies. Some scenarios of putative sequence homology require more transformations (however weighted) on a given tree than others. In essence, that the congruence principle existed not only between topologies, but within, and applied to the determination of characters themselves as well at to the disposition of their states. The idea that even putative homology schemes could vary among trees was termed *dynamic homology* (Wheeler, 2001b) to distinguish it from those schemes that were invariant over topologies (such as many complex anatomical attributes, for example, forelimbs in tetrapods). Such dynamic homologies are not limited to molecular data, but can be found in other sequences such as developmental events, call songs in insects, and even "serial" homology of anatomical features in segmented organisms (Schulmeister and Wheeler, 2004; Robillard et al., 2006; Ramírez, 2007).

Salvatore "Lucky" Luciano (1897–1962)

One of the hallmarks of dynamic homology is the lack of transitivity among sequence elements in taxa (Fig. 7.4). While a forelimb in taxon A, B, and C is comparable, that of "the third Guanine residue in taxon A" will have multiple potential comparable elements in other taxa. This was formalized in the definition of static and dynamic characters (Varón et al., 2010):

> *Static homology characters.* Let A and B be two states of a character.
> A correspondence between the elements in A and B is a relation

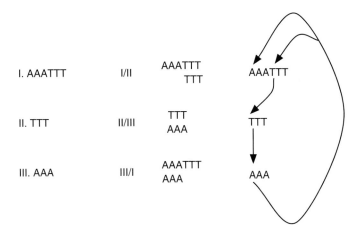

Figure 7.4: Non-transitive nature of the relationship among dynamic homology sequence elements (after Wheeler, 2001b).

between them. We define static homology characters as those in which for every element in A there is at most one corresponding element in B, and the correspondence relations are transitive (*i.e.* let $a \in A$, $b \in B$, and $c \in C$ be elements of different states, where a corresponds to b, and b corresponds to c; then a and c must also correspond to each other). Corresponding elements with the same value match the notion of primary homology (DePinna, 1991).

Dynamic homology characters. We define as dynamic homology characters (Wheeler, 2001b) the complement of their static homology counterparts: for some pair of states A and B, there exists an element $a \in A$ that has more than one corresponding element in B, or the correspondences are not transitive. Dynamic homology characters typically have states that may have different cardinalities, and no putative homology statements among the state elements. These characters formalize the multiple possibilities in the assignment of correspondences (primary homologies) between the elements in a pair of states, which can only be inferred from a transformation series linking the states, and the distance function of choice.

With dynamic homology, the principle of connections of Geoffroy Saint-Hilaire drives the optimality function, and the test is still one of congruence (if more generally). Position, composition, and congruence are simultaneously and quantitatively optimized and tested (as in Brady, 1985).

This leads to a further modification of the homology definition:

Features are homologous when their origins can be traced to a unique transformation on the branch of a cladogram leading to their most recent common ancestor (Wheeler et al., 2006a).

In the end, the central issue is whether aspects of organisms have single or multiple origins. If single, they are homologous; if not, not.

7.7   Exercises

1. Discuss the example of wings in birds, bats, pterosaurs, and moths in terms of the homology definitions discussed above. Under what conditions are these structures homologous?

2. What are the issues involved in the homology of therapod hind limb digits? Discuss these digit homology scenarios in terms of positional, developmental, and dynamic homology frameworks.

3. "Eyes" are broadly distributed, but not uniformly present, in Metazoa. Discuss homology in this context. Are the "eyes" of cephalopods and vertebrates homologous? How would the underlying genetic mechanisms affect this determination?

Chapter 8

Sequence Alignment

8.1 Background

Traditional analysis of sequence data (Chapters 9, 10, 11, and 12) requires an *a priori* scheme of correspondences among the observed sequence elements. Given that sequence data can vary in length, a process is required to convert the variable-length sequence data into constant-length aligned data suitable for this form of character-based optimization. Sequence alignment (or Multiple Sequence Alignment—MSA when there are more that two sequences) is this process.

Sequences can be composed of a variety of objects, from familiar nucleic acid and protein data to observations less frequently treated as sequences such as developmental, behavioral, and chromosomal data. As pre-aligned characters, MSA can be performed under a variety of criteria. This section will focus on cost minimization (= parsimony), and later sections will take up others (Chapters 11 and 12). MSA is not a necessary prerequisite for phylogenetic analysis (Sect. 10.6), but it is a common one.

8.2 "Informal" Alignment

The discussion here is limited to alignment procedures that are precisely defined both in methodology and objective. There will be no discussion of "by-eye" alignment or subjective hand adjustments made to alignments generated with software tools. Such operations, to the extent that they are ill-defined and not based in optimizing objective measures of quality, are better discussed elsewhere.

8.3 Sequences

Sequences are linear arrays of elements. As such, they contain two sorts of information—the type of each element and its position relative to other elements. The positional information can come from any one-dimensional ordering such as body-axis, time and position on a chromosome or within a protein (Fig. 8.1).

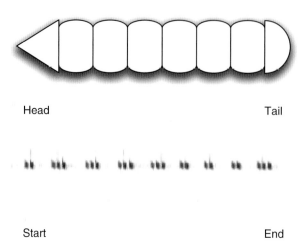

5'-GCAAAAATAACGAT-3'

Head   Tail

Start   End

ND1-16SrDNA-12SrDNA-CytB-ND6

Figure 8.1: Example sequences. Top: Segment of heteropteran 18S rRNA (Wheeler et al., 1993); Middle Top: Generalized arthropod body plan; Middle Bottom: Orthopteran stridulation (Robillard et al., 2006); Bottom: Fragment of mtDNA gene order.

Given that organisms (in general) have defined morphologies (with a body-axis) and undergo growth and development (time-axis), this type of data is quite commonly encountered. When discussing sequences, there are three aspects that must be defined: the sequence alphabet, the transformations it may suffer, and an objective distance function to quantify change.

8.3.1 Alphabets

The set of possible elements that may appear in a sequence defines its alphabet (Σ). Familiar alphabets include those for DNA ($\Sigma = \{A, C, G, T, gap\}$), Morse code ($\Sigma = \{dit, dah, short\ gap, middle\ gap, long\ gap\}$), and arthropod segment appendages ($\Sigma = \{no\ appendages, antennae, legs, mouthparts\}$). Locus synteny maps (such as mtDNA gene order) are sequences of annotated loci with the genes themselves as alphabet elements (*e.g.* 18S rRNA). The set of developmental events in the lifetime of an organism presents another alphabet. A special character representing an empty string of characters (λ)—not an element itself—is often used to represent "gaps" or indels since there are no actual elements present. The gap symbol ("-") may be used interchangeably with λ. Any given sequence is an instantiation of possible elements and is represented as Σ^*.

8.3.2 Transformations

Sequences of elements can undergo several modes of transformation. Four commonly encountered types are substitution, insertion–deletion, inversion, and move (Fig. 8.2). Transformations can occur in complex combinations (*e.g.* *Inversion* + *Move*) making their reconstruction in systematic analysis complex.

(a) $x_0, x_1, x_2, x_3, x_4, x_5 \rightarrow x_0, x_1, x'_2, x'_3, x_4, x_5$

(b) $x_0, x_1, x_2, x_3, x_4, x_5 \rightarrow x_0, x_2, x_3, x_4, x_6, x_5$

(c) $x_0, x_1, x_2, x_3, x_4, x_5 \rightarrow x_0, x_1, \bar{x}_4, \bar{x}_3, \bar{x}_2, x_5$

(d) $x_0, x_1, x_2, x_3, x_4, x_5 \rightarrow x_0, x_3, x_4, x_5, x_1, x_2,$

Figure 8.2: Sequence transformations. (a) Substitution; (b) Insertion–Deletion; (c) Inversion; and (d) Move.

8.3.3 Distances

It is always convenient and often necessary to specify an objective distance function between sequences. These distances are some function (differentially weighted or not) of the transformations involved in converting, or *editing*, one sequence into another. Such edit distances can include substitutions only (*e.g.* Hamming, 1950, or Manhattan) for equal length sequences, or more complex combinations of moves and indels. Distances are usually required to be metric and follow the triangle inequality, among other constraints (Eq. 8.1, see Sect. 5.4).

$$
\begin{aligned}
\forall x \quad d(x,x) &= 0 \\
\forall x, y; x \neq y \quad d(x,y) &> 0 \\
\forall x, y \quad d(x,y) &= d(y,x) \\
\forall x, y, z \quad d(x,y) &\leq d(x,z) + d(z,y)
\end{aligned}
\tag{8.1}
$$

Metricity can be applied at the level of distances among elements or entire sequences. Wheeler (1993) argued that non-metric element distances (as in among indels and nucleotides) would lead to nonsensical results (*e.g.* empty sequence medians; Fig. 8.3).

8.4 Pairwise String Matching

Alignment of sequence pairs is the foundation of all more elaborate procedures. The problem, simply stated, is to create the series of correspondences between

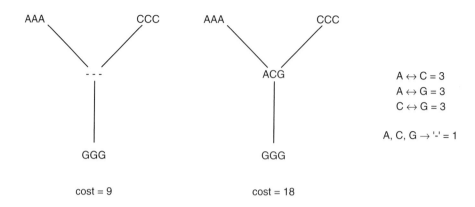

cost = 9   cost = 18

Figure 8.3: Nonsensical sequence median with non-metric indel costs. Since indels cost less than one-half nucleotide substitutions, the lowest cost median will always be an empty (all gap) sequence. All 27 possible medians of length 3 with A, C, and G will yield a cost of 18.

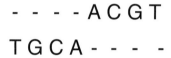

Figure 8.4: Trivial alignment of cost zero when indels have cost = 0 and substitutions cost > 0.

the nucleotides in two sequences via the insertion of gaps, such that the edit cost (the weighted sum of all events—insertions, deletions, nucleotide substitutions—required to convert one sequence into another) between the sequences is minimized (or some other function optimized). Non-zero costs must be assigned to each type of event, or trivial, zero-cost alignments can result (*e.g.* indels costing zero and an alignment that places each nucleotide opposite a gap; Fig. 8.4).

The first algorithmic solution to this form of string-matching problem was proposed by Needleman and Wunsch (1970) and is used throughout most alignment procedures (See Gusfield, 1997, for more extensive discussion).

The procedure follows a dynamic programming approach (Bellman, 1953) by solving a series of small, dependent sub-problems that implicitly examine all possible alignments (Eq. 8.2; Torres et al., 2003).

$$f(n,m) = \sum_{k=0}^{\min(n,m)} 2^k \binom{m}{k} \binom{n}{k} \tag{8.2}$$

$$f(n,m) \approx (1+\sqrt{2})^{2n} \text{ with } n = m$$

Richard Bellman
(1920–1984)

There are two components to the procedure. The first determines the cost of the best alignment (or alignments—there may be multiple solutions) by applying the recursion in Equation 8.3.

$$cost[i][j] = \min \begin{cases} cost[i-1][j-1] + \sigma_{i,j} & \text{match/mismatch} \\ cost[i-1][j] + \sigma_{indel} & \text{insertion} \\ cost[i][j-1] + \sigma_{indel} & \text{deletion} \end{cases} \quad (8.3)$$

This is often referred to as the "wavefront" update (Alg. 8.1). The second is the "traceback" (Alg. 8.2), which yields the alignment itself (more complex examples

---

Algorithm 8.1: PairwiseSequenceAlignmentCost

Data: Input strings X and Y of lengths $|X|$ and $|Y|$
Data: Element distance matrix σ of pairwise substitution costs between
all elements in Σ and λ (indel)
Result: The minimum pairwise alignment cost.
Initialize first row and column of matrices;
$direction[0][0] \leftarrow$ '$\searrow$';
$cost[0][0] \leftarrow 0$;
$length[0][0] \leftarrow 0$;
for $i = 1$ **to** $|X|$ **do**
$\quad cost[i][0] \leftarrow cost[i-1][0] + \sigma_{X_i,\lambda}$;
$\quad direction[i][0] \leftarrow$ '$\rightarrow$';
$\quad length[i][0] \leftarrow length[i-1][0] + 1$;
end
for $j = 1$ **to** $|Y|$ **do**
$\quad cost[0][j] \leftarrow cost[0][j-1] + \sigma_{Y_j,\lambda}$;
$\quad direction[0][j] \leftarrow$ '$\downarrow$';
$\quad length[0][j] \leftarrow length[0][j-1] + 1$;
end
Update remainder of matrices $cost$, $direction$, and $length$;
for $i = 1$ **to** $|X|$ **do**
$\quad$ **for** $j = 1$ **to** $|Y|$ **do**
$\quad\quad ins \leftarrow cost[i-1][j] + \sigma_{X_i,\lambda}$;
$\quad\quad del \leftarrow cost[i][j-1] + \sigma_{Y_j,\lambda}$;
$\quad\quad sub \leftarrow cost[i-1][j-1] + \sigma_{X_i,Y_j}$;
$\quad\quad cost[i][j] \leftarrow \min(ins, del, sub)$;
$\quad\quad$ **if** $cost[i][j] = ins$ **then**
$\quad\quad\quad direction[i][j] \leftarrow$ '$\rightarrow$';
$\quad\quad\quad length[i][j] \leftarrow length[i-1][j] + 1$;
$\quad\quad$ **else if** $cost[i][j] = del$ **then**
$\quad\quad\quad direction[i][j] \leftarrow$ '$\downarrow$';
$\quad\quad\quad length[i][j] \leftarrow length[i][j-1] + 1$;
$\quad\quad$ **else**
$\quad\quad\quad direction[i][j] \leftarrow$ '$\searrow$';
$\quad\quad\quad length[i][j] \leftarrow length[i-1][j-1] + 1$;
$\quad\quad$ **end**
$\quad$ **end**
end
return $cost[|X|][|Y|]$

Algorithm 8.2: PairwiseSequenceAlignmentTraceback

Data: Strings X and Y of Algorithm 8.1
Data: *direction* matrix of Algorithm 8.1
Data: *length* matrix of Algorithm 8.1
Result: X' and Y' contain aligned sequences of X and Y via inclusion of
   gaps
$alignCounter \leftarrow length\,[|X|]\,[|Y|];$
$xCounter \leftarrow |X|;$
$yCounter \leftarrow |Y|;$
while $xCounter \geq 0$ *and* $yCounter \geq 0$ *and* $alignCounter \geq 0$ **do**
  if $direction\,[i]\,[j] = \text{'}\rightarrow\text{'}$ **then**
   $X'\,[alignCounter] \leftarrow X\,[xCounter];$
   $Y'\,[alignCounter] \leftarrow GAP;$
   $xCounter \leftarrow xCounter - 1;$
   $alignCounter \leftarrow alignCounter - 1;$
  else if $direction\,[i]\,[j] = \text{'}\downarrow\text{'}$ **then**
   $X'\,[alignCounter] \leftarrow GAP;$
   $Y'\,[alignCounter] \leftarrow Y\,[yCounter];$
   $yCounter \leftarrow yCounter - 1;$
   $alignCounter \leftarrow alignCounter - 1;$
  else
   $X'\,[alignCounter] \leftarrow X\,[xCounter];$
   $Y'\,[alignCounter] \leftarrow Y\,[yCounter];$
   $xCounter \leftarrow xCounter - 1;$
   $yCounter \leftarrow yCounter - 1;$
   $alignCounter \leftarrow alignCounter - 1;$
  end
end

---

can be found in Phillips et al. (2000)). Needleman and Wunsch described a maximization of identity algorithm, whereas a minimization of difference is presented here. The underlying principles are unchanged.

The first part of the algorithm fills a matrix *cost* of size $(n+1) \times (m+1)$ to align a pair of sequences X and Y of lengths n and m respectively. Each cell (i, j) is the cost of aligning the first i characters of a with the first j characters of b (*i.e.* aligning $x_1 \cdots x_i$ and $y_1 \cdots y_j$). Each value is calculated using the previously aligned subsequences—that is the cost of cell (i, j) will be for indel and substitution cost matrix σ.

The additional first row and column (the reason for the $+1$ in the matrix dimensions) represents the alignment of a sequence with an empty string, that is, initial gaps. Each decision minimum is recorded, to follow the path that leads to the cost of aligning X and Y, that is, the cost in cell (n, m).

In order to create the actual alignment between the sequences, a traceback step is performed that proceeds back up and to the left of the matrix, keeping track of the optimal indels and substitutions performed in the matrix update operations (Alg. 8.2). The minimum cost path is followed back, where the best move is diagonal if the nucleotides of the sequences correspond, and the left and up moves signify indels.

8.4.1 An Example

Consider two sequences "ACGT" and "AGCT" and alignment parameters of nucleotide substitution cost equal to 1 ($\sigma_{substitution}$) and indel cost equal to 10 (σ_{indel}) (Fig. 8.5).

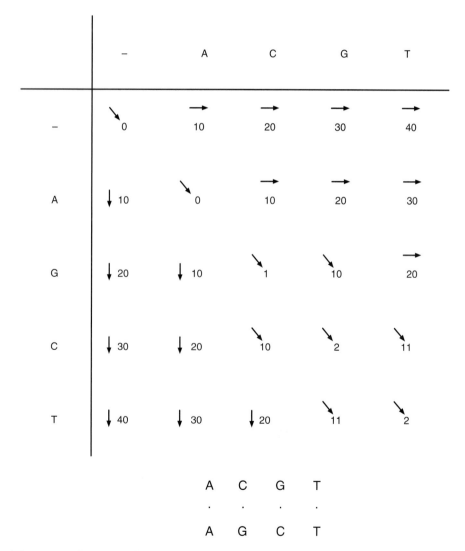

Figure 8.5: Pairwise alignment cost matrix for $\sigma_{substitution} = 1$ and $\sigma_{indel} = 10$. Arrows denote optimal path to each cell.

The minimal cost alignment for these sequences (ACGT and AGCT) with the cost regime indels = 10, substitutions = 1 is two, with two base substitutions implied between the sequences (C↔G, and G↔C).

If a complementary cost scenario is specified, *e.g.* indels = 1 and substitutions = 10, a different optimal solution is found (Fig. 8.6). In this case as well, the minimum cost is two, but no substitutions are implied—only indels (2). Furthermore, there are two equally optimal solutions differing in the placement of the gaps. This ambiguity comes from the equally costly paths found at matrix element 3,3 (of 0,0 to 4,4). The non-unique nature of such solutions is a frequent

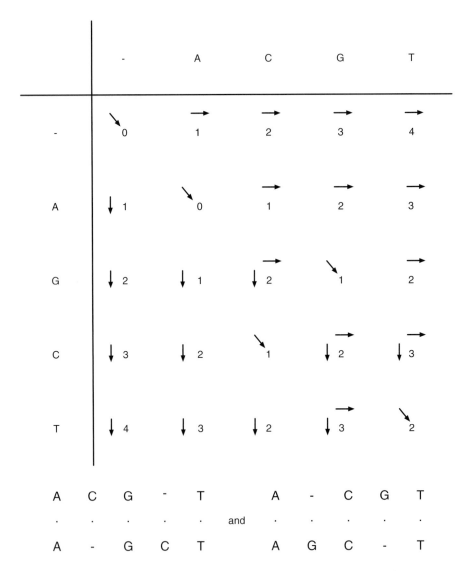

Figure 8.6: Pairwise alignment cost matrix for $\sigma_{substitution} = 10$ and $\sigma_{indel} = 1$. Arrows denote optimal path to each cell.

property of alignments and can have dramatic effects on phylogenetic conclusions (Wheeler, 1994).

The Needleman–Wunsch procedure loops through the length of both input sequences in their entirety, hence has a time (and space) complexity of $O(nm)$ or more simply $O(n^2)$.

8.4.2 Reducing Complexity

The n^2 matrix can be quite large for long sequences, and when they are highly similar, as in most systematic analyses, there is a great deal of the matrix that plays no role in the optimal pairwise alignment. This observation motivated Ukkonen (1985) to improve on the quadratic time complexity of the basic algorithm. Ukkonen showed that when sequences were similar, only a central diagonal (whose width depended on that similarity) was required. The more similar the sequences, the tighter the bound and the faster the alignment.

Algorithm 8.3 shows the outline of the Ukkonen method. In essence, the distance between the sequences (*threshold*) is converted into the number of cells away from the diagonal that are required to ensure a correct result (*barrier*)— minimally the difference in lengths of the input sequences. The alignment proceeds, limited by these barriers (Fig. 8.7). If the cost of the resulting alignment

Esko Ukkonen

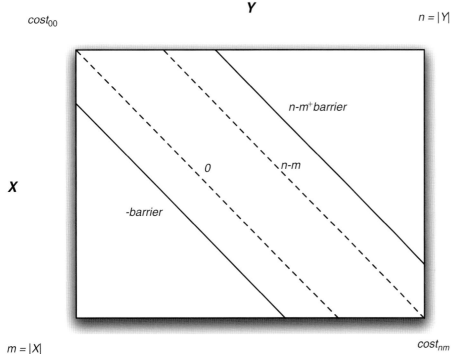

Figure 8.7: Ukkonen barriers in *cost* matrix to align sequences X and Y as in Algorithm 8.3.

Algorithm 8.3: PairwiseSequenceAlignmentUkkonen

Data: Input strings X and Y of lengths $|X|$ and $|Y|$ with $|Y| \geq |X|$

Data: Element distance matrix σ of pairwise substitution costs between all elements in Σ and λ (indel)

Result: The minimum pairwise alignment cost.

$threshold \leftarrow (|Y| - |X| + 1) \cdot \sigma_{indel}$;

$barrier \leftarrow \lfloor \left(\frac{1}{2} \left(\frac{threshold}{\sigma_{indel}} - (|Y| - |X|) \right) \right) \rfloor$;

while $threshold \geq cost[|X|][|Y|]$ **do**

$\quad$ **for** $i = 0$ **to** $|X|$ **do**

$\quad\quad$ **for** $j = max(0, i - barrier)$ **to** $min(i + barrier + (|Y| - |X|), |Y|)$ **do**

$\quad\quad\quad$.

$\quad\quad\quad$ · As in Algorithm 8.1

$\quad\quad\quad$.

$\quad\quad$ **end**

$\quad$ **end**

$\quad$ **if** $threshold \leq cost[|X|][|Y|]$ **then**

$\quad\quad$ $threshold \leftarrow 2 \cdot threshold$;

$\quad\quad$ $barrier \leftarrow \lfloor \left(\frac{1}{2} \left(\frac{threshold}{\sigma_{indel}} - (|Y| - |X|) \right) \right) \rfloor$;

$\quad$ **end**

end

return $cost[|X|][|Y|]$

---

is greater than or equal to the *threshold* value, then the barriers limited the alignment and optimality is not assured. If this occurs, the threshold is doubled and the procedure repeated until the threshold is greater than the cost of the alignment. The time complexity of this Ukkonen's algorithm is $O(nd)$, where d is the distance between the sequences. For related sequences, this can be a huge speed-up.

8.4.3 Other Indel Weights

The Needleman–Wunsch algorithm operates on indel functions where the cost of a gap (w_k) is a linear function of its length (k), $w_k = kw_1$ (w_1 is the cost of a gap of length 1). Other, more complex indel cost functions have been proposed and analyzed, usually resulting in greater time complexity as the cost for additional flexibility or realism.

Non-Linear

Waterman et al. (1976) examined a large number of possible sequence metrics and described algorithms for $w_k \leq kw_1$ with time complexity $O(n^3)$. The reason for this additional complexity has to do with the non-independent nature of the indels. Briefly, in addition to the *cost* matrix of the basic algorithm, two additional matrices are required (P and Q) that track the cost of indels of varying length (Eq. 8.4). These matrices are recursed through repeatedly for each element in $cost$, greatly adding to execution time.

$$
\begin{aligned}
cost_{i,j} &= \min\left(cost_{i-1,j-1} + \sigma_{X_i,Y_j}, P_{i,j}, Q_{i,j}\right) & (8.4)\\
P_{i,j} &= \min_{1 \le k \le i}\left(cost_{i-k,j} + w_k\right)\\
Q_{i,j} &= \min_{1 \le k \le j}\left(cost_{i,j-k} + w_k\right)
\end{aligned}
$$

Affine

A special case of $w_k \le kw_1$, where gaps have the cost function $w_k = uk + v$ ($u, v \ge 0$), is often referred to as an "affine" function. Gotoh (1982) was able to show that for this restricted situation, multiple traversals through P and Q were not required (Eq. 8.5), yielding a time complexity of $O(n^2)$ (but with a constant factor of 3 over atomic indels).

$$
\begin{aligned}
cost_{i,j} &= \min\left(cost_{i-1,j-1} + \sigma_{X_i,Y_j}, P'_{i,j}, Q'_{i,j}\right) & (8.5)\\
P'_{i,j} &= \min\left(cost_{i-k,j} + w_1, P'_{i-1,j} + u\right)\\
Q'_{i,j} &= \min\left(cost_{i,j-k} + w_1, Q'_{i,j-1} + u\right)
\end{aligned}
$$

8.5 Multiple Sequence Alignment

Other than techniques that operate on pairwise distances (Chapter 9), most sequence-based tree reconstruction methods begin with a multiple sequence alignment (MSA). This is not necessary (Sections 10.6 and 11.5), but is often the case. There are potentially huge numbers of MSAs (for m sequences of lengths $n_1, \ldots, n_m$; Eq. 8.6; Table 8.1; Slowinski, 1998),

$$
f(n_1, \ldots, n_m) = \sum_{N=\max(n_1,\ldots,n_m)}^{\sum_{k=1}^{m} n_k} \sum_{i=0}^{N} (-1)^i \binom{N}{i} \prod_{j=1}^{m} \binom{N-i}{N-n_j-i} \qquad (8.6)
$$

and the problems involved in constructing them are legion and non-obvious.

m	2	3	4	5
$n=$ 1	3	13	75	541
2	13	409	23917	2244361
3	63	16081	10681263	14638756721
4	321	699121	5552351121	117629959485121
5	1683	32193253	3147728203035	1.05×10^{18}
10	8097453	9850349744182729	3.32×10^{26}	1.35×10^{38}

Table 8.1: Number of multiple sequence alignments for *very* small data sets (m sequences of length n).

David Sankoff

Lusheng Wang

The first problem to present itself is how to assess the quality of a given MSA, with each option implying different meaning to the MSA. There are three basic approaches: "sum-of-pairs" (SP), "consensus," and "tree alignment" (Gusfield, 1997). Each of these criteria assumes metricity in sequence distances (stochastic model-based versions of these optimality criteria are discussed in Chapters 11 and 12).

SP–alignment (Carrillo and Lipman, 1988) seeks to minimize the pairwise distance (substitutions and indels, potentially weighted) summed over all sequence pairs in the MSA. This distance is calculated based on the multiply aligned sequences (as opposed to their individual pairwise alignments). To create a multiple alignment, pairwise alignment can be generalized in a straightforward fashion to align more than two sequences. The matrix would have an axis for each sequence (k sequences would require k dimensions), and there would be 2^{k-1} paths to each cell representing all the possible combinations of gaps and substitutions possible (seven in the case of three sequences). These two factors add enormously to the calculations ($O\left(n^k 2^{k-1}\right)$), making multidimensional alignments unattainable for real data sets. Furthermore, Wang and Jiang (1994) have shown that the exact SP–problem is NP–complete, so solutions for non-trivial data sets are unlikely ever to be identified. Gusfield (1997) defined a method, lifted alignment, with a guaranteed bound of not more than twice the optimal SP–score (assuming metric element distance). MUSCLE (Edgar, 2004b,a) and SAGA (Notredame and Higgins, 1996) are SP–alignment programs.

Consensus alignment methods seek to create an MSA such that the summed distances between the aligned sequences and a consensus sequence is minimized. This sequence can be a central sequence in a "star" (a Steiner sequence) or one created by the most frequent element in each aligned position (plurality character). Again, assuming element metricity, a guaranteed bound of a factor of two can be found for consensus as for SP, and by the same "lifting" procedure (Gusfield, 1997). Consensus MSAs can also be scored using the Shannon entropy measure (Shannon, 1950). The score (s_{MSA}) is based on treating each multiply aligned column as independent and calculating the logarithm of the probability of the number of elements in the column calculated from their frequency in each aligned position (Eq. 8.7 with n_{ij} of element j in aligned position i).

$$pr_i = \prod_{j=1}^{elements} \text{freq}_{ij}^{n_{ij}} \qquad (8.7)$$

$$s_{MSA} = - \sum_{i=1}^{columns} \sum_{j=1}^{elements} n_{ij} \log pr_{ij}$$

Tree alignment (Sankoff, 1975) places sequences to be aligned on the leaves of a tree, seeking to minimize the pairwise distances along the edges. This tree is the phylogenetic tree sought in systematic analysis, hence is the ideal form of MSA in systematics.

8.5.1 The Tree Alignment Problem

The problem of determining the minimum cost of a tree ($T = (V, E)$) for a set of terminal (leaf, $L \subseteq V$) sequences and a specified distance function (d) is known as the Tree Alignment Problem (TAP; Sankoff, 1975). The key operation is the construction of vertex sequences (medians) such that the pairwise distance between vertices (V) along the edges (E) is minimal (Fig. 8.8). If the distance is Hamming (hence sequences equal in length), the calculation of this cost is straightforward. If, however, the set of allowed transformations includes indels, the optimization (*i.e.* finding the minimum cost) is NP–hard (Wang and Jiang, 1994) and cannot (unless P = NP) be guaranteed. This is the situation we are faced with in molecular sequence data, and the problem in which we are most interested. The combinatorial complexity of TAP comes from the exponential number of possible sequence medians at the internal vertices. There may be, in fact, an exponential number of minimum cost solutions. MALIGN (Wheeler and Gladstein, 1998) and POY (Wheeler et al., 2005; Varón et al., 2008, 2010) are Tree–alignment programs.

Tao Jiang

8.5.2 Trees and Alignment

The TAP was originally described in terms of identifying the alignment, given a tree, that is of minimal cost. Each median assignment, however, has an associated alignment ("Implied Alignment"; Wheeler, 2003a), hence there are, potentially, an exponential number of minimum cost alignments for any given tree. Even a tiny data set (for 10 sequences of length 5)—an unrealistically

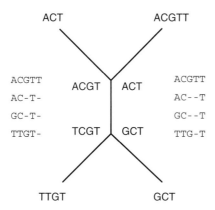

Figure 8.8: Example tree alignment. Either set of medians (ACGT, TCGT; ACT, GCT) yields a cost of 6 transformations (indels and substitutions). The two alignments also result in a tree cost of 6.

small and well-behaved case—presents 1.35×10^{38} possible homology schemes (Slowinski, 1998). This problem, for a single cladogram, is as hard as the more familiar tree search problem itself.

8.5.3   Exact Solutions

Sankoff and Cedergren (1983) proposed a recursive, exact solution to TAP with $O(n^k)$ (k sequences of length n) time complexity based on the recursion in Equation 8.8. Clearly, this was not likely to be useful for non-trivial data sets given the k-dimensional sequence alignment. Wheeler (2003c) proposed another exact solution, also using dynamic programming. In this case, the time complexity is exponential in sequence length ($O(k \cdot 4^{2n})$) due to the explicit enumeration of possible medians. Both methods could be improved by a branch-and-bound type procedure, perhaps based on some generalized form of Ukkonen (1985).

$$d_{i,j,\dots,k} = \min_{\delta_1 + \dots + \delta_n \neq 0} \left(\begin{array}{l} d_{i-\delta_1,\dots,k-\delta_N} + \min_{x_1,\dots,x_m} \\ \times \text{ (cost of } 2^n \text{ vertex state assignments)} \end{array} \right) \quad (8.8)$$

$$\text{where } \delta_i = 0 \text{ if gap, 1 otherwise}$$

8.5.4   Polynomial Time Approximate Schemes

Of theoretical use, but almost no practical utility due to the large time complexity, Polynomial Time Approximate Schemes (PTAS) are methods with known bounded behavior for TAP (Wang et al., 1996; Wang and Gusfield, 1997; Wang et al., 2000). These methods use a combination of "lifted" alignment (Sect. 10.9.3) and exact $O(n^k)$ solutions to create algorithms that achieve provable boundedness, at an increasing cost of time complexity. Currently (Wang et al., 2000), the best procedure is $O(kdn^5)$ (for k sequences of length n on a tree with depth d) for a bound of 1.5 over the optimal solution. Clearly, an $O(n^5)$ procedure will not be useful for real data sets of thousands of nucleotides, but this does provide a benchmark to evaluate the bounds and time complexity of other heuristic procedures.

8.5.5   Heuristic Multiple Sequence Alignment

Current heuristic procedures are similar in that many attempt to render multiple alignment tractable by breaking down simultaneous n–dimensional alignments into a series of manageable pairwise alignments related by a "guide tree" (in the parlance of Feng and Doolittle, 1987). These differ in the techniques used to generate the guide tree and conduct the pairwise alignments at the guide tree nodes ("profile alignment"). Furthermore, the procedures may or may not be explicitly linked to optimality criteria. Below several commonly (and some less commonly) used methods are discussed. This discussion is by no means exhaustive and restricts itself to the algorithmic procedures as opposed to execution time efficiency. More complete lists can be found at `http://pbil.univ-lyon1.fr/alignment.html` and additional comparisons in Phillips et al. (2000).

Additionally, methods that deal with simultaneous tree reconstruction and alignment (*e.g.* POY, Wheeler, 1996; Varón et al., 2008, Hein et al., 2003; BAli-Phy, Redelings and Suchard, 2005; Suchard and Redelings, 2006) will be discussed in other sections (Chapters 10, 11 and 12). Hidden Markov Model methods fall outside the metric tree-alignment paradigm and will be discussed with other likelihood and Bayesian techniques (Chapters 11 and 12).

8.5.6   Implementations

Common Operations

Most, but not all, MSA implementations share core components. These include guide trees, progressive alignment, profile alignment, refinement, and optimality (scoring) criteria. These components may vary in detail, but most MSA methods can be constructed from combinations of these fundamental operations (Fig. 8.9).

Guide Trees are directed graphs that specify an order of pairwise alignments to build up a multiple alignment. Each vertex on the guide tree signifies the

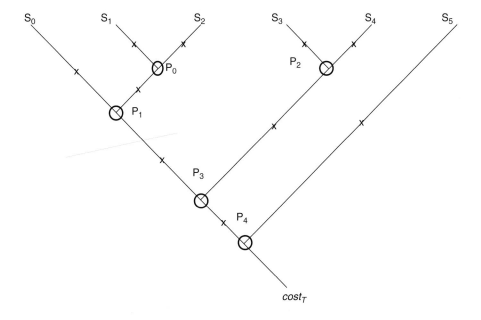

Figure 8.9: Common components of heuristic MSA implementations. Observed sequences $(S_0, \ldots, S_5)$ are placed as leaves of a guide tree T and profile alignments $(P_0, \ldots, P_4)$ created at each non-leaf vertex as the process moves progressively down the tree (leaves to root, post-order) resulting in a full MSA at P_4 with cost $cost_T$. Each edge may be revisited in a refinement step by deleting each edge in turn ("x") and recreating it based on pairwise alignment of its two connected vertices.

alignment of the two child vertex sequences (usually $O(n^2)$ or so). The tree is traversed post-order (leaves to root) adding up alignments until the root is reached with the complete MSA. Since the guide tree determines alignment order, and there are so many possibilities (Eq. 2.1), the choice of guide tree is an important aspect of MSA implementation. Usually, an initial guide tree is constructed from pairwise distances via Unweighted Pair Group Method using arithmetic Averages (UPGMA) (Sokal and Michener, 1958), Fitch–Margoliash (Fitch and Margoliash, 1967), Neighbor-Joining (Saitou and Nei, 1987), or an other distance-based technique (Chapter 9), since distances can be based on pairwise alignments. The initial guide tree may be revised based on an initial MSA with this second tree constructed via the same distance-based techniques.

Progressive Alignment is the alignment process based on the guide tree (Feng and Doolittle, 1987). As mentioned above, each vertex on the guide tree requires a pairwise alignment, hence the overall complexity for m sequences is $O(mn^2)$ for each MSA. For vertices whose descendants are not exclusively leaves, alignment techniques are required to align sequences to alignments and alignments to other alignments. In order for the heuristic to be useful, these must be roughly quadratic, hence sub-alignments are aligned by the same basic pairwise technique, but with the entire aligned *column* acting as the single element in standard procedures. This may be done on partial alignments, or "profiles," which are MSAs reduced to single strings (Fig. 8.10).

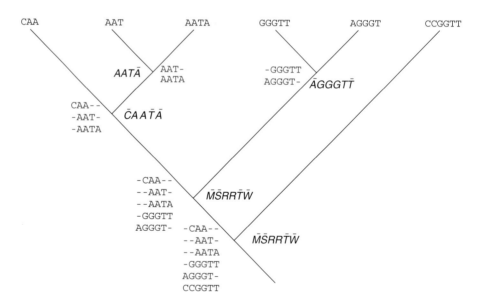

Figure 8.10: Progressive alignment of Feng and Doolittle (1987) where pairwise alignments are performed via post-order tree traversal at non-leaf vertices to create partial MSAs following "once a gap, always a gap" (left of vertices) or profile sequences (right of vertices) of IUPAC symbols for each column with a "bar" if a gap is also present in that aligned column.

Profile alignment. Sequence profiles are used during progressive alignment to allow the alignment of two partial MSAs (with l and m sequences of length n) or a sequence and an MSA in quadratic rather than $O(n^{lm})$ time.

Refinement improves the initial MSA through realignment along guide tree edges, recombining candidate alignments, or modifying the guide tree itself. Usually, refinement proceeds until no improvement (by optimality score) can be found.

Optimality Criterion, or score, is the objective function used to measure MSA quality. As mentioned above, frequently encountered measures are sum-of-pairs (SP), Shannon entropy, and phylogenetic tree cost.

MSA Software

Below are described a few (there are a large number) MSA implementations. This list is not intended to be exhaustive, but illustrative (Fig. 8.11).

CLUSTAL (Higgins and Sharp, 1988; Thompson et al., 1994) creates a single multiple alignment based on a single guide tree. A Neighbor-Joining (NJ) tree (Saitou and Nei, 1987) is calculated from the pairwise alignments. Internal profile sequences are consensus sequences of their descendent partial alignments. When profiles are aligned at internal vertices, the average distance (potentially weighted) between elements in each profile is used in pairwise alignment. There is no optimality value associated with a CLUSTAL alignment (CLUSTAL 2.0, however, has additional features).

TREEALIGN (Hein, 1989a,b) also produces a single multiple alignment based on a single guide tree, but by keeping all potential sequence medians in an "alignment graph." The method is potentially very strong in solving the Tree Alignment Problem, but due to the very large number of medians for real data sets, does not scale beyond small collections of relatively short sequences.

MALIGN (Wheeler and Gladstein, 1998) uses multiple guide trees to generate a diversity of multiple sequence alignments, choosing the "best" on the basis of the parsimony score (indels included) of the most parsimonious cladogram derived from that alignment. Hence, MALIGN is a Tree Alignment program. Guide trees are "searched" (using standard tree refinement procedures such as branch-swapping and randomization) and multiple alignments created for each candidate guide tree. Partial multiple alignments are pairwise aligned at guide tree vertices based on the union of elements in aligned columns as profiles. Each alignment is used as the basis for a heuristic cladogram search (indels weighted and included). The cost of the most parsimonious cladogram is attached to the alignment as its optimality score. MALIGN will output multiple multiple-alignments if they are equally optimal.

POY (Wheeler et al., 2005; Varón et al., 2008, 2010) explicitly attempts to solve the Tree Alignment Problem. As such it is not an alignment program *per se*, but complete phylogenetic tree search software. POY can output an *Implied Alignment* (Wheeler, 2003a) for a given tree or trees.

SAGA (Notredame and Higgins, 1996) takes a different approach, eschewing progressive alignment and guide trees altogether in favor of a Genetical-Algorithm (Holland, 1975) approach to directly improve the MSA SP–score.

SAGA generates an initial set (generation) of MSAs via a random gap padding process. These MSAs are evaluated via SP, and undergo generations of breeding, mutation, recombination, and selection. This cycle is continued until the MSA population stabilizes.

MUSCLE (Edgar, 2004b,a) seeks to minimize the SP–score, and adds guide tree edge (*i.e.* iterative) refinement to the basic progressive alignment paradigm. Like CLUSTAL, MUSCLE creates an initial guide tree, but uses a rapid UP-GMA or NJ approach. Progressive alignment is then employed using the log-expectation (LE) score. This initial MSA is used to create a second guide tree again with UPGMA or NJ but using Kimura (1983) distance, and a second MSA via the LE score and progressive alignment. At this point, MUSCLE performs an edge-based refinement. In each step, an edge is removed from the guide tree following the procedure of Hirosawa et al. (1995), and profiles created at the incident vertices as if they were roots of two separate trees. These are aligned, calculating the SP–score. The process is repeated until no further improvements are found.

MAFFT (Katoh et al., 2002) operates in much the same way that MUSCLE does. MAFFT differs in using a Fourier transform method to both identify common sequence motifs (reducing pairwise sequence alignment time complexity) and for profile and pairwise alignments. MAFFT uses NJ to create guide trees.

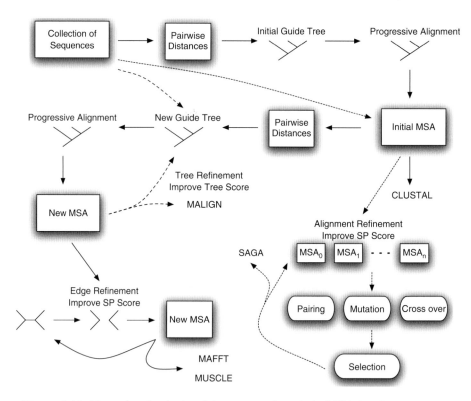

Figure 8.11: Procedural relationships among heuristic MSA implementations.

MAFFT also goes through two rounds of initial guide tree and MSA generation via progressive alignment, followed by edge refinement (Hirosawa et al., 1995) to improve a weighted SP–score.

DIALIGN (Morgenstern et al., 1996; Morgenstern, 1999) differs from other methods in looking for alignments of contiguous gap-free fragments of DNA that may have mismatches. This contrasts with the approach that attempts to globally align each position in a sequence. No gap penalty is employed. The idea behind this method is to create complete alignments by stitching together locally similar sequences that may be separated by highly divergent regions. An optimal alignment is one that maximizes the weighted sum of the matches in the smaller segments. Alignments can be compared on this basis. This method makes no reference to cladograms or trees whatsoever.

COFFEE (Notredame et al., 1998, 2000) behaves as a "wrapper" or meta-algorithm, using a genetic algorithm to optimize multiple alignments based on consistency with the pairwise alignments of the same sequences. Any pairwise alignment procedure can be used under the COFFEE optimality function.

8.5.7 Structural Alignment

Many of the most common molecular sequences submitted to MSA are those of structural RNA such as transfer RNAs (tRNAs) and ribosomal RNAs (rRNAs). These sequences are not translated into protein and interact as mature molecules in protein synthesis. These interactions depend on higher order structure created by the internal nucleotide pairing of the single-stranded RNA. At its most simple level, RNA secondary structure is centered around single-stranded "loop" regions and double-stranded "stem" or "helix" regions (Fig. 8.12). For all but the smallest RNAs, secondary structures are reconstructed computationally-based on the minimization of the combined free energy of stem and loop regions. As with the primary sequence data, the secondary structures provide historical information and can inform alignment and tree reconstruction. Sankoff (1985) presented an algorithm to simultaneously optimize the MSA, sequence medians, and sequence folding energy. These three problems can be solved by dynamic programming and the Sankoff algorithm does so in $O(n^{3m})$ time and $O(n^{2m})$ space for m sequences of length n [1]. The solution is based on an enhanced optimality criterion for the tree alignment based on the structural folding energy. The cost of the alignment, C_A, is equal to the traditional parsimony cost summed over edge distances, D_T, added to the energy of the median sequence structures, $E(s)$. In this way, there is an explicit trade-off between pure tree alignment cost and the energy of the reconstructed structures—a true joint solution. There are limits to this procedure, the most prominent of which is that the structures need to be broadly similar. This is to allow for the correspondences and indels of structural elements (subsequences). Additionally, some forms of 'pseudo-knotting' (see below) are forbidden and these may occur in some RNA structures.

---

[1] This is not to be confused with the informal "by-eye" manual alignment procedures said to be structural. These are clearly *ad hoc* in nature (*e.g.* Kjer, 2004).

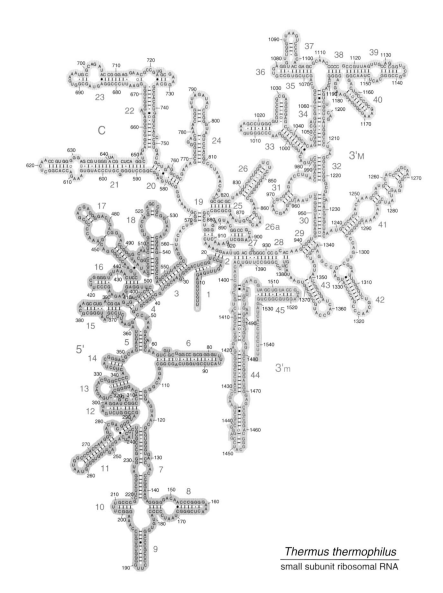

Figure 8.12: Secondary structure model for *Thermus thermophilus*. `http://rna.ucsc.edu/rnacenter/ribosome_images.html`. See Plate 8.12 for the color figure.

Sankoff Algorithm

The Sankoff (1985) algorithm is based on three recursions: alignment, folding, and median sequence reconstruction. These are combined to solve the joint problem simultaneously.

- Alignment—The core sequence alignment step in this context is slightly generalized from that discussed above (Eq. 8.3) in that the cost is defined for *subsequences*. For sequences $a = \{a_i, \ldots, a_m\}$ and $a = \{b_1, \ldots, b_n\}$ and $1 \le i \le j \le m$ and $1 \le h \le k \le n$, $D(i, j; h, k)$ is the minimum alignment cost between partial sequences $a_i, \ldots, a_j$ and $b_h, \ldots, b_k$. D can be calculated via the recursion relationships:

$$D(i, j; h, k) = \min \begin{cases} D(i, j; h, k-1) + y & \\ D(i, j-1; h, k-1) + x & \text{if } a_j \ne b_k \\ D(i, j-1; h, k-1) & \text{if } a_j = b_k \\ D(i, j-1; h, k) + y & \\ \\ D(i+1, j; h, k) + y & \\ D(i+1, j; h+1, k) + x & \text{if } a_i \ne b_h \\ D(i+1, j; h+1, k) & \text{if } a_i - b_h \\ D(i, j; h+1, k) + y & \end{cases} \quad (8.9)$$

with base indel cost y and substitution cost x, and initial conditions:

$$D(i, i; h, h) = \begin{cases} x & \text{if } a_i \ne b_h \\ 0 & \text{if } a_i = b_h \end{cases} \quad (8.10)$$

This will have a time complexity of $O(m^2 n^2)$ or approximately $O(n^4)$ when $m \sim n$. In previous sections, we dealt with the special case of whole sequences or $D(1, m; 1, n)$, which then has quadratic time complexity only requiring the upper four terms in Equation 8.9.

- Folding—The central idea behind secondary structure reconstruction is the minimization of the Gibbs Free Energy of the structure. This is determined by the internal pairing of nucleotides and the geometry of loop and stem regions. The structure S of a sequence a is defined as a set of pairs (i, j) that creates bonds between the nucleotides with an experimentally determined energy $e(s)$. These pairs must satisfy the "knot constraint" where if $i \le i' \le j \le j'$, (i, j) and (i', j') cannot be distinct elements of S (Fig. 8.13).

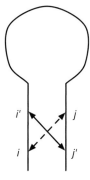

Figure 8.13: "Knot" in secondary structure between structure pairs (i, j) and (i', j') where $i \le i' \le j \le j'$.

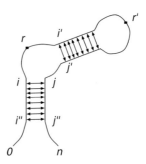

Figure 8.14: r is accessible from (i, j) while r' is not. The k-loop closed by (i, j) would be all z where $i < z < i'$ and $j' < z < j$. Those in position z' where $0 \leq z' < i''$ and $j'' < z' \leq n$ are external.

Several additional terms need to be defined before setting up the folding recursion: *accessible*, *k-loop*, and *external* (Fig. 8.14). Given $(i, j) \in S$ and $i < r < j$, if there is no $(i', j') \in S$ where $i < i' < r < j' < j$ then r is accessible from (i, j). The pair (p, q) is accessible if both p and q are. The k-loop closed by (i, j) comprises the $u \geq 0$ unpaired positions and the $k - 1 \geq 0$ accessible pairs. The k-loops correspond to hairpins ($k = 1$), bulges, and interior loops ($k = 2, u > 0$), stacked pairs ($k = 2, u = 0$), and multiple loops ($k \geq 3$).

The free energy of a given loop can be approximated by the function

$$e(s) = A + (k - 1)P + uQ \tag{8.11}$$

with the experimentally determined A, P, Q and structural u, k from above.

These are the components of the recurrence relationship to determine $F(i, j)$, the minimum free energy for the secondary structure S on the partial sequence $i, \ldots, j$.

$$F(i, j) = \min \begin{cases} C(i, j) \\ \min_{i \leq h < j}\{F(i, h) + F(h + 1, j)\} \end{cases} \tag{8.12}$$

$$C(i, j) = \min \begin{cases} e(s), \text{ hairpin closed by } (i, j) \\ \min\{e(s) + C(p, q)\}, \text{ 2-loop closed by } (i, j) \\ \qquad \text{and } (p, q) \text{ accessible} \\ \min_{i < h < j-1}\{G(i + 1, h) + G(h + 1, j - 1) + A\} \end{cases}$$

$$G(i, j) = \min \begin{cases} C(i, j) + P \\ \\ \min_{i \leq h < j} \min \begin{cases} G(i, h) + (j - h)Q \\ G(i, h) + G(h + 1, j) \\ (h - i + 1)Q + G(h + 1, j) \end{cases} \end{cases}$$

These have initial conditions $F(i, i) = 0$, $C(i, i) = \infty$, and $G(i, i) = \infty$. The minimum free energy of the entire structure would then be $F(1, n)$.

- Medians—Median sequences (= hypothetical ancestors or HTUs) are constructed to be the lowest cost (= most parsimonious) assignments given a tree and the alignment as in Section 8.5.1. The alignment cost D from above (Eq. 8.10) is generalized to be the cost over the edges of the phylogenetic tree between vertices (Eq. 8.13).

$$D\overrightarrow{(i, j)} = \min\nolimits_{not\forall j'_r = j_r} \{D\overrightarrow{(i', j')} + \gamma\overrightarrow{(a_j(j - j'))}\} \qquad (8.13)$$

In this situation, the medians are derived from a fixed tree and alignment, hence can be determined in $O(nm)$ for n sequences of length m.

In combining the three above elements, there are two points worth keeping in mind. The first is that the indel and match/substitution costs of the alignment operations need to be set in such a way to be comparable with the free energy values calculated for the structure. Only in this way can these values be summed to determine the overall optimality value. The second is that the secondary structures are required to maintain a great deal of higher-level identity. Individual loops must be inserted or deleted *in toto*, and loops may only correspond with individual loops in other sequences. A single loop in one sequence cannot correspond to multiple sequences in another.

When we combine these elements, we are presented with a rather complex recursion (Eq. 8.14) for $F\overrightarrow{(i, j)}$, the minimum cost alignment and secondary structure for N leaf sequences on a tree.

$$F\overrightarrow{(i, j)} = \min \begin{cases} C\overrightarrow{(i, j)} + \gamma(a_{i_1}^{(1)}, \ldots, a_{i_N}^{(N)}) + \gamma(a_{j_1}^{(1)}, \ldots, a_{j_N}^{(N)}) \\ \min_{i_r \leq h_r < j_r} \{F\overrightarrow{(i, h)} + F\overrightarrow{(h + 1, j)}\} \\ D\overrightarrow{(i, j)} \end{cases} \qquad (8.14)$$

$$C\overrightarrow{(i, j)} = \min \begin{cases} \sum e(s_r) + D\overrightarrow{(i + 1, j - 1)} \\ \min_{not\forall (p_r, q_r) = (i_r, j_r)} \\ \qquad \times \{\sum e(s_r) + C(p, q) + D\overrightarrow{(i + 1, p)} + D\overrightarrow{(q, j - 1)}\} \\ \min_{i_r < h_r < j_r - 1} \{G\overrightarrow{(i + 1, h)} + G\overrightarrow{(h + 1, j - 1)} + NA\} \end{cases}$$

$$G\overrightarrow{(i, j)} = \min \begin{cases} C(i, j) + NP + \gamma(a_{i_1}^{(1)}, \ldots, a_{i_N}^{(N)}) + \gamma(a_{j_1}^{(1)}, \ldots, a_{j_N}^{(N)}) \\ \min_{i_r \leq h_r < j_r} \min \begin{cases} G\overrightarrow{(i, h)} + \sum (j_r - h_r)Q + D\overrightarrow{(h + 1, j)} \\ G\overrightarrow{(i, h)} + G\overrightarrow{(h + 1, j)} \\ \sum (h_r - i_r + 1)Q + G\overrightarrow{(h + 1, j)} + D\overrightarrow{(i, h)} \end{cases} \end{cases}$$

These have initial conditions $\overrightarrow{F(i,i)} = 0$, $\overrightarrow{C(i,i)} = \infty$, and $\overrightarrow{G(i,i)} = \infty$ if any $i_r = j_r$.

The overall time complexity of the Sankoff procedure is high—$O(n^6)$ for two sequences of length n, and the MSA case even more daunting, with exponential dependency on the number of taxa (N), $O(n^{3N})$.

Implementations

The basic approach of most implementations of structural alignment is the Sankoff algorithm. In order to make the problem tractable, however, simplifying restrictions are made. Several of these approaches are compared in Gardner and Giegerich (2004).

FOLDALIGN (Gorodkin et al., 1997) reduces the pairwise alignment complexity by not allowing branching structures. In doing so, the time complexity of aligning two sequences is reduced by a quadratic factor to $O(n^4)$. The lack of branching in the structures effectively limits the operation to relatively short sequences. A further reduction in the exponential factor comes from the use of progressive alignment (for N sequences, $O(N)$) after determining the all-pairs costs ($O(N^2)$). The result is a procedure akin to CLUSTAL, but with a structural component.

DYNALIGN (Mathews and Turner, 2002) is limited to pairwise alignment, but is free of the structural restrictions in FOLDALIGN. DYNALIGN limits the distance between aligned nucleotide positions to m, thereby reducing the pairwise Sankoff time complexity from $O(n^6)$ to $O(m^3n^3)$, with m^3 as a constant factor.

PMCOMP/PMMULTI (Hofacker et al., 2004) uses a probabilistic pairing model to calculate initial pairwise alignments and then uses progressive alignment to create the full multiple alignment. In the progressive alignment step, the same restriction on distances between subsequence alignments ($D(i,j;h,k)$ above) used in DYNALIGN is applied, reducing time complexity from $O(n^6)$ to $O(n^4)$. A further factor of N sequences is added for the progressive step, hence, the overall time complexity is $O(n^4N)$.

RNAcast (Reeder and Giegerich, 2005) is based on a "consensus shape" approach where the reconstructed structures are limited to those common to a body of sequences. This reduces the space of possible structures and the specific structures of sequences are optimized within this space. RNAcast does *not* align the input sequences, but creates a set of consensus structural elements.

RNASalsa (Stocsits et al., 2009) is a meta-algorithm implementing a refinement procedure based on input alignments and structural models. An aligned set of sequences are input with structural information demarcated for one of the input sequences. RNASalsa first generalizes the single input structure to the other sequences in the input set based on the input alignment. Second, refined structural reconstructions are performed on the input sequences (using the same probability structure codes in PMCOMP above), given the constraints based on the generalized input structure. The third and final step progressively aligns the sequences based on the refined structural models. Given the reliance on input alignment and structural information, the main time complexity factors are in the structural refinements ($O(n^4)$ for n nucleotides) and progressive alignment ($O(N)$ for N taxa).

Comparisons—What is a Good Alignment?

There are myriad comparisons and comparison methodologies that have been published (*e.g.* Notredame, 2002; Batzoglou, 2005; Wheeler, 2007a) and I have no desire to create another one here. Comparisons center on two sorts of analyses: 1) whether one procedure or another is more similar to a "real" or simulated alignment (Ogden and Rosenberg, 2007), and 2) whether competing methods optimize an objective function (such as SP or tree-length) more effectively or in a more timely fashion (Boujenfa et al., 2008; Wheeler and Giribet, 2009).

It seems most unlikely that we will ever encounter a "real" alignment in nature, hence comparison of procedures that are at least nominally optimization-based to simulated, "known" alignment seems fruitless, but see Ogden and Rosenberg (2007) for a different view. Even comparison on the basis of objective functions can be difficult, since the various implementations do not attempt to optimize precisely the same functions.

As far as systematic analysis is concerned, the central goal is to identify optimal trees. Those methods that yield better solutions to this problem, no matter what optimality criterion one chooses, are to be favored. Most likely, the best way a method can do this effectively is to directly attempt to optimize tree costs—the Tree Alignment Problem itself. It is worth remembering that even for a single tree, there may be an exponential number of equally costly alignments. This reality is rarely explored, and its impact on the potentially equally large number of tree solutions is largely unexamined. Multiple alignment methods offer one set of heuristics to this problem; in later sections we will encounter others.

8.6 Exercises

1. Align the two sequences "ACGTTA" and "TCTA" by hand using the cost scenarios where all transformations are equally costly.

2. What is the best method to choose a MSA? Should this be optimality-based?

3. Choose a small data set and run it through several MSA programs. How do they differ? Change indel and gap extension parameters. How do the alignments change?

4. Are MSA columns independent?

5. What are some implications for phylogenetic analysis of affine indel models?

6. Is hand ("by-eye") alignment adequate for phylogenetic analysis?

7. Are there any cases where it would be advisable/defensible to discard sequence data?

Part III

Optimality Criteria

Chapter 9

Optimality Criteria–Distance

Distance analysis, as opposed to character-based analysis, reconstructs phylogenetic relationships based on the pairwise overall distances between taxa. This approach has a long history, beginning with the first computational approaches to systematics and the early use of molecular biological data to create phylogenetic trees. This form of phylogenetic analysis is characterized by fast algorithms (generally) and statistically motivated statements about tree topology, edge weights, and correctness. Distances are highly amenable to simulation tests as well. For these reasons, they have been very popular in mathematical treatments. Unfortunately, distance analysis is also characterized by a necessary loss of information (*i.e.* multiple character data sets generating identical dissimilarity matrices) and departure from optimizing a set of specific character changes. For these reasons, distances have not been as popular in empirical analyses.

9.1 Why Distance?

Phenetic techniques are inherently distance-based (whether similarity or dissimilarity) given that they are constructed from *overall* similarity as opposed to specific primitive or derived character change. The entire foundation of phenetic classification rests on distance quantified similarity (Chapter 1). Michener and Sokal (1957) defined their initial clustering procedures on a data set of solitary bees (Megachilidae) comprising 97 taxa and 122 characters. These resulted in 11,834 individual observations, which were converted into 4656 pairwise distances (actually similarities in their example) and analyzed on that basis. These data did not require distance analysis; it was thrust upon them. Other forms of data, especially that from molecular sequences such as DNA or protein, have been transformed and treated this way.

Systematics: A Course of Lectures, First Edition. Ward C. Wheeler.
© 2012 Ward C. Wheeler. Published 2012 by Blackwell Publishing Ltd.

Several forms of data are observed as distances. Although not practiced much today, both immunological distances (Micro-complement fixation) (Sarich and Wilson, 1967) and DNA:DNA hybridization (Sibley and Ahlquist, 1984, 1990) were extremely popular in the 1970s and 1980s. These data expressed dissimilarity directly and required appropriate analytical tools (also based on dissimilarity) to erect phylogenetic hypotheses.

Vincent Sarich

9.1.1 Benefits

The most commonly used distance methods are, in general, much faster than character-based methods (such as parsimony or likelihood). Least-squared methods (such as Fitch and Margoliash (FM), 1967) have a time complexity of at least $O(n^4)$ for n taxa, and Neighbor-Joining (Saitou and Nei, 1987) $O(n^3)$. These may seem time consuming but, compared to NP–hard tree search, these time complexities are quite manageable. Yet, this comes not from a factor unique to distance data themselves (general distance tree optimization is NP–hard—see below) but from the application of methodologies (such as Neighbor-Joining) that are low complexity heuristics of the full combinatorial problem.

A further benefit (although not all would agree) is the ability to adjust or "correct" observed dissimilarities for multiple hits via stochastic character (usually DNA) transformation models (*e.g.* Nei and Li, 1979)—although likelihood methods possess this ability as well. The objective of these model-based transformations is to estimate the "true" number of changes between leaf taxa. Inherent in such models are ideas of variance and expectation that can be used to make statistical statements of asymptotic behavior and expected data requirements for correct resolution of trees (Erdös et al., 1997).

Other methods establish the validity of edges (branches) of trees through analysis of their implied quartets (four-taxon statements), leading to procedures with high probability (given a correct distance model) of returning the "correct" tree (Buneman, 1971).

Allan Wilson
(1934–1991)

9.1.2 Drawbacks

There are three main drawbacks to distance analysis: the loss of information, the treatment of heterogeneous data is unclear, and the absence of optimized specific character transformation events.

Clearly, if sequences of thousands of nucleotides are recoded as simple distances, considerable information is lost (*e.g.* Penny, 1982). This is an obvious point, usually countered by the vision of nucleotides as individual random samples drawn from a universe of possibilities defined by the parameters of the system. The concept of nucleotides as historically unique events clashes with this view.

Charles Sibley
(1917–1998)

Many, if not most, modern analyses combine data from different sources (Kluge, 1989). Although it might seem reasonable to create distances among homogeneous character types, how would one combine 10 nucleotide changes, three gene rearrangements, a shift in development, and a change in body tagmosis into a single dissimilarity value (even if one could somehow normalize the numbers)?

Finally, if we are interested in specific transformation events, distance-based trees can only serve as a heuristic for the global solution to the event based (*i.e.* character) scenario. The global solution to optimal edge transformations cannot be created from a series of potentially contradictory pairwise statements. Only direct, global analysis of character data can achieve this goal.

9.2   Distance Functions

Raw distance values can be transformed via evolutionary models to attempt to account for hidden evolutionary change. A simple example is the back mutation (multiple hits) from A $\rightarrow$ C $\rightarrow$ A. Although two changes have occurred, the observed distance would be 0. The goal of the transformation is to account for the actual number of changes (observed and unobserved) between the sequences. Such a distance would be *additive* (see below) and would allow correct tree reconstruction with high probability. This is illustrated by a simple homogeneous evolutionary model—Jukes–Cantor (Jukes and Cantor, 1969). After some time interval, the expected distance between two DNA sequences will hit its maximum at 0.75 $(1 - [4 \cdot (\frac{1}{4})^2])$. After that point, even with continuous change, there will be no observable difference in overall dissimilarity. The Jukes–Cantor model would correct for this and extrapolate to higher levels of dissimilarity, hopefully restoring additivity to the distance values. There are many such models, based on a variety of assumptions of sequence evolution (up to General-Time-Reversible in complexity, Lanave et al., 1984).

9.2.1   Metricity

One immediate issue of note is that once stochastic models are used to transform observed distances, these distances are no longer metric (Sect. 5.4). In order to account for multiple hits and back mutations, there must be a non-zero probability that two identical sequences have undergone change. Hence, the identity condition will be violated ($d(i, i) > 0$, if $\mu t > 0$). Furthermore, the triangle inequality itself may be violated if the sequence dissimilarities are near saturation (0.75 for Jukes–Cantor). As an example, posit three sequences s_0, s_1, and s_2 with sequence dissimilarities $d(s_0, s_1) = 0.375$, $d(s_1, s_2) = 0.375$, and $d(s_0, s_1) = 0.75$. The Jukes–Cantor distance ($D_{ij} = -\frac{3}{4} \log(1 - \frac{4}{3} d_{ij})$) between s_0 and s_2 would be unbounded, the other two finite—hence violating the triangle inequality ($D(s_0, s_2) > D(s_0, s_1) + D(s_1, s_2)$). Even though the dissimilarities themselves are additive, the corrected distances are clearly non-metric.

9.3   Ultrametric Trees

Let us suppose that we have a tree, $T = (V, E)$, that exactly reflects the "true" branching pattern of a set of leaf taxa ($L \subset V$) in that the non-leaf nodes, $V \setminus L$, are labeled with the time since divergence of its descendent vertices, which must be strictly increasing along all paths from leaves to root (Fig. 9.1). Such a tree

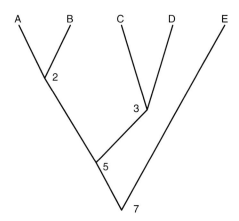

Figure 9.1: An ultrametric tree for the data in Table 9.1.

	A	B	C	D	E
A	0	2	3	5	7
B		0	3	5	7
C			0	3	7
D				0	7
E					0

Table 9.1: Ultrametric distances for ultrametric tree in Figure 9.1.

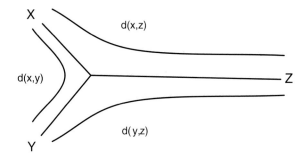

Figure 9.2: Three point condition of ultrametric distances of Buneman (1971).

is an *ultrametric* tree. A true tree would be an ultrametric tree. It would be nice to have such an object, but such trees tend not to present themselves. What we are likely to have are data, in this case distance data. A data matrix, D, defines an *ultrametric distance* if and only if the matrix is metric, and for x, y, and z, of distances $d(x,y)$, $d(x,z)$, and $d(y,z)$ two of these must be the maximum value (hence equal; Table 9.1). This is referred to as the *three point condition* (Buneman, 1971) (Eq. 9.1; Fig. 9.2). Matrix D is also referred to as an ultrametric matrix.

Any three points x, y, z can be renamed such that:

$$d(x, y) \leq d(x, z) = d(z, y) \tag{9.1}$$

If D is ultrametric, then it has a unique ultrametric tree T (and this tree can be found in $O(n^2)$ time). The inverse is also true; if a tree T is ultrametric, then it must have an ultrametric matrix D (Gusfield, 1997).

Recall the conditions of metric distances (Eqs. 5.5 and 9.2).

$$
\begin{aligned}
\forall x \quad d(x, x) &= 0 \\
\forall x, y; x \neq y \quad d(x, y) &> 0 \\
\forall x, y; \quad d(x, y) &= d(y, x) \\
\forall x, y, z \quad d(x, y) &\leq d(x, z) + d(z, y)
\end{aligned}
\tag{9.2}
$$

An ultrametric distance adds a further condition (Eq. 9.3) derived from the three point condition (above).

$$\forall x, y, z \quad d(x, y) \leq \max(d(x, z), d(z, y)) \tag{9.3}$$

This corresponds to the behavior of distances under a rigid molecular clock model (Zuckerkandl and Pauling, 1962). In addition to other restrictions, all leaf taxa must be equidistant from the root.

Ultrametric distances are an ideal case of distance data that allow rapid, "correct" tree reconstruction. Unfortunately, real data are never (to my knowledge) ultrametric.

9.4   Additive Trees

A weaker condition than ultrametricity, is *additivity*. An additive tree possesses edge weights (branch lengths) that directly match the observed distances among taxa. That is, the edge weights summed over the path between two leaves equals the distance between them (Fig. 9.3, Table 9.2). It follows that all ultrametric trees are additive, but not all additive trees are ultrametric. Furthermore, an

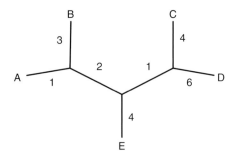

Figure 9.3: An additive tree (edges not drawn to scale).

	A	B	C	D	E
A	0	4	8	10	7
B		0	10	12	9
C			0	10	9
D				0	11
E					0

Table 9.2: Additive distances for additive tree in Figure 9.3.

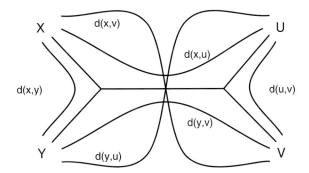

Figure 9.4: Four point condition of additive metric distances of Buneman (1971).

additive tree is characterized by the *four point condition* (Fig. 9.4; Eq. 9.4; Buneman, 1971).

Any four points x, y, u, v can be renamed such that:

$$d(x, y) + d(u, v) \leq d(x, u) + d(y, v) = d(x, v) + d(y, u) \tag{9.4}$$

There are two important implications of additive distances that bear on tree reconstruction: the Farris transform, and Buneman trees.

9.4.1 Farris Transform

Farris et al. (1970) defined a transform of additive data into ultrametric values (in $O(n^2)$ time), which can then be reconstructed by $O(n^2)$ methods. Completely non-intuitive (at least to me), the Farris transform reveals deep connections between additive and ultrametric distances (Fig. 9.5). Due to its general importance (it pops up in many areas of mathematics—Dress et al., 2007), the transform was given its patronym by Bandelt (1990).

The Farris transform of observed, additive d_{ij} to ultrametric d'_{ij} (Eq. 9.5) adds a constant factor δ to distances recalculated from fixed leaf x (originally the most distant, presumed out-taxon yielding $\delta \geq \max d_{xy}$) to create an advancement index, which is the vertex labeling of ultrametric trees.

$$\text{if } x \neq y \quad d'_{i,j} = \delta + \frac{1}{2}(d_{i,j} - d_{x,i} - d_{xj}) \tag{9.5}$$
$$\text{if } x = y \quad d'_{i,j} = 0$$

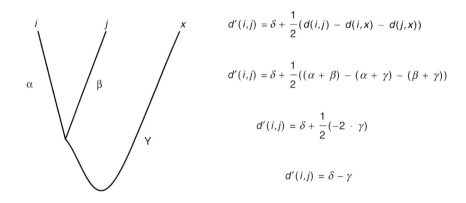

$$d'(i,j) = \delta + \frac{1}{2}(d(i,j) - d(i,x) - d(j,x))$$

$$d'(i,j) = \delta + \frac{1}{2}((\alpha + \beta) - (\alpha + \gamma) - (\beta + \gamma))$$

$$d'(i,j) = \delta + \frac{1}{2}(-2 \cdot \gamma)$$

$$d'(i,j) = \delta - \gamma$$

Figure 9.5: Farris transform of a triplet.

Original Observed Distances

	A	B	C	D	E
A	0	4	8	10	7
B		0	10	12	9
C			0	10	9
D				0	11
E					0

Transformed Distances

	A	B	C	D	E
A		3	6	12	5
B			6	12	5
C				12	6
D					12

Figure 9.6: Farris transform of additive metric distances of Table 9.2 with $\delta = 12$ (Farris et al., 1970).

Peter Buneman

The new ultrametric matrix of advancement indices can then be used for ultrametric tree reconstruction (Figs. 9.6 and 9.7).

9.4.2   Buneman Trees

Based on the four point condition, Buneman (1971, 1974) developed a procedure to identify splits (= edges) that were pairwise compatible. The central concept is that a split on a tree divides the tree on an edge into two subtrees containing sets of leaf taxa. The split in Figure 9.4 would yield $xy|uv$. A value, $\beta(xy|uv)$, can be defined (Eq. 9.6) based on the four point condition (Fig. 9.8).

$$\beta(xy|uv) = \min[d(x,u) + d(y,v), d(x,v) + d(y,u)] - (d(x,y) + d(u,v)) \quad (9.6)$$

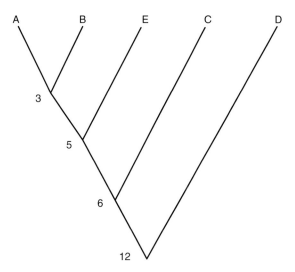

Figure 9.7: Ultrametric tree derived from transformed matrix of Table 9.2 and Figure 9.6.

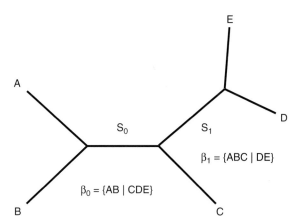

Figure 9.8: Buneman Tree (Buneman, 1971).

Equation 9.6 can be generalized to splits on larger trees, defining the *Buneman Index* β_s (Eq. 9.7).

$$\beta_s \text{ of split } \{A, B\} = \frac{1}{2} \min \beta(xy|uv) \tag{9.7}$$
$$\forall x, y \in A$$
$$\forall u, v \in B$$

All splits with $\beta_s > 0$ are pairwise compatible, hence imply a tree. Furthermore, the β_s values are the weights of the tree edges. Given the quartets that must be

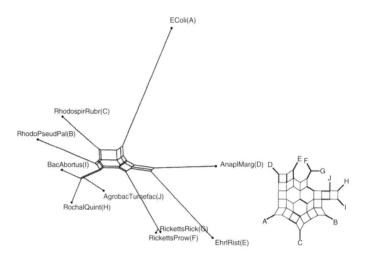

Figure 9.9: "Splits" Tree (Bandelt and Dress, 1992b) with splits drawn showing relative support on left, and pattern of splits on right.

evaluated, the construction of a Buneman tree is $O(n^4)$. A point of note here is that the Buneman tree is based on properties of *additive* distances. The result is not based on optimality in any way, but the compatibility of supported splits.

Due to the strict requirements of Buneman trees, they frequently have few supported edges. To deal with the unresolved nature of Buneman trees, the method of "Split Decomposition" (Bandelt and Dress, 1992a,b) was developed. In this method of representing distance data, the relative Buneman index of each edge—and alternate quartet resolutions—are represented simultaneously as axes of a box, the major axis proportional to the dominant split, the minor the residual. A clear, well-Buneman-supported edge would be a narrow box; an ambiguous one would be square (Fig. 9.9).

9.5   General Distances

The two previous sections dealt with distances exhibiting desirable, but rare to non-existent (at least for real data) properties. The remainder of this discussion centers on methods that do not require ultrametricity or additivity (Table 9.3), but may require metricity to correctly reconstruct distances with high probability. These methods attempt to optimize an objective function (even if they were not described that way initially). The general problem, like other tree-searching operations, is NP–complete (Day, 1987), hence no method short of explicit or implicit enumeration of solutions can guarantee an optimal result. In this sense, the series of $O(n^2)$, $O(n^3)$, and $O(n^4)$ methods below are low-complexity

	A	B	C	D	E
A	0	2.4	7.9	10.1	3.5
B		0	7.6	9.1	10.5
C			0	8.1	10.1
D				0	11.2
E					0

Table 9.3: Non-additive distances.

heuristic algorithms. The methods that were based on explicit optimality criteria (Percent Standard Deviation, minimum length, and minimum evolution) certainly implied NP–hard optimizations from their conceptualization. UPGMA and Neighbor-Joining were more procedural tree-building techniques, which (at least initially) had no optimality objective.

9.5.1 Phenetic Clustering

Clustering algorithms are the heart of phenetic techniques. First enunciated by Michener and Sokal (1957), a large collection of procedures developed through variation in the agglomeration steps. The most (historically) popular algorithm is the Unweighted Pair Group Method using arithmetic Averages (UPGMA).

UPGMA

UPGMA proceeds by defining clusters of most similar leaf taxa and joining them in a cluster, and then adding others in decreasing order of similarity. The level at which taxa are joined is called the "linkage level" (Fig. 9.10). As each cluster is created, the distance between that cluster and all other leaf taxa (or clusters) has to be calculated to proceed. The distance between two clusters, c_i and c_j

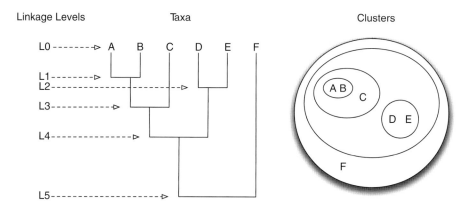

Figure 9.10: UPGMA clustering (Michener and Sokal, 1957).

is simply the average pairwise distances, d, between the members if c_i and c_j (Eq. 9.8).

$$d_{c_i,c_j} = (|c_i| \cdot |c_j|)^{-1} \sum_{\forall a \in c_i, \forall b \in c_j} d_{a,b} \qquad (9.8)$$

Using this relation, clusters are constructed:

1. Assign each leaf to a cluster with linkage level 0.

2. Choose one pair of clusters with minimum distance d_{ij}.

3. Create a new cluster $c_k = c_i \cup c_j$ at linkage level $d_{ij}/2$.

4. Remove clusters c_i and c_j.

5. Calculate distance between c_k and remaining clusters (Eq. 9.8).

6. Repeat until all leaves are added.

UPGMA is (by inspection) an $O(n^2)$ algorithm (Alg. 9.1).

Although not employed recently in general biological systematics, the medical literature continues to employ this method (Fig. 9.11, Achtman et al., 2001).

The UPGMA tree derived from the distances of Table 9.3 is shown in Figure 9.12.

Algorithm 9.1: UPGMA

Data: Input pairwise distance matrix, D, for a set of taxa, L.
Result: UPGMA Tree, $T = (V, E)$ and linkage levels for each vertex.
$V \leftarrow L$;
$E \leftarrow \varnothing$;
$clusterNumber = |L|$;
for $clusterNumber = 0$ **to** $|V| - 1$ **do**
  $\mid$ $linkageLevel_{clusterNumber} \leftarrow 0$;
end
while $\nexists$ *a cluster with all leaf nodes* **do**
  Choose smallest distance in D
  $(i, j) \leftarrow \min D$;
  Create vertex and edges;
  $V \leftarrow V \cup v_{clusterNumber}$;
  $E \leftarrow E \cup (v_{clusterNumber}, i)$;
  $E \leftarrow E \cup (v_{clusterNumber}, j)$;
  Remove v_i and v_j distances from D;
  $linkageLevel_{clusterNumber} \leftarrow D_{i,j}/2$;
  Add $v_{clusterNumber}$ distances to remaining v_k in D via Eq. 9.8;
  $clusterNumber \leftarrow clusterNumber + 1$;
end
$T \leftarrow (V, E)$;
return *(T, linkageLevel)*

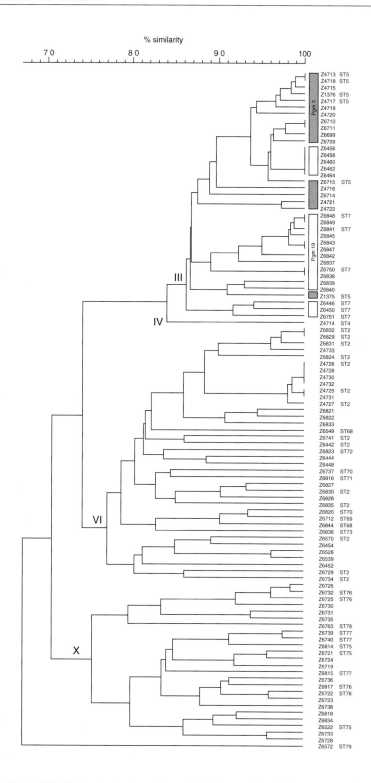

Figure 9.11: UPGMA clustering of meningitis strains (Achtman et al., 2001).

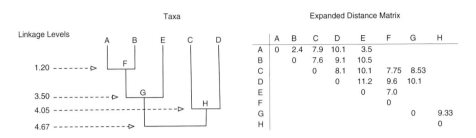

Figure 9.12: UPGMA clustering of distances of Table 9.3 with matrix expanded to include added vertices.

Variations

Many permutations of this process were defined depending on whether the cluster distances were calculated from the average, minimum, or maximum of the leaf distances; or weighting the linkage levels based on recency of addition (Sneath and Sokal, 1973).

- Single linkage clustering: $d(x, u) \leftarrow min(d(y, u), d(z, u))$.

- Maximal linkage clustering: $d(x, u) \leftarrow max(d(y, u), d(z, u))$.

- Weighted by size of subgroup, WPGMA (later arrivers given more weight $(1/2)^n$).

Cophenetic Correlation Coefficient

Although UPGMA (and relatives) were not defined with an optimality criterion in mind, Sokal and Rohlf (1962) defined the Cophenetic Correlation Coefficient (CPCC) to compare phenograms. This was the simple least-squares regression of the linkage levels, $l_{i,j}$ (average value $\bar{l}$), determined by clustering to the observed distances, $d_{i,j}$ (average value $\bar{d}$), (Eq. 9.9). Farris (1969) made great use of the CPCC in his criticism of phenetic methods.

$$CPCC = \frac{\sum_{i<j}(d_{i,j} - \bar{d}) \cdot (l_{i,j} - \bar{l})}{\sqrt{\left(\sum_{i<j}(d_{i,j} - \bar{d})^2\right) \cdot \left(\sum_{i<j}(l_{i,j} - \bar{l})^2\right)}} \quad (9.9)$$

9.5.2 Percent Standard Deviation

Walter Fitch
(1929–2011)

Fitch and Margoliash (1967) (Fig. 9.13) and Cavalli-Sforza and Edwards (1967) independently were the first to define an optimality criterion for their tree construction method, Percent Standard Deviation (PSD in Fitch and Margoliash parlance). The PSD value was calculated based on a given tree, by estimating branch lengths ($D_{u,v}$) from observed (usually model transformed) distances (d_{ij}) via a least-squares approach (Eq. 9.10). The original method built up trees

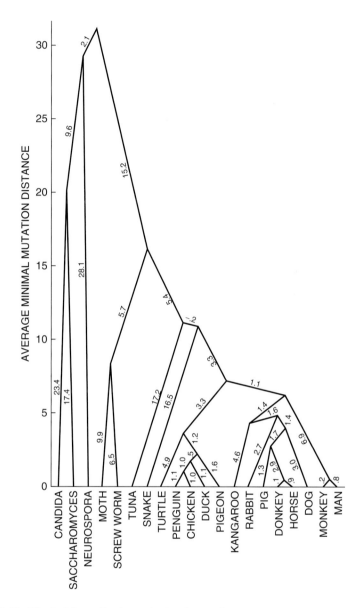

Figure 9.13: Fitch–Margoliash analysis of cytochrome c protein sequences (Fitch and Margoliash, 1967). Note the negative branch length on the edge leading to tetrapods.

by comparing triples, choosing those that implied the lowest average distance between pairs of groups compared to a third (*i.e.* A and B in ((A, B) C)). The three branches in the triples were estimated by *average* distances among members of the three groups. After an initial tree was constructed, others were constructed and compared to others based on PSD.

$$PSD_{T=(V,E)} = \sum_{e \in E} \frac{(d_{uv} - D_{uv})^2}{d_{uv}^2} \qquad (9.10)$$

Later use of the PSD criterion minimized the value over trees as a familiar tree search, and with equally familiar NP–hard complexity (Day, 1987). The procedure (tree-building heuristic) described by Fitch and Margoliash (1967) has a time complexity of $O(n^4)$, but was later improved to $O(n^3)$ (Desper and Gascuel, 2007).

Estimating Branch Lengths

Several methods, including those that minimize PSD (above) and ME (below), estimate edge weights (branch lengths) by least-squares. This is based on a statistical view of the distribution of distances and the desire to create edge weights that conform to their statistical expectation (summarized by Felsenstein, 2004). The process is relatively simple in that each observed distance corresponds to a series of edges on a given tree. With n leaf taxa and $2n - 3$ edges, there are $(n^2 - n)/2$ linear equations created by the sum of edge weights linking each pair of leaves. Given that there are more equations than unknowns, the system is overdetermined and can be solved using elementary linear algebra.

For the tree in Figure 9.14, there are 10 equations and 7 variables (Eq. 9.11).

$$
\begin{aligned}
d(A, B) &= e_0 + e_1 \qquad\qquad (9.11)\\
d(A, C) &= e_0 + e_2 + e_4 + e_5 \\
d(A, D) &= e_0 + e_2 + e_4 + e_6 \\
d(A, E) &= e_0 + e_2 + e_3 \\
d(B, C) &= e_1 + e_2 + e_4 + e_5 \\
d(B, D) &= e_1 + e_2 + e_4 + e_6 \\
d(B, E) &= e_1 + e_2 + e_3 \\
d(C, D) &= e_5 + e_6 \\
d(C, E) &= e_5 + e_4 + e_3 \\
d(D, E) &= e_6 + e_4 + e_3
\end{aligned}
$$

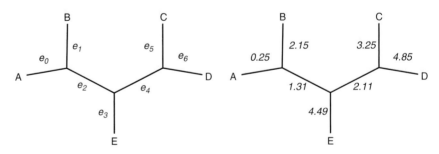

Figure 9.14: Tree of five leaves and seven edges. Edges labeled on left and estimated (distances of Table 9.3) by least-squares on right. The tree has a length (ME) of 18.41 and PSD of 3.24.

The main time-consuming step in solving for the edge weights comes in inverting the matrix representation of Equation 9.11 with $O(n^3)$ complexity. If these operations take place during a search, however, where similar trees have been solved earlier, significant economies can be realized (Bryant and Waddell, 1998).

9.5.3 Minimizing Length

Minimum Evolution

Kenneth Kidd

An alternate notion of minimization—evolutionary tree length—was proposed by Kidd and Sgaramella-Zonta (1971). These authors proposed to use a least-squares method, as in Fitch and Margoliash (1967) and Cavalli-Sforza and Edwards (1967), to calculate edge weights (branch lengths), but to use the linear sum of the edge lengths (Minimum Evolution; ME) as an optimality criterion—in the spirit of parsimony (Eq. 9.12, Fig. 9.14).

$$ME = \sum_{e \in E} D_{uv} \qquad (9.12)$$

The ME criterion was adopted by Neighbor-Joining (NJ) (Saitou and Nei, 1987; Rzhetsky and Nei, 1993) as an optimality criterion for NJ, and others (e.g. Desper and Gascuel, 2007) have employed ME as an optimality criterion in standard tree searching (Fig. 9.15).

Pauplin (2000) derived an $O(n^2)$ method to calculate the total length of a distance tree without the necessity of determining edge weights. This fast

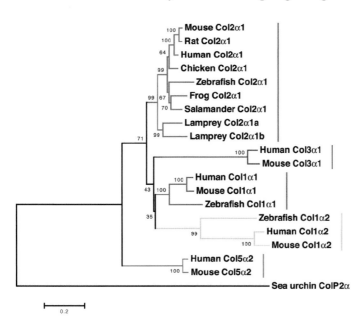

Figure 9.15: Minimum evolution phylogeny for fibril A collagen proteins (Zhang et al., 2006).

ME method has been used to improve the execution time of a variety of distance tree-building heuristics (*e.g.* Balanced Minimum Evolution; Desper and Gascuel, 2002).

Distance Wagner

After his (Farris, 1970) character-based method for calculating Wagner trees, Farris (1972) modified his $O(n^3)$ (cubed because the closest taxon is chosen at each point; if the taxon addition order is determined *a priori*, the process is $O(n^2)$) procedure for distances. The key modification to allow the use of distances comes in the calculation of the distances between median (*i.e.* internal or HTU) nodes and leaf (or OTU) nodes. Farris used a greatest-lower-bound approach based on metric distances (specifically the triangle inequality) as opposed to the least-squares method of Fitch and Margoliash (1967) and Kidd and Sgaramella-Zonta (1971). The distance from new node F, is created by adding node C to edge (G, H) to node Z not yet placed on the tree (Eq. 9.13, Fig. 9.16). Taxa were added to the growing tree by minimizing the patristic (*i.e.* edge path) distance between the leaf to be added and a specific edge on the tree (Eq. 9.14, Alg. 9.2).

1. Begin by choosing the closest pair of taxa (min $d_{G,H}$).

2. Add the taxon C closest (via Eq. 9.14) to the pair (G, H) (or to the closest edge of the tree if the tree has > 2 taxa).

3. Create a new vertex F with distances to (C, G, H) calculated via Eq. 9.14, and distances to taxa yet to be added (Z) via Eq. 9.13.

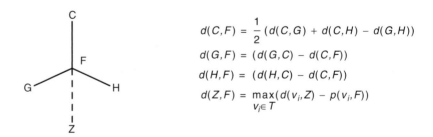

$$d(C,(G,H)) = \frac{1}{2}(d(C,G) + d(C,H) - d(G,H))$$

$$d(C,F) = \frac{1}{2}(d(C,G) + d(C,H) - d(G,H))$$
$$d(G,F) = (d(G,C) - d(C,F))$$
$$d(H,F) = (d(H,C) - d(C,F))$$
$$d(Z,F) = \max_{v_i \in T}(d(v_i,Z) - p(v_i,F))$$

Figure 9.16: Distance Wagner procedure to create edge weights between leaves and non-leaf vertices (upper), and distances between non-leaf vertices and leaves (Farris, 1972). The distances d are observed and p are patristic, previously calculated edges.

Algorithm 9.2: DistanceWagner

Data: Input pairwise distance matrix, D, for a set of taxa, L.
Result: Distance Wagner Tree, $T = (V, E)$.
$V \leftarrow L$;
$E \leftarrow \varnothing$;
Choose two starting taxa
$E \leftarrow (v_0, v_1)$;
While there are leaf nodes to add
while $\exists v_k \notin T$ **do**

  Choose smallest distance between next leaf and each edge in tree
  $(i, j) \leftarrow \min_{i,j} d(v_k, e_{i,j}) = \frac{1}{2}(d(v_k, v_i) + d(v_k, v_j) - d(v_i, v_j))$;
  Remove $e_{i,j}$
  $E \leftarrow E \setminus e_{i,j}$;
  Create vertex and edges;
  $V \leftarrow V \cup v_{HTU}$;
  $E \leftarrow E \cup (v_{HTU}, i)$;
  $E \leftarrow E \cup (v_{HTU}, j)$;
  $E \leftarrow E \cup (v_{HTU}, v_k)$;
  Update tree
  $T \leftarrow (V, E)$;
  Determine distances from v_{HTU} to v_i, v_j and v_k
  $d(v_k, v_{HTU}) = \frac{1}{2}(d(v_k, v_i) + d(v_k, v_j) - d(v_i, v_j))$;
  $d(v_i, v_{HTU}) = d(v_i, v_k) - d(v_k, v_{HTU})$;
  $d(v_j, v_{HTU}) = d(v_j, v_k) - d(v_k, v_{HTU})$;
  Determine distances from v_k to remaining nodes
  $\forall l \notin T, d(v_l, v_{HTU}) = \max_{v_p \in i,j,k}(d(v_p, v_l) - p(v_p, v_{HTU}))$;
  $HTU \leftarrow HTU + 1$;

end
$T \leftarrow (V, E)$;
return T

---

4. Repeat until all taxa have been added to the tree.

$$d(Z, F) = \max_{v_i \in T}(d(v_i, Z) - p(v_i, F)) \tag{9.13}$$

$$d(C, (G, H)) = \frac{1}{2}(d(C, G) + d(C, H) - d(G, H)) \tag{9.14}$$

Although not conspicuous in its initial description, the Distance Wagner technique strives to minimize total patristic distance over the tree (Fig. 9.17).

Distance Wagner trees can contain negative branch lengths and are often longer (sum of estimated edge weights) than those reconstructed by ME or FM. Both of these effects come from the application of the triangle inequality to the distances. If metricity is violated, negative branches may occur. The triangle inequality also requires that patristic distances on the tree cannot be *less* than those observed, placing a lower bound on edge weights not found in ME or FM.

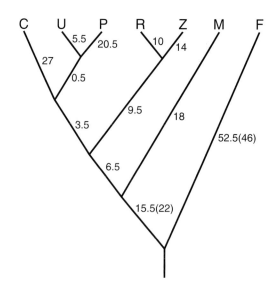

Figure 9.17: Distance Wagner tree for seven carnivores from Farris (1972).

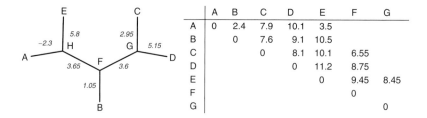

	A	B	C	D	E	F	G
A	0	2.4	7.9	10.1	3.5		
B		0	7.6	9.1	10.5		
C			0	8.1	10.1	6.55	
D				0	11.2	8.75	
E					0	9.45	8.45
F						0	
G							0

Figure 9.18: Distance Wagner tree and expanded distances of Table 9.3. Tree length is 19.9. ME tree length is 18.43 and PSD is 3.70.

These effects can be seen using the example data of Table 9.3 (Fig. 9.18), where the edge connecting leaf A and non-leaf vertex H is -2.3. The overall length of this tree is 19.9 compared to 18.43 for this tree using ME branch estimation (PSD = 3.70).

Neighbor-Joining

Neighbor-Joining (NJ; Saitou and Nei, 1987) is a very popular method of distance analysis that was originally described without any particular optimality criterion. Subsequently, Rzhetsky and Nei (1993) published a justification of

NJ in terms of minimum evolution. NJ is guaranteed to give the correct tree if the distances are additive. Of course, in this unlikely event, with additive distances, we can apply the Farris Transform, convert the distance matrix to an ultrametric (in $O(n^2)$ time), and reconstruct the tree exactly in $O(n^2)$ time (Gusfield, 1997) as opposed to $O(n^3)$ for NJ (below).

The NJ algorithm defines a normalized distance between the leaf taxa that adjusts the distance value for each pair of leaf taxa by their average distance to other leaves. For observed d_{ij} of leaf set L, the normalized distance, D_{ij}, is given in Equation 9.15.

$$D_{ij} = d_{ij} - (r_i + r_j)) \qquad (9.15)$$

$$r_i = (|L| - 2)^{-1} \sum_{\forall k \in L} d_{ik}$$

$$r_j = (|L| - 2)^{-1} \sum_{\forall k \in L} d_{jk}$$

NJ begins with a "star" phylogeny and proceeds by joining pairs of "neighbors" based on Equation 9.15, creating new nodes that connect the joined taxa to the remaining star (Fig. 9.19). The overall time complexity to build a tree is $O(n^3)$ ($n = |L|$; Alg. 9.3).

1. Define a Tree T, of all leaf taxa L, initially a "star".

2. Choose a pair of leaf taxa, i, j, for which D_{ij} (Eq. 9.15) is minimal.

3. Create new node k, set $d_{kl} = \frac{1}{2}(d_{il} + d_{jl} - d_{ij})$ for all l in L.

4. Add edges $d_{ik} = \frac{1}{2}(d_{ij} + r_i - r_j)$ and $d_{jk} = d_{ij} - d_{ik}$.

5. Add edge from l to T, remove i and j from L.

6. Repeat until all leaf taxa are added.

The NJ tree of Table 9.3 is shown in Figure 9.20. The length of this tree is 18.11, but without the least-squares edge weight estimates of ME. NJ has been used in the vast majority of distance-based analyses and in empirical studies, and is nearly ubiquitous in viral systematics (Fig. 9.21).

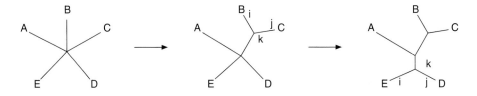

Figure 9.19: Neighbor-Joining procedure (Saitou and Nei, 1987).

Algorithm 9.3: NeighborJoining

Data: Input pairwise distance matrix, D, for a set of taxa, L.
Result: Neighbor-Joining Tree, $T = (V, E)$.
$E \leftarrow \varnothing$;
$V \leftarrow L \cup v_{|L|}$;
Create "star" edges
for $i = 0$ **to** $|L| - 1$ **do**
  $\mid$ $E \leftarrow E \cup (v_{|L|}, v_i)$;
end
While there are nodes to create
while $|V| < 2 \cdot |L| - 2$ **do**
  Choose smallest distance between pair of vertices
  $(i, j) \leftarrow \min_{i,j} D(v_i, v_j)$ (Eq. 9.15);
  Destroy $e_{v_i, v_{|L|}}$ and $e_{v_j, v_{|L|}}$
  $E \leftarrow E \setminus (e_{v_i, v_{|L|}} \cup e_{v_j, v_{|L|}})$;
  Create vertex and edges;
  $V \leftarrow V \cup v_{HTU}$;
  $E \leftarrow E \cup (v_{HTU}, v_i)$;
  $E \leftarrow E \cup (v_{HTU}, v_j)$;
  $E \leftarrow E \cup (v_{HTU}, v_{|L|})$;
  Determine new edge distances
  $d(v_i, v_{HTU}) \leftarrow \frac{1}{2}(d_{ij} + r_i - r_j)$;
  $d(v_j, v_{HTU}) \leftarrow (d_{ij} - d(v_i, v_{HTU}))$;
  Determine new edge distances to leaf vertices
  $\forall l \in L, d_{l,HTU} = \frac{1}{2}(d_{il} + d_{jl} - d_{ij})$;
  $HTU \leftarrow HTU + 1$;
end
$T \leftarrow (V, E)$;
return T

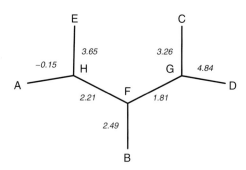

Figure 9.20: Neighbor-Joining tree of example data Table 9.3. The total tree length (by NJ edge estimation) is 18.11.

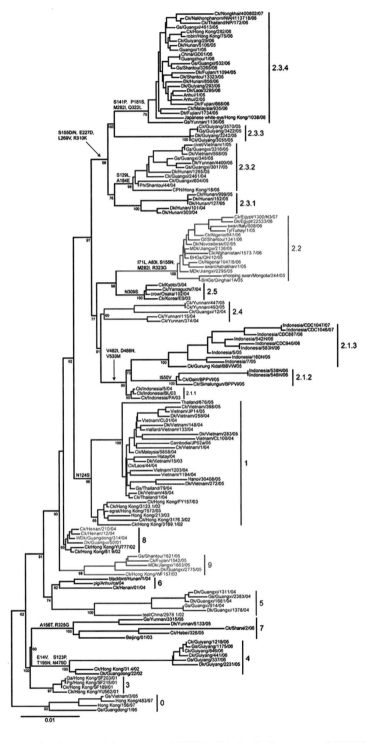

Figure 9.21: Neighbor-Joining tree of H5N1 "Avian" flu virus of WHO/OIE/ FAO H5N1 Evolution Working Group http://www.cdc.gov/eid/content/14/ 7/e1-G2.htm. See Plate 9.21 for the color Figure.

Desper and Gascuel (2002) proposed a method analogous to the NJ procedure by requiring the weight of sister taxa (the two descendent edges incident on a vertex) to be equal as opposed to being weighted by the number of leaves in each subtree. This "Balanced Minimum Evolution" procedure uses the same optimality value—ME—but with lower time complexity than that for NJ ($O(n^2 \log n)$).

ME and Tree Search

The NJ algorithm can be viewed as a heuristic solution to the general ME problem. If this point of view is taken, standard tree search procedures can be employed in an effort to optimize ME and identify the optimal tree. As mentioned before, this will be an NP–hard optimization, as for any other optimality criterion.

9.6 Comparisons

There are two means of comparing the various distance approaches: optimality criterion and tree-building algorithm. Three optimality criteria (at least by the methods) have been discussed here: Cophenetic Correlation Coefficient, Percent Standard Deviation, and Minimum Evolution. The choice among these is more one of aesthetic than any empirical means since a tree may be optimal for one criterion or another. Any argument among them must be made on the relative merits of the criteria themselves.

One point of comparison is important to make clear in the two forms of minimum evolution, that of the Distance Wagner (DW) and Minimum Evolution proper (ME). As mentioned above, the total DW tree lengths will always be greater than or equal to that yielded from ME. This is due to the different objectives of the two length criteria. DW seeks to minimize overall tree length given the constraints of metricity (Sect. 9.2.1 above), specifically the triangle inequality. ME, on the other hand, seeks to create tree lengths based on the statistical expectation of distances. As a result of this, ME can reconstruct branch lengths that yield patristic distances between taxa that are impossibly low—less than that observed in the input matrix. In the examples based on Table 9.3, the ME paths between AC, AD, BE, and EC are all less than that observed. They are impossibly short. Of the 10 pairwise paths, four are too short, four are longer than observed, and two are exactly on, coincident with the objective of creating edge lengths as expected amount of evolution. The DW does not have this problem, but does display a negative branch length on AH. This negative value indicates that the observed distances are non-metric (similarly, negative branch lengths in NJ trees are indicative of non-additive distances (Fig. 9.20). Negative branches frequently occur in real and simulated data sets (Gascuel, 1997), especially those that have been transformed by evolutionary models.

When considering the tree-building procedures themselves (UPGMA, Distance Wagner, Neighbor-Joining, *etc.*), they must be evaluated on efficacy (of optimizing some criterion) and efficiency (time complexity). Given the reality of the overall NP–hard optimization, these tree-building procedures are best viewed as starting points for refinement procedures discussed later (Chapter 14).

9.7 Exercises

1. Given the additive tree in Figure 9.22, produce the ultrametric tree via the Farris transform.

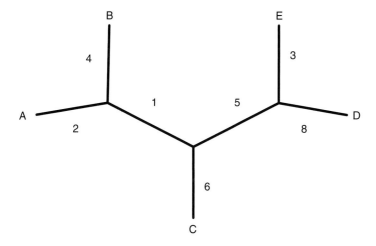

Figure 9.22: Convert this additive tree to an ultrametric via the Farris transform.

2. Given the following symmetrical distance matrix (Table 9.4), create the UPGMA tree.

	A	B	C	D	E
A	0	3.5	8.2	9.1	4.3
B		0	8.9	11.2	9.3
C			0	7.9	9.9
D				0	11.6
E					0

Table 9.4: A non-additive distance matrix.

3. Given the symmetrical distance matrix (Table 9.4), create the Distance Wagner tree.

4. Given the symmetrical distance matrix (Table 9.4), create the NJ tree.

5. What are the implications for edge weights of the differing methods used by Distance Wagner and least-squares to estimate them?

Chapter 10

Optimality Criteria–Parsimony

The operation of assigning states to non-leaf vertices is often referred to as *optimization*. This operation implies a set of transformations between adjacent (ancestor–descendant) vertices, and that assignment is also referred to as optimization. This chapter discusses the operations involved in both these forms of character optimization under the parsimony criterion. Median (vertex) states will be assigned to minimize overall tree cost without reference to a stochastic model of character change or evolution—although there may be non-homogenous transformation cost scenarios.

As per usual, we define a tree, $T = (V, E)$, and a transformation cost matrix σ that specifies the cost of transformations among the elements (character states) of the data set. The purpose of the operation is to determine the minimum cost of a tree given a transformation cost matrix and data $(T(\sigma, D))$. There are two general types of characters discussed here: static and dynamic *sensu* Wheeler (2001b). Static characters are those whose correspondences (observations in leaf taxa) are fixed and invariant over alternate trees, while the correspondences of dynamic characters may vary with tree topology such that overall cost is minimized. The static character types include additive, non-additive, and matrix (alternately referred to as general or Sankoff), while unaligned sequence and chromosomal characters are referred to as dynamic homology characters. A computational distinction between these classes of characters can be drawn based on the time complexity of their optimization. Static characters can be optimized on a given tree in polynomial time, whereas the optimization of dynamic characters is, in general, NP–hard.

10.1   Perfect Phylogeny

The most simple (and potentially satisfying case) is a matrix of n leaf taxa and m binary observations, or features, for each leaf. If the observations are binary, $(0, 1)$, we can test to see if the character columns are in either of two relationships. Consider two characters: the set A_i of taxa with a 1 for a character i (0 can always be set to the out or root state) and a similarly constructed set A_j for some character j, where $|A_i| \leq |A_j|$. If:

$$\forall i, j \in m \text{ character set}$$
$$\text{if } A_i \cap A_j = \varnothing \text{ or}$$
$$A_i \subseteq A_j \tag{10.1}$$

then there is a perfect phylogeny. If:

$$\forall i, j \in m \text{ character set}$$
$$\forall a, b \in n \text{ leaf taxa}$$
$$A_{ia} = 1, A_{ib} = 0 \text{ and } A_{ja} = 0, A_{ib} = 1 \tag{10.2}$$

then there is not (this can be tested in $O(nm)$ time; Gusfield, 1997). If the data are "perfect," derived distances will be ultrametric, and there will be $O(n^2)$ time reconstruction of a unique tree (Fig. 10.1). The problem can be generalized to r-states characters. If the characters are additive (Sect. 10.2.1), there is a polynomial time algorithm to solve for the tree; if non-additive (Sect. 10.2.2), the problem is NP–hard for $r > 2$ (Steel, 1992). Of course, these perfect scenarios are nearly non-existent in empirical situations. Homoplasy happens.

10.2   Static Homology Characters

There are three classes of static characters each with exact, polynomial optimization procedures: Additive, Non-Additive, and Matrix (or Sankoff). Due to

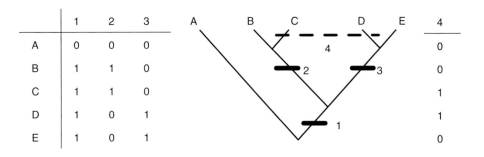

	1	2	3		4
A	0	0	0		0
B	1	1	0		0
C	1	1	0		1
D	1	0	1		1
E	1	0	1		0

Figure 10.1: A perfect phylogeny with the characters and states on the left and the tree in the center. If character 4 were added, the problem is no longer perfect.

their low complexity, procedures to reduce more difficult to analyze data types to these static types are common (*e.g.* developmental "event-pairing," Bininda-Emonds et al., 2002). The entire multiple sequence alignment enterprise (Chapter 8) can be seen as an effort to convert dynamic sequence characters to more tractable static characters.

The algorithms presented in this section have two major components, an initial post-order traversal (down-pass) that establishes the cost (in terms of weighted events) of the tree for the data, and a second pre-order (up-pass) to establish the median (non-leaf vertex) state (character element) assignments (Wheeler et al., 2006a).

10.2.1   Additive Characters

Additive (or ordered) characters (Farris, 1970) are those with transformation costs determined by the difference in their state index (Eq. 10.3). Each successive index represents an increasingly restrictive homology statement. State 3 implies all features inherent in state 2, which in turn implies all in state 1 and so forth (Fig. 10.2).

$$\text{for states } \{s_0, s_1, \ldots, s_{n-1}\}$$
$$\sigma_{s_i, s_j} = |i - j| \tag{10.3}$$

As part of the optimization procedure, we first define the concept of an *interval*. The interval $[a, b]$ contains all the states numerically intermediate between, and including, the lower (a) and higher (b). As an example, $[0, 3]$ would contain 0, 1, 2, and 3. The singleton state a would imply the interval $[a, a]$. P denotes the preliminary state of a node, A the final state of ancestor of that node, F its final state, and L and R the preliminary states of the left and right descendants

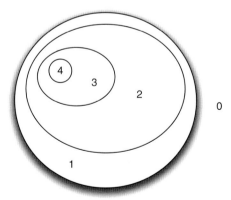

Figure 10.2: Additive character states with "4" implying "3", which implies "2" and so on. State "0" would signify the absence of all variations in the feature.

Algorithm 10.1: AdditiveDownPass

Data: Tree, $T = (V, E)$, with leaf taxa $L \subset V$

Data: Character set, $c \in \mathbb{N}$ for V. The down-pass or *preliminary* states
   are c^p

Data: Vertex, v, of T, initially the root. If $v \notin L$ then v has two
   descendants v_{left} and v_{right}

Result: Return the minimum cost of c on T

Initialize tree cost to 0

$cost \leftarrow 0$;

Preliminary state of leaf taxa are their observed states

for $v \in L$ **do**

 | $c_v^p \leftarrow L_v$;

end

v is not a leaf taxon

if $v \notin L$ **then**

  if *not set* $c_{v_{left}}^p$ **then**

   | $AdditiveDown(T, c, v_{left})$;

  end

  if *not set* $c_{v_{right}}^p$ **then**

   | $AdditiveDown(T, c, v_{right})$;

  end

  $c_v^p \leftarrow c_{v_{left}}^p \cap c_{v_{right}}^p$;

  No intersection of descendent states

  if $c_v^p = \varnothing$ **then**

    Smallest closed interval between descendent states

    $c_v^p \leftarrow sci\{c_{v_{left}}^p, c_{v_{right}}^p\}$;

    Cost of closest elements in $c_{v_{left}}^p$ and $c_{v_{right}}^p$

    $l = $ closest element of $c_{v_{left}}^p$ in $c_{v_{right}}^p$;

    $r = $ closest element of $c_{v_{right}}^p$ in $c_{v_{left}}^p$;

    $cost \leftarrow cost + |l - r|$;

  end

end

return $cost$

---

of the node. For intervals $I_1 = [a, b]$ and $I_2 = [a', b']$, the closest state in I_1 to I_2 is a' if $a' > b$ or b' if $b' < a$. The smallest closed interval between I_1 and I_2 is defined as $[b, a']$ if $a' > b$ and $[b', a]$ if $b' < a$; and the largest closed interval as $[\min(a, a'), \max(b, b')]$. In order to determine the cost of optimizing character c on tree T, a post-order traversal of the tree is performed (Alg. 10.1). This down-pass is that of Farris (1970).

1. Begin at leaves (post-order). The preliminary states of the leaves are their observed states.

2. Choose a non-leaf node whose descendants have known preliminary states. Test for overlap between descendants by taking the intersection of the

intervals $([a, b] \cap [a', b']) = [a', b])$ of its left and right descendants (preliminary $P = L \cap R$).

3. If the intervals overlap $(P \neq \varnothing)$, assign P to the node.

4. If no overlap $(P = \varnothing)$, take the smallest closed interval between them $(P = [b, a']$ if $a' > b$ and $[b', a]$ if $b' < a)$ and increment the cost by the minimum difference between the largest state in L and smallest in R, or the reverse, whichever is smaller $(\min (b' - a, a' - b))$.

5. Continue to the root, setting the preliminary states of each vertex.

After the down-pass, a pre-order up-pass is performed (Alg. 10.2) to establish the set of median states at each non-leaf vertex that is compatible with the

---

Algorithm 10.2: AdditiveUpPass

Data: Tree, $T = (V, E)$, with leaf taxa $L \subset V$

Data: Character set, $c \in \mathbb{N}$ for V initialized by Algorithm 10.1.

Data: Preliminary states are c^p, *final* or up-pass states are c^f.

Data: Vertex, v, of T, initially the root. If $v \notin L$ then v has two descendants v_{left} and v_{right}.

Result: Require that c contain the set of all states consistent with the tree cost

leaf or root;

if $v \in L$ *or* $v = root$ **then**

$\quad \lfloor \ c_v^f \leftarrow c_v^p;$

if $v \notin L$ **then**

$\quad$ **if** $c_v^p \cap c_{v_{parent}}^f = c_{v_{parent}}^f$ **then**

$\quad \quad \mid \ c_v^f \leftarrow c_{v_{parent}}^f;$

$\quad$ **else if** $(c_{v_{left}}^p \cup c_{v_{right}}^p) \cap c_{v_{parent}}^f \neq \varnothing$ **then**

$\quad \quad \quad X = (c_{v_{left}}^p \cup c_{v_{right}}^p \cup c_v^p) \cap c_{v_{parent}}^f;$

$\quad \quad$ **if** $X \cap c_v^p \neq \varnothing$ **then**

$\quad \quad \quad \mid \ c_v^f \leftarrow X;$

$\quad \quad$ **else**

$\quad \quad \quad \mid$ Largest closed interval between X and c_v^p

$\quad \quad \quad \lfloor \ c_v^f \leftarrow lci\{X, c_v^p\};$

$\quad$ **else**

$\quad \quad \mid$ Largest closed interval between $\{c_v^p$ and $c_{v_{parent}}^f\}$

$\quad \quad$ and $\{(c_{v_{left}}^p \cup c_{v_{right}}^p)$ and $c_{v_{parent}}^f\}$

$\quad \quad$ $c_v^f \leftarrow$

$\quad \quad \lfloor \ lci\{\left[c_v^p \text{ closest to } c_{v_{parent}}^f\right], \left[\left(c_{v_{left}}^p \cup c_{v_{right}}^p\right) \text{ closest to } c_{v_{parent}}^f\right]\}$

$\quad$ Recurse up the tree until all $V \notin L$ are updated

$\quad$ $AdditiveUp(T, c, v_{left});$

$\quad \lfloor \ AdditiveUp(T, c, v_{right});$

minimum cost. The original (Farris, 1970) up-pass yields a single median of a potential set. Algorithm 10.2 is that of Goloboff (1993a) (Fig. 10.3), which yields the complete set of parsimonious assignments. The intersection and union operations are on the intervals of the character sets defined above.

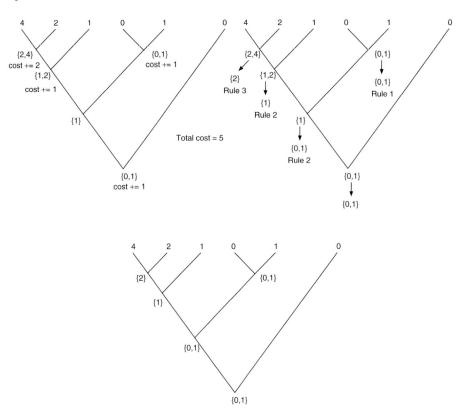

Figure 10.3: Additive character optimization. Down-pass of Farris (1970) (top left), up-pass of Goloboff (1993a) (top right), and final states (lower center).

1. Begin at root (pre-order), root $F = P$.

2. Move up the left and right descendants of the node.

3. Rule 1: If the overlap between the preliminary state (interval) of the node and its ancestor is equal to the ancestral state, the final state of the node is that intersection (if $A \cap P = A$ then $F = A$).

4. Rule 2: (If Rule 1 does not apply and $((L \cup R) \cap A \neq \varnothing)$.) If the union of the preliminary states of the two descendants of the current node has state(s) in common with the ancestor of the current node, then define $X = ((L \cup R \cup P) \cap A)$. If $X \cap P \neq \varnothing$. Then, the final states of the node are equal to X ($F = X$). Otherwise ($X \cap P = \varnothing$), the final states are the set (largest closed interval) of X and the state in P closest to X.

5. Rule 3: (If Rule 1 and 2 do not apply) Then, the final states (F) of the current node is the largest closed interval between the state in P closest to A and the state in ($L \cup R$) closest to A.

6. Continue to leaves (for leaves $F = P$).

Continuous Characters

Continuous (*i.e.* real valued) characters, such as measurements of length, or ratios can be optimized via this same additive procedure simply by multiplying the values by some large constant factor representing the precision of the measurement and truncating the result. The "real" values become integerized and can be optimized as above. The character weight would need to be scaled down to the same degree to maintain parity with other characters and to remove any dependency on units. This approach was taken by Goloboff et al. (2006) and is implemented in TNT (Goloboff et al., 2003) and POY4 (Varón et al., 2008).

10.2.2 Non-Additive Characters

When character transformation costs are constant[1] between all state pairs (Eq. 10.4), those characters are said to be non-additive or unordered (Fitch, 1971) (Fig. 10.4).

$$\text{for states } \{s_0, s_1, \ldots, s_{n-1}\}$$
$$\text{and constant } k$$
$$\forall_{i,j} \ \sigma_{s_i,s_j} = k \tag{10.4}$$

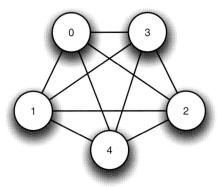

Figure 10.4: Non-additive character states with all possible transformations equal in cost.

---

[1]In practice, all transformation costs are set to unity (1) since the constant term k can be applied as a character weight after the optimization process. The presentation here is for clarity in comparison with matrix characters (below).

Algorithm 10.3: NonAdditiveDownPass

Data: Tree, $T = (V, E)$, with leaf taxa L

Data: Character set, $c \in \mathbb{N}$ for V. The down-pass or *preliminary* states are c^p

Data: Vertex, v, of T, initially the root. If $v \notin L$ then v has two descendants v_{left} and v_{right}

Data: Constant k cost of character transformation

Result: Return the minimum cost of c on T

Initialize tree cost to 0;

$cost \leftarrow 0$;

Preliminary state of leaf taxa are their observed states

for $v \in L$ **do**

    $\lfloor \quad c_v^p \leftarrow L_v$;

v is not a leaf taxon

if $v \notin L$ **then**

    if *not set* $c_{v_{left}}^p$ **then**

        $\lfloor \quad NonAdditiveDown(T, c, v_{left}, k)$;

    if *not set* $c_{v_{right}}^p$ **then**

        $\lfloor \quad NonAdditiveDown(T, c, v_{right}, k)$;

    $c_v^p \leftarrow c_{v_{left}}^p \cap c_{v_{right}}^p$;

    No intersection of daughter states;

    if $c_v^p = \varnothing$ **then**

        $c_v^p \leftarrow c_{v_{left}}^p \cup c_{v_{right}}^p$;

        $cost \leftarrow cost + k$;

return *cost*

The down-pass of the non-additive optimization procedure is similar to that for additive (Alg. 10.3) with the exception that in the place of intervals, the sets of characters in non-additive are not filled in between maximum and minimum states. The union and intersection operations act as standard set operations. Additionally, the incremental cost of character change is calculated not as the distance between the intervals of the parent states:

$$cost \leftarrow cost + |c_{v_{left}}^p - c_{v_{right}}^p| \tag{10.5}$$

but by the constant k (Eq. 10.4). The down-pass procedure is identical to that of Farris (1970) (using standard set notation), although he defined it only for binary characters. The up-pass is due to Fitch (1971).

$$cost \leftarrow cost + k \tag{10.6}$$

1. Begin at leaves (post-order).

2. Choose a non-leaf node whose descendants have known preliminary states. Test for overlap between descendants (preliminary $P = L \cap R$).

3. If no overlap ($P = \varnothing$), take the union (combination) of their states ($P = L \cup R$) and increment the cost by k.

4. Move to the ancestor of the node. Continue to root.

The up-pass is also similar to that of additive (Alg. 10.4) (Fig. 10.5):

1. Begin at root (pre-order), root final state = root preliminary state ($F = P$).

2. Move up the left and right descendants of the node.

3. Rule 1: If the overlap between the preliminary state, P, of the node and its ancestor, A, is equal to A, (if $A \cap P = A$) then the final states, F, are equal to that of the ancestor ($F = A$).

4. Rule 2: (If Rule 1 does not apply) If the union of preliminary states of the two descendants of the current node (L and R) are equal to the preliminary states of the current node ($P = L \cup R$), then $F = P \cup A$.

5. Rule 3: (If Rule 1 and 2 do not apply) Then the final state is the union of the preliminary state set, augmented by states that are common to the ancestor and either of its descendants ($F = P \cup (L \cap A) \cup (R \cap A)$).

6. Continue to leaves (leaf final state = leaf preliminary state, $F = P$).

Algorithm 10.4: NonAdditiveUpPass

Data: Tree, $T = (V, E)$, with leaf taxa L

Data: Character set, $c \in \mathbb{N}$ for V initialized by Algorithm 10.3.

Data: Vertex, v, of T, initially the root. If $v \notin L$ then v has two descendants v_{left} and v_{right}. If $v \neq root$ then
$$v = (v_{parent})_{left} \text{ or } (v_{parent})_{right}$$
The down-pass or *preliminary* states are c^p, *final* or up-pass states are c^f

Result: Require that c contain the set of all states consistent with the tree cost

v is leaf or root

if $v \in L$ or $v = root$ **then**
  $\mid \quad c_v^f \leftarrow c_v^p;$

v is not a leaf taxon

else if $v \notin L$ **then**
  $\mid \quad$ **if** $c_{v_{parent}}^f \cap c_v^p = c_{v_{parent}}^f$ **then**
  $\mid \quad \mid \quad c_v^f \leftarrow c_{v_{parent}}^f;$
  $\mid \quad$ **else if** $c_{v_{left}}^p \cup c_{v_{right}}^p = c_v^p$ **then**
  $\mid \quad \mid \quad c_v^f \leftarrow c_v^p \cup c_{v_{parent}}^f;$
  $\mid \quad$ **else**
  $\mid \quad \mid \quad c_v^f \leftarrow c_v^p \cup \left(c_{v_{parent}}^f \cap c_{v_{left}}^p \right) \cup \left(c_{v_{parent}}^f \cap c_{v_{right}}^p \right);$
  $\mid \quad$ Recurse up the tree until all $V \notin L$ are updated
  $\mid \quad NonAdditiveUpPass(T, c, v_{left});$
  $\mid \quad NonAdditiveUpPass(T, c, v_{right});$

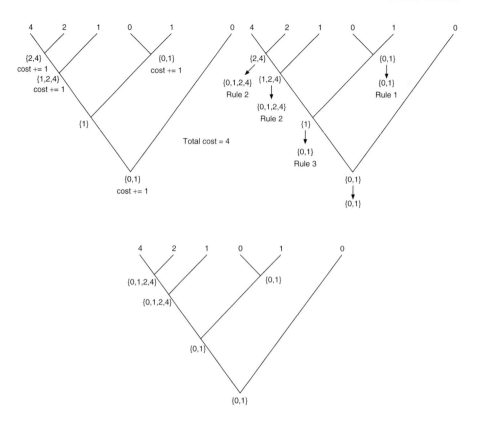

Figure 10.5: Non-additive character optimization Fitch (1971) down-pass (top left), up-pass (top right), and final states (lower center).

10.2.3   Matrix Characters

Characters with arbitrary transformation costs (Eq. 10.7) were first described with their optimization procedure by Sankoff (1975). Unlike the procedures for additive and non-additive characters above, matrix characters require a different approach using dynamic programming to determine the cost and final median state sets for the tree (Alg. 10.5).

$$\text{for states } \{s_0, s_1, \ldots, s_{n-1}\}$$
$$\sigma_{s_i, s_j} = k_{ij} \tag{10.7}$$

As with pairwise sequence alignment (Chapter 8), there are two phases to the dynamic programming algorithm here. The first determines the cost of the tree (here a post-order down-pass), and this is followed by a pre-order trace back step to determine the median state assignments (Fig. 10.6).

    The up-pass of the Sankoff–Rousseau algorithm determines the final state sets for each node by tracing the $from^{left}$ and $from^{right}$ values. If these are true

---

Algorithm 10.5: MatrixDownPass

Data: Tree, $T = (V, E)$, with leaf taxa L

Data: Character set, $c \in \mathbb{N}$ for V with k states

Data: Vertex, v, of T, initially the root. If $v \notin L$ then v has two descendants v_{left} and v_{right}. If $v \neq root$ then $v = (v_{parent})_{left}$ or $(v_{parent})_{right}$

Data: $v.cost$ is an array of size k such that $v.cost_i$ $(0 \leq i < k)$ is the total cost of the subtree rooted at v with state v_k

Data: If $v \in L$ then $v.cost_i \leftarrow 0$ if the observed state of v is i, otherwise $v.cost_i \leftarrow \infty$

Result: Cost of T

v is not a leaf taxon

if $v \notin L$ **then**

    $MatrixDownPass(T, v_{left}, \sigma)$;

    $MatrixDownPass(T, v_{right}, \sigma)$;

    for $i = 0$ **to** $k - 1$ **do**

        $v.cost_i \leftarrow \infty$;

        $from_i^{left} \leftarrow$ false;

        $from_i^{right} \leftarrow$ false;

    for $i = 0$ **to** $k - 1$ **do**

        $min_{left} \leftarrow \infty$;

        $min_{right} \leftarrow \infty$;

        for $j = 0$ **to** $k - 1$ **do**

            if $v_{left}.cost_j + \sigma_{ij} < min_{left}$ **then**

                $min_{left} \leftarrow v_{left}.cost_j + \sigma_{ij}$;

            if $v_{right}.cost_j + \sigma_{ij} < min_{right}$ **then**

                $min_{right} \leftarrow v_{right}.cost_j + \sigma_{ij}$;

        $v.cost_i \leftarrow min_{left} + min_{right}$;

        for $j = 0$ **to** $k - 1$ **do**

            if $v_{left}.cost_j + \sigma_{ij} = min_{left}$ **then**

                $from_j^{left} \leftarrow$ true;

            if $v_{right}.cost_j + \sigma_{ij} = min_{right}$ **then**

                $from_j^{right} \leftarrow$ true;

return $\min v_{root}.cost_i$ over i

---

for a state i, for a left or right vertex (respectively), then those states are included in the final state set for that vertex. The trace back proceeds analogously to that of the Needleman–Wunsch algorithm (Alg. 8.1) in sequence alignment.

The time complexity for the additive and non-additive procedures are linear in the number of taxa n and in character states k, hence $(O(nk))$, whereas the Sankoff–Rousseau algorithm is quadratic in character states, $(O(nk^2))$. Clearly, both additive and non-additive optimization are special cases of the general

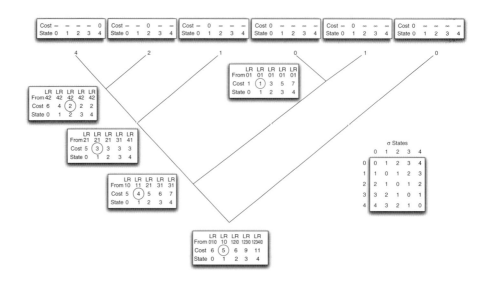

Figure 10.6: Matrix character optimized using the Sankoff–Rousseau algorithm (Sankoff, 1975). The states with their costs circled are the final state assignments determined during the up-pass traceback.

matrix type, with non-diagonal elements either all 1 (non-additive) or the difference in matrix indices (additive).

Although the matrix character optimization algorithm does not require metricity, biologically odd results may occur otherwise. As an example, an additional state could be added to an existing set, with very low transformation cost to all other elements $(\sigma_{k,0} < \frac{1}{2} \min \sigma_{i,j})$. The median state at all internal vertices $(V \setminus L)$ would then be this new state for all trees, no matter what the leaf conditions were.

10.3   Missing Data

Missing data are always undesirable, not only because of lack of information, but also because of how they are optimized on trees. The primary requirement of the optimization of a missing observation is that it should not affect the choice of optimal tree, implying that such a non-observation must never contribute cost to a tree. In order for this to be the case, optimization routines (such as Additive, Non-Additive, and matrix methods) treat missing values as a maximally polymorphic state (in order to ensure that the missing datum adds no cost to the tree)—in essence, that all states are present in that non-observation. Recalling the down-pass of these operations, it is clear that the missing value will conform to the first non-missing observation. This is because of the intersection operations ($\cap$) between the states. This operation is seeking the common information

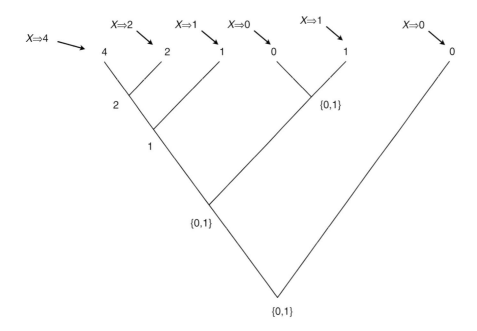

Figure 10.7: Missing character X will assume different states depending on where it is placed on the tree.

between the two character sets. If one is maximally polymorphic (has all states), then the intersection of that state and any other will be that other set. Any other scenario would add cost to the optimization, which is impossible. Unfortunately, unlike unambiguous observations, the missing data will conform to different observations on different trees as it encounters various other taxa first, assuming (in essence) a variety of character states simultaneously (Fig. 10.7). This curious and perhaps undesirable behavior is at the core of the discussions of Platnick et al. (1991) and Nixon and Davis (1991). Unfortunately, as bad as this seems, this treatment of missing data—like democracy—is terrible, except for everything else.

It is worth noting that autapomorphic data can behave as if they were missing. A unique state of a non-additive character will be lost and conform (as with missing data) to the first non-unique state it encounters (Fig. 10.8). The same situation occurs with an additive character but yields different behavior. Due to the different cost scenario of additive characters, the placement of a unique state can affect tree cost (Fig. 10.9). This is due to the fact that a unique state will either be basal, intermediate, or derived compared to other states. Hence, its homology implications and costs are not uniform.

The behavior of missing data cannot be predicted *a priori* in any general way; their pathological effects (if any) are specific to individual analyses.

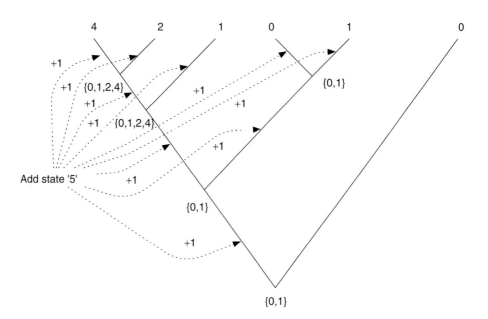

Figure 10.8: Autapomorphic non-additive character will act as missing data, except for adding a single step, no matter where it is placed.

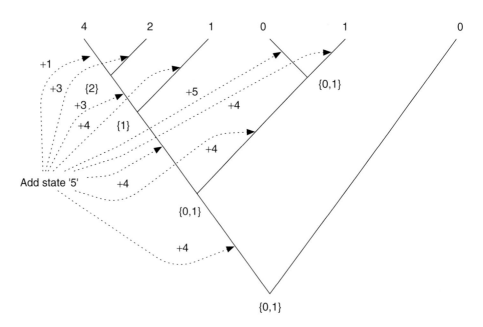

Figure 10.9: Autapomorphic additive character will assume different states depending on where it is placed on the tree, adding from one to five steps to the tree depending on its placement.

10.4   Edge Transformation Assignments

Frequently, median vertex assignments are not unique. In the example of Figure 10.5, all internal vertices are ambiguously optimized. It may seem odd that the tree cost itself can be set so precisely, when the states that cause it cannot. Often, investigators wish to examine branch lengths or localize transformations to a particular edge on a tree. If the vertex assignments are unique, this is straightforward. When ambiguity exists, this may be impossible. As a means of examining change, we often make arbitrary (if repeatable) choices (or random for that matter) in resolving vertex ambiguity and examine the impact on edge weights.

One option is to resolve states such that transformations occur as closely to the root as possible, and a second is to refrain from changes as long as possible, pushing change to the leaf tips of the tree (Fig. 10.10). These two scenarios have been called "ACCTRANS" and "DELTRANS" in PAUP (Swofford, 1993, 2002). It is important to recognize that these are only two possibilities (and not extremes either for anything other than the root and pendant edges). All transformations (and there may be an exponentially large number of unique scenarios) can be generated easily (if tiresomely) via recursively visiting each vertex, assigning each potential state and moving on to the next vertex up the tree towards the leaf nodes. This approach has been used to examine the diversity of optimization scenarios in discussions of evolutionary trends and repeated patterns (*e.g.* Maddison, 1995).

Given that the weights of edges (branch lengths) may vary tremendously with these different optimizations, the question of when to collapse or resolve branches arises immediately.

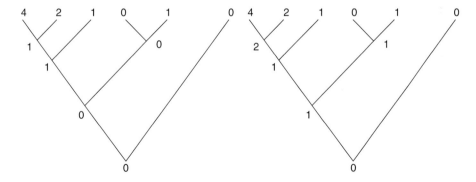

Figure 10.10: Alternate vertex character optimizations of Figure 10.5. The left is "delayed" change and the right "accelerated." The tree cost is unaffected.

Jon Coddington

10.5   Collapsing Branches

In general, systematists would like the graphs they present to reflect or represent all those clades and only those clades that are supported. "Supported," in this context, usually means unambiguous character change between the parent of a vertex (node) and itself (but for other ideas of support see Chapter 15). Consider the additive optimization of Figure 10.3. Several vertices (those closest to the root) are ambiguous, while those further away are uniquely optimized. Coddington and Scharff (1994) enumerated four options as to how to handle this issue.

1. Collapse if the *minimum* edge weight is zero.

2. Collapse if the edge weight is zero on an arbitrary tree (from all possibilities).

3. Collapse if the *maximum* edge weight is zero.

4. Discard trees that must contain zero-length branches.

If we restrict ourselves to options 1 and 3 (2 and 4 seem—to me, at least—to be arbitrary and unreasonable), then the question is whether to collapse branches (edges) because they *can* be zero (case 1) or *must* be zero (case 3) (Fig. 10.11). Over time, different phylogenetic software implementations have approached this problem in different ways. Most have presented options (whether or not users were aware of them) to choose alternate collapsing regimes. At this point in time, the consensual opinion is to be conservative and collapse an edge if its minimum weight is zero. In this way, the investigator can be sure that there is at least some character support for each resolved node on a tree.

10.6   Dynamic Homology

This section treats sequences (in their general form) as characters and presents methods to create minimum cost scenarios on trees directly. As such, these

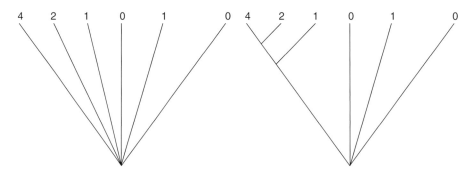

Figure 10.11: Alternate branch collapsing scenarios of the optimizations of Figure 10.5. Left is case 1 and right case 3 of Coddington and Scharff (1994).

ideas are both in opposition and close cousins to the multiple sequence alignment methods of Chapter 8. The sections of that chapter concerning sequences (Sect. 8.3) and pairwise-string matching (Sect. 8.4) are introductory to the concepts discussed here.

10.7 Dynamic and Static Homology

Dynamic homology (Wheeler, 2001a,b) is based on the idea that statements of homology cannot exist outside of a specific cladogram. Two features are homologous if and only if their origin can be traced back to a specific change on a specific branch of a specific cladogram (Fig. 10.12). This definition not only does not require primary or putative homology, but does not allow it (*sensu* DePinna, 1991). Since each cladogram may have its own set of correspondences among features, any "putative" or "primary" homology statements would be equally bound to that cladogram.

The tasks of homology determination and cladogram searching are inseparable. An optimal cladogram is one that minimizes the costs of all transformations among attributes of the organisms under study, allowing for all possible correspondences among variants.

This manner of thinking leaves static homology characters unaffected. Their *a priori* fixed correspondences are a special case of the more general dynamic homology scenario.

Following the treatment in Varón et al. (2010), we can define A and B as two states of a character (such as a sequence). A relationship exists between the elements in A and B that is a *correspondence*. A static homology character

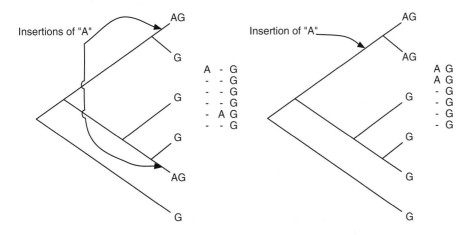

Figure 10.12: The feature "A" has non-unique origins on the left, hence is not homologous. "A" has a single origin on the right, hence the "A"s are homologous. The Implied Alignments (Wheeler, 2003a) are shown to the right of the cladograms.

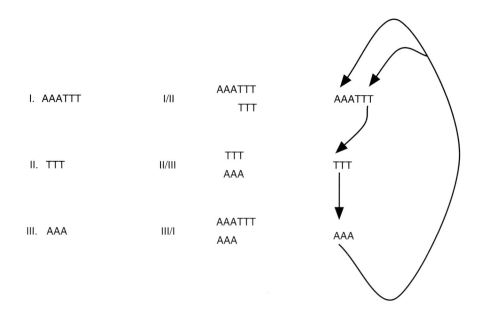

Figure 10.13: Three simple sequences characters showing multiple element correspondences and non-transitivity (Wheeler, 2001a).

is one in which, for each element in A, there is at most one element in B, and these correspondences are transitive (*i.e.* if $a \in A$ corresponds to $b \in B$, and b corresponds to $c \in C$, then a and c must also correspond). A dynamic homology character would be the complement of this definition. For some pair of states A and B, there exists an element $a \in A$ that corresponds to more than one element in B, or the correspondences are not transitive (Fig. 10.13).

10.8   Sequences as Characters

Unlike static characters where individual observations constitute states, sequence character states are extended strings. Individual observations (*e.g.* nucleotides in a DNA sequence) are part of an array of states whose optimizations are interrelated. The optimization of a given additive (Sect. 10.2.1) or non-additive (Sect. 10.2.2) feature takes place independently of all others (if on the same cladogram). Those in a sequence character are determined by the lowest cost correspondence among all observations in that sequence. In short, static homology elements are independent, those of dynamic homology are not. This is the root cause of the complexity of their optimization.

A sequence character, then, is the entire contiguous sequence. If the sequence is a genetic sequence of a particular locus, then that locus is the character. If the sequence is the set of developmental steps involved in the origin of an anatomical feature, then all those steps are components of the character. In the case of a nucleotide sequence character, we may or may not specify individual locus identities (annotations) of sequences (hence, sequences of loci). This is a static homology statement at the locus level, and the sequence characters of the various loci would be optimized independently. This is, however, only a special case. The general problem would involve all nucleotides in a genome and all events that relate them. Annotation, or identification of loci is not a necessary input of dynamic homology analysis (but may be a result; Wheeler, 2007b). If locus annotations are known, the gene homologies would be static and allow for more rapid optimization.

10.9 The Tree Alignment Problem on Trees

The Tree Alignment Problem (TAP; Sect. 8.5.1) was originally described in the context of tree optimization (Sankoff, 1975), but later discussed in terms of multiple sequence alignment (Sankoff and Cedergren, 1983). The basic problem is the same—given a tree $(T = (V, E))$, a set of sequences $(L \subseteq V)$, and a metric distance function (d), determine the median sequence assignments for the internal vertices $(V \setminus L)$ such that the overall tree cost $(\sum_{u,v \in E} d(u, v))$ is minimal. As mentioned earlier, this optimization problem has been shown to be NP–hard (Wang and Jiang, 1994). The TAP is the parsimony problem for dynamic homology sequence characters.

10.9.1 Exact Solutions

As mentioned above, exact solutions will be unavailable for non-trivial data sets. Sankoff and Cedergren (1983) proposed a k-dimensional recursive procedure (for k sequences) that used the tree topology to determine the alignment cost for each cell in cost matrix (Eq. 8.8). The time complexity of this procedure is $[O\left((2^k - 1) \cdot n^k\right)$ for k sequences of length $n]$. Wheeler (2003c) proposed a similarly time-complex approach, but based on a different dynamic-programming model. The idea behind this procedure was to use the matrix character optimization of Sankoff (1975) with the potential sequences as states. The matrix of character state transformations (σ) is determined by the pairwise edit cost (Sect. 8.4) between sequences. Given the large number of possible sequences (from minimum length n to maximum $2n$ $N_{sequences} = \sum_{i=n}^{i=2n} i^4$), the time complexity for an exact solution would be $O(n \cdot N_{sequences}^2)$ (after an initial $\binom{N_{sequences}}{2}$ set-up operation to determine the pairwise costs). This leads to an unexplored, but potentially useful heuristic procedure where a heuristic sequence set (akin to a heuristic tree search set) could be generated where $N_{heuristic} \ll N_{sequences}$ with time complexity $O(kN_{heuristic}^2)$ (Sect. 10.9.3)(Wheeler, 2003c).

10.9.2 Heuristic Solutions

The central problem in heuristic TAP procedures is the assignment of the internal $(V \setminus L)$ vertices. Multiple sequence alignment (MSA) can be viewed as a TAP heuristic, constructing median sequences by static optimization (usually

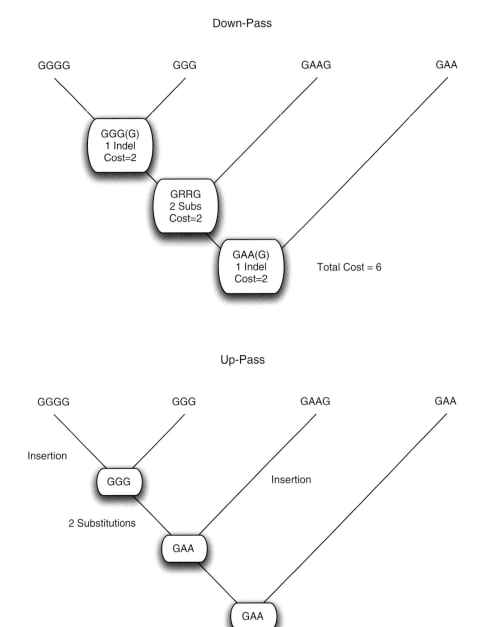

Figure 10.14: Direct Optimization (Wheeler, 1996) down-pass (top) and up-pass (bottom). In this example, all substitutions cost 1 and indels 2.

non-additive, or matrix) of the independent leaf characters (*i.e.* columns) of the MSA. Given the exponential complexity nature of the problem, this approach, as all others, will be an upper bound for non-trivial data sets.

Other heuristics operate directly on the tree, without recourse to MSA. These are presented roughly in increasing order of time complexity. Unfortunately, two of the most commonly used methods do not have known guaranteed bounds (Direct Optimization, Iterative Improvement) but guaranteed time complexity.

Direct Optimization

Direct Optimization (DO; Wheeler, 1996) is currently the TAP heuristic most commonly used in empirical studies (Liu et al., 2009). DO creates a set of minimum cost medians at each internal vertex $(V \setminus L)$ on a tree by pairwise comparison of the two descendants of that vertex. A preliminary median is created from the minimum cost pairwise alignment (Sect. 8.4) by choosing those sequence elements (including "gap" or *indel* element) nearest to the corresponding descendent elements given a metric objective cost function d. All positions consisting solely of *indel* elements (implying that corresponding elements in the descendent vertices contain *indel* elements) are deleted from the candidate median (since sequences do not contain gaps).

The procedure continues down the tree in a post-order traversal, summing up the cost of each median (from the pair wise comparison of vertices) until the root is reached. As with other optimization procedures, an up-pass is then performed to assign final median assignment to each vertex (Fig. 10.14). The DO algorithm is similar to the first pass of pairwise alignment (Alg. 8.1). Both the first-pass and traceback (Alg. 10.6, 10.7) are augmented by a matrix of elements σ' containing those elements closest to the input elements based on the costs specified in σ (*i.e.* elements and combinations of elements and their nearest element or elements).

10.9.3 Lifted Alignments, Fixed-States, and Search-Based Heuristics

This category of heuristics is based on inputting pre-specified sequences. Lifted alignments (Wang and Gusfield, 1997; Wang et al., 2000) rely on using a subset of the leaf (observed) sequences to assign to the remaining vertex sequences. In a lifted alignment, the assignment of a vertex sequence must be from the set of sequences that are descendants of that node. A *uniform* lifted alignment chooses either the left or right descendant consistently for each assignment at that level (Fig. 10.15). The powerful result derived from lifted alignments is that they have a guaranteed bound. The average cost of the 2^d (tree depth d) uniform-lifted alignments will be no greater than twice the global minimum. As such, uniform-lifted alignments are not likely to be useful for empirical analysis,

Daniel Gusfield

Algorithm 10.6: DirectOptimizationFirstPass

Data: Input strings X and Y of lengths $|X|$ and $|Y|$

Data: Element cost matrix σ' of elements nearest to all pairs of elements in Σ and λ (indel) based on element cost matrix σ of Algorithm 8.1

Result: Median cost.

Initialize first row and column of matrices;

$direction\,[0]\,[0] \leftarrow$ '$\searrow$';

$cost\,[0]\,[0] \leftarrow 0$;

$length\,[0]\,[0] \leftarrow 0$;

for $i = 1$ **to** $|X|$ **do**

    $cost\,[i]\,[0] \leftarrow cost\,[i-1]\,[0] + \sigma'_{X_i,\lambda}$;

    $direction\,[i]\,[0] \leftarrow$ '$\rightarrow$';

    $length\,[i]\,[0] \leftarrow length\,[i-1]\,[0] + 1$;

end

for $j = 1$ **to** $|Y|$ **do**

    $cost\,[0]\,[j] \leftarrow cost\,[0]\,[j-1] + \sigma'_{Y_j,\lambda}$;

    $direction\,[0]\,[j] \leftarrow$ '$\downarrow$';

    $length\,[0]\,[j] \leftarrow length\,[0]\,[j-1] + 1$;

end

Update remainder of matrices $cost$, $direction$, and $length$;

for $i = 1$ **to** $|X|$ **do**

    for $j = 1$ **to** $|Y|$ **do**

        $ins \leftarrow cost\,[i-1]\,[j] + \sigma'_{X_i,\lambda}$;

        $del \leftarrow cost\,[i]\,[j-1] + \sigma'_{Y_j,\lambda}$;

        $sub \leftarrow cost\,[i-1]\,[j-1] + \sigma'_{X_i,Y_j}$;

        $cost\,[i]\,[j] \leftarrow \min\,(ins, del, sub)$;

        if $cost\,[i]\,[j] = ins$ **then**

            $direction\,[i]\,[j] \leftarrow$ '$\rightarrow$';

            $length\,[i]\,[j] \leftarrow length\,[i-1]\,[j] + 1$;

        else if $cost\,[i]\,[j] = del$ **then**

            $direction\,[i]\,[j] \leftarrow$ '$\downarrow$';

            $length\,[i]\,[j] \leftarrow length\,[i]\,[j-1] + 1$;

        else

            $direction\,[i]\,[j] \leftarrow$ '$\searrow$';

            $length\,[i]\,[j] \leftarrow length\,[i-1]\,[j-1] + 1$;

        end

    end

end

return $cost\,[|X|]\,[|Y|]$

Algorithm 10.7: DirectOptimizationTraceback

Data: Strings X and Y of Algorithm 10.6

Data: *direction* matrix of Algorithm 10.6

Data: *length* matrix of Algorithm 10.6

Data: Element cost matrix σ' of elements nearest to all pairs of elements
in Σ and λ (indel) based on element cost matrix σ of Algorithm 8.1

Data: Median sequence M' between input sequences X and Y of
minimum cost.

$alignCounter \leftarrow length\,[|X|]\,[|Y|]$;

$medianCounter \leftarrow 0$;

$xCounter \leftarrow |X|$;

$yCounter \leftarrow |Y|$;

while $xCounter \geq 0$ *and* $yCounter \geq 0$ *and* $alignCounter \geq 0$ **do**

    if $direction\,[i]\,[j] = ins$ **then**

        $M[alignCounter] \leftarrow \sigma'\,(X\,[xCounter]\,,GAP)$;

        $xCounter \leftarrow xCounter - 1$;

        $alignCounter \leftarrow alignCounter - 1$;

    else if $direction\,[i]\,[j] = del$ **then**

        $M[alignCounter] \leftarrow \sigma'\,(GAP, Y[yCounter])$;

        $yCounter \leftarrow yCounter - 1$;

        $alignCounter \leftarrow alignCounter - 1$;

    else

        $M[alignCounter] \leftarrow \sigma'\,(X\,[xCounter]\,,Y[yCounter])$;

        $xCounter \leftarrow xCounter - 1$;

        $yCounter \leftarrow yCounter - 1$;

        $alignCounter \leftarrow alignCounter - 1$;

Remove GAP only positions;

for $i = 0$ *to* $length\,[|X|]\,[|Y|] - 1$ **do**

    if $M\,[i] \neq GAP$ **then**

        $M'\,[medianCounter] \leftarrow M\,[i]$;

        $medianCounter \leftarrow medianCounter + 1$;

Median sequence M' without GAP elements;

return M'

---

but do give us some idea of the lower bound on tree costs. Polynomial Time
Approximation Schemes (PTAS) for the TAP (Wang et al., 2000) make use of a
combined exact and uniform-lifted assignment to create guaranteed performance
bounds in polynomial time (Table 10.1).

Fixed-States optimization (FS; Wheeler, 1999) is not strictly speaking a
lifted alignment since the vertex assignments are not limited to their descendants, but may be drawn from any of the leaf sequences (Fig. 10.16). Since FS
allows a superset of vertex assignments, the bound for FS can be no greater
than the factor of two for lifted alignment (Wang and Gusfield, 1997). FS
can be accomplished via the Sankoff–Rousseau algorithm for matrix characters

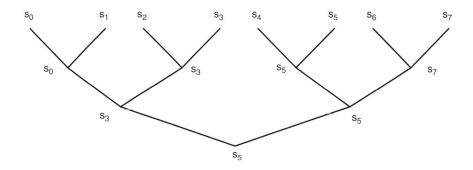

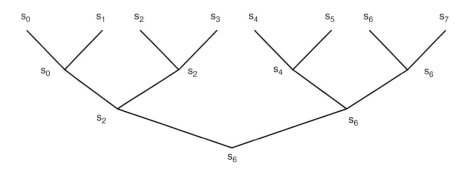

Figure 10.15: Lifted assignment (Wang and Gusfield, 1997) (top) and uniform-lifted assignment (Wang et al., 2000) (bottom).

Running time	$O(kdn^3)$	$O(kdn^4)$	$O(kdn^5)$	$O(kdn^6)$	$O(kdn^7)$	$O(kdn^8)$	$O(kdn^9)$
T_{PTAS}/T_{min}	1.67	1.57	1.50	1.47	1.44	1.42	1.40

Table 10.1: PTAS time complexity for k sequences of length n and tree depth d (Wang et al., 2000).

(Alg. 10.5), with the observed sequences as the states, and their transformation costs determined by their pairwise edit cost (Alg. 8.1). Since the number of states is equal to the number of leaves (k), the time complexity of FS is $O(k^3)$ for each tree, which is usually much smaller than that of DO ($O(kn^2)$ for sequence length n) but can grow to be worse for very large data sets ($k > n$).

Search-based optimization (SBO; Wheeler, 2003c), as mentioned above, expands the possible vertex assignments to non-leaf sequences spanning the divide between FS and an exact solution. SBO allows unobserved sequences to

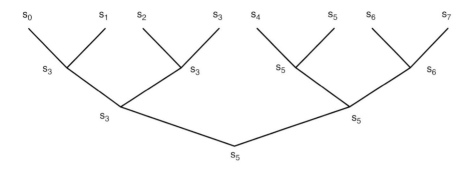

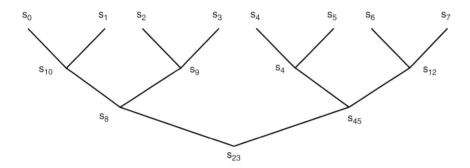

Figure 10.16: Fixed-States Optimization (Wheeler, 1999) (top) and Search-Based Optimization (Wheeler, 2003b) (bottom). Note that the Fixed-States assignments need not be descendent sequences, and that Search-Based assignments may be outside the leaf set.

be assigned to non-leaf vertices to further improve tree cost. If k additional potential sequence medians are added to the n observed, the time complexity of tree evaluation will be $O((n + k)^2)$ (via Alg. 10.5) after an initial $O((n + k)^2 m^2)$ (for sequence length m) edit cost determination. It is unclear whether heuristic sequence sets can be identified that are sufficiently compact to make SBO a useful tree heuristic.

10.9.4   Iterative Improvement

Iterative improvement is a process of improving vertex median assignments by creating an exact median for each vertex in turn from its three neighbors (Sankoff and Cedergren, 1983; Wheeler, 2003b) via a three-dimensional DO. This procedure can be used after any initial median assignment (*e.g.* DO, lifted,

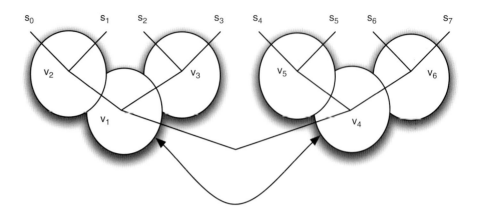

Figure 10.17: Iterative improvement (Sankoff and Cedergren, 1983; Wheeler, 2003b). Vertices $v_1, \ldots, v_6$ are each recalculated from their three neighbors. Note that v_1 and v_4 are neighbors.

or "nearest" leaf). Since the alteration of one vertex may affect its three neighbors, the process is repeated, iteratively, over all internal vertices until the vertex set is stable, or the tree cost has reached a stable minimum (Fig. 10.17). The time complexity will depend on the number of taxa, k, and the cube of the length of the sequences, m, for $O(km^3)$. Experience shows that tree optimality is often improved this way, but at high (factor of m) cost (Frost et al., 2006).

10.10   Performance of Heuristic Solutions

There are only a few comparisons of TAP heuristic effectiveness. In general, worst-case effectiveness should follow worst-case time complexity, hence the order: lifted assignment, FS, DO, IA, SBO, to exact solutions. Although as yet unstudied, the average case quality of bounds should improve as computational effort increases. An area of great interest is in the comparison of MSA as a TAP heuristic to the methods described here (specifically DO as an $O(kn^2)$ method similar in complexity to progressive alignment). A handful of limited studies (*e.g.* Wheeler, 2007a; Wheeler and Giribet, 2009) have shown the superiority (in terms of TAP cost) of DO (via POY; Wheeler et al., 2005; Varón et al., 2008) to MSA (via CLUSTAL; Thompson et al., 1994) in simulated and real data. More general studies of stronger alignment methods (*e.g.* MUSCLE, MAFFT) and broader sequence problem sets have yet to reveal themselves.

10.11   Parameter Sensitivity

A topic of great importance to TAP analysis is parameter sensitivity. At its most basic, the relative costs of indels and substitutions must be specified (unless analyses are limited to areas of exact matching). The objective distance functions that relate sequences to one another via edit costs are based on element edit costs (σ here) among all pairs of sequence elements and the empty (λ) or gap element. As shown for MSA earlier (Sect. 8.4), the specific values of these parameters are likely to have a large effect on the correspondences among sequence elements in MSA and TAP. The repercussions of the choice of cost model will continue through to the TAP cost and then through tree search playing a role in the solution to the General Tree Alignment Problem (GTAP; TAP with tree search to identify the optimal tree). The exploration of this effect has been termed *sensitivity analysis* (Wheeler, 1995).

10.11.1 Sensitivity Analysis

Unfortunately, at least in the world of parsimony, there is no method to determine the "true" values of these necessary parameters via observation in nature, and no way to choose one set over another purely on their optimality values (since the numbers are not comparable). As a result, Wheeler (1995) suggested sensitivity analysis to track the influence of parameter variation on phylogenetic results (Fig. 10.18). The idea was to repeat analyses using different sets of parameters (indel costs, transition–transversion ratios *etc.*) and compare the end results on the basis of congruence and stability among data sets. While not an impediment to theoretical studies, empirical studies are driven by the desire for specific results. Wheeler (1995) suggested two approaches. The first was to identify those taxonomic groups (= vertices) that were largely or completely robust to parameter variation. These groups are relatively insensitive to the unmeasurable assumptions of analysis (Fig. 10.19). Unfortunately, as there can be huge variation in parameters, few aspects of tree topology are entirely free of this sort of dependence.

The second was to identify those areas in parameter space that minimized incongruence among characters using an incongruence metric (Wheeler, 1995; Wheeler et al., 2006b) (Fig. 10.20). Studies that report sensitivity results (*e.g.* Giribet and Edgecombe, 2006; Murienne et al., 2008) usually present both of these modes of analysis.

One solution to the problem of sensitivity has been proposed by Grant and Kluge (2003, 2005) and that is to treat all transformations as equally costly in all circumstances. The form of parsimony advocated by Grant and Kluge is one of minimizing transformations. This interpretive choice was regarded as somewhat arbitrary by Giribet and Wheeler (2007) since this is only one of a very large set of possible cost regimes. Furthermore, this completely homogeneous cost regime could well result in non-metric transformation costs when extended to high-order changes involving the insertion and deletion of sequence blocks.

10.12 Implied Alignment

Although DO-based trees (or lifted, FS, and SBO for that matter) are not based on multiple sequence alignment, and vertex medians may vary in length, it is

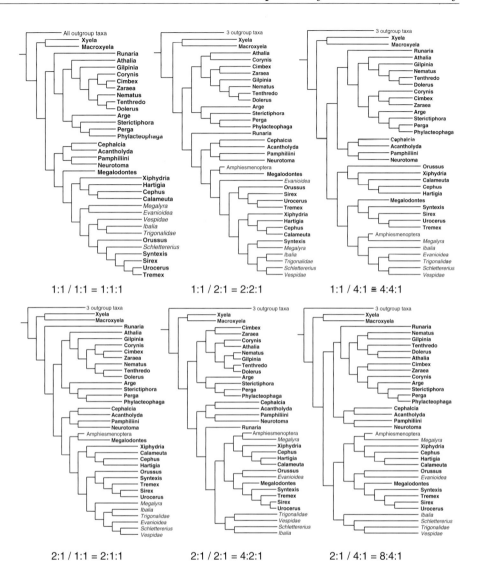

Figure 10.18: A sampling of sawfly trees, based on morphological and molecular data of Schulmeister et al. (2002), showing the effects of parameter set variation. The notation signifieds indel:transition cost ratio / transversion:transition cost ratio = indel:transversion:transition cost ratio.

possible to trace the correspondences between vertex sequence elements over the tree. The tracing would follow a pre-order path from root to leaves keeping track of the chain or element correspondences between ancestor and descendant. These traces would link all the leaf sequences through their hypothetically ancestral medians. Where there were deletions (or areas basal to insertions), gap

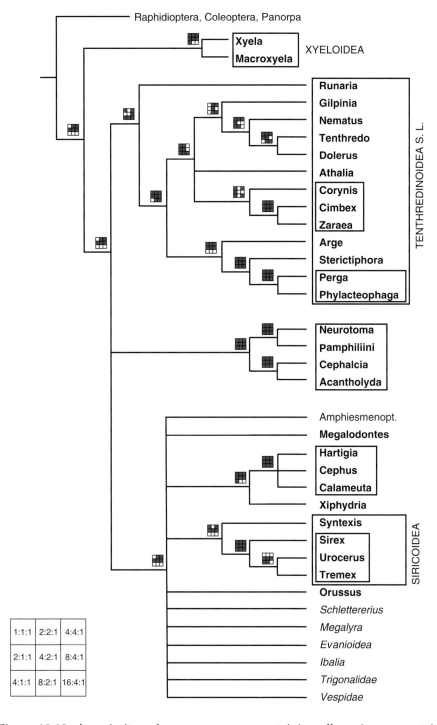

Figure 10.19: A majority rule consensus tree containing all vertices present in more than half of the analyses performed under different parameter sets (Schulmeister et al., 2002). "Navajo rugs" are displayed at the vertices.

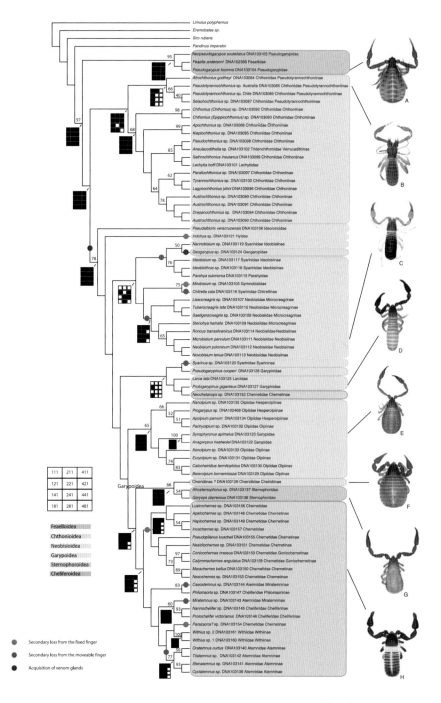

Figure 10.20: Pseudoscorpion analysis of Murienne et al. (2008). The base tree is that which minimized incongruence among multiple molecular loci. The "Navajo rugs" show the presence or absence of each vertex in parameter space. See Plate 10.20 for the color Figure.

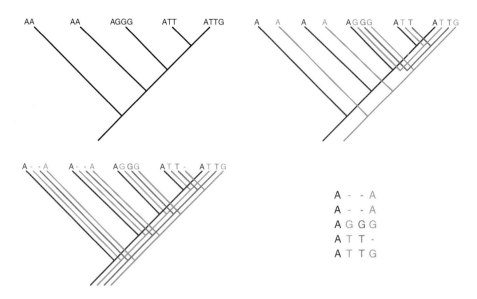

Figure 10.21: Implied alignment (Wheeler, 2003a) of five sequences: AA, AA, AGGG, ATT, and ATTG. The original optimized tree is shown on the upper left; the implied traces upper right; implied traces with traces extended and gap characters filled in lower left; and the final implied alignment in the lower right. See Plate 10.21 for the color Figure.

characters ('-') are to be placed. The weighted sum of the transformations along these traces would exactly match the tree cost. When displayed in 5' to 3' order, these correspondences among leaf elements (with medians removed) would look a great deal like a multiple sequence alignment. It would be the alignment *implied* by the tree and cost parameters. It would be an *implied alignment* (Fig. 10.21) (Wheeler, 2003a). Implied alignments are different from traditional multiple alignments in that the column identity (= putative homology) varies with, and is linked irrevocably to, the tree topology. Sequence elements (usually nucleotide, but they could be anything in theory) involving non-homologous (*i.e.* multiple origin) insertions may appear to be "mis-aligned" when inspected visually. This is due, however, to their non-homology (heuristic effects aside). They cannot align since they cannot be homologous. If such elements had a single origin (such as if they were present in a monophyletic group), they would line up (Fig. 10.12).

One nice feature of implied alignments is that they exactly reflect the sequence transformations that have occurred on the tree. This allows verification of the TAP heuristic cost in that the implied alignment can then be diagnosed as a series of static characters. Given the same tree topology and cost regime (*e.g.* indel cost), either mode of analysis should return the same cost. This aspect of implied alignments leads to a useful TAP search heuristic. Since TAP optimization heuristics such as DO are quadratic in sequence length, and static characters linear, the implied alignment can be used to search for optimal trees much more rapidly than DO can. Of course, the implied alignment potentially would change with each tree topology, but an alternation between DO and implied alignment (*Static Approximation*; Wheeler, 2003a) can be a fast and effective tool for initial tree search on unaligned sequence data (Varón et al., 2010).

10.13 Rearrangement

In addition to the insertion–deletion and substitution modes of sequence change, other transformations are possible. These are placed in the general category of *moves* or *rearrangement*. Chromosomes can be viewed as sequences with gene regions (loci) as sequence elements. Variation in gene order, orientation, and complement have been used as historical information since the first multi-gene mitochondrial sequences became available. Initially, these were informal analyses (*e.g.* Boore et al., 1995, Fig. 10.22), but more recently, they have been based on explicit, optimality-based algorithmic procedures (Moret et al., 2002). Most frequently, sequence and gene order information have been treated independently. Ideally, however, these data would be optimized simultaneously.

10.13.1 Sequence Characters with Moves

The straightforward, if daunting, simultaneous option is to extend sequence edit algorithms to include moves of blocks of sequences from one position to another (Eq. 10.8). A naïve analysis would suggest complexity for the general case of

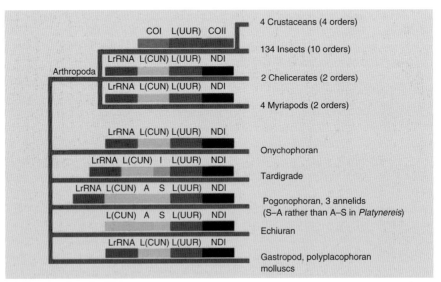

Figure 10.22: Mitochondrial gene order variation in protostome taxa (Boore et al., 1998). See Plate 10.22 for the color Figure.

$O(n^5)$ with order n potential sequence blocks moving to order n positions, and order n such moves between two sequences. Clearly, the general case will be very difficult. Cormode and Muthukrishnan (2002) studied a special case of approximate edit cost calculation yielding a sub-quadratic time complexity of $O(\log n \log^* n)$.

Substitution at i with e:
$$S_0 \dots S_{n-1} \rightarrow S_0 \dots S_{i-1}, e, S_{i+1}, \dots S_{n-1} \qquad (10.8)$$
Deletion at i:
$$S_0 \dots S_{n-1} \rightarrow S_0 \dots S_{i-1}, S_{i+1}, \dots S_{n-1}$$
Insertion at i of e:
$$S_0 \dots S_{n-1} \rightarrow S_0 \dots S_{i-1}, e, S_i, \dots S_{n-1}$$
Move with $0 \le i \le j \le k \le n-1$:
$$S_0 \dots S_{n-1} \rightarrow S_0 \dots S_{i-1}, S_k \dots S_{k-1}, S_i, \dots, S_{j-1},$$
$$S_k, \dots, S_{n-1}$$

Given the inherent complexity of this sequence-based analysis, moves are generally studied at the gene locus level, as rearrangements of high-order elements as opposed to nucleotides themselves.

10.13.2   Gene Order Rearrangement

Chromosomal gene synteny patterns can undergo a variety of transformations including moves, inversions, insertions, and deletions (Fig. 10.23). There are two components required for systematic analysis of gene order data. First, an edit distance function must be defined, and second, median genomes must be reconstructed. This is a brief summary of several approaches to rearrangement analysis.

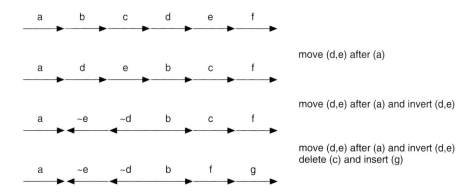

Figure 10.23: A sequence of elements $[a, \dots, g]$ (top) transformed via move (upper center), inversion and move (lower center), and inversion, move, insertion, and deletion (bottom). The complement of element x is represented by $\sim x$.

Distances

Gene order distance functions measure the edit cost between synteny maps based on specific classes of transformation events. These vary from the linear time Break point distance to NP–hard inversion distance. Each of these functions is based on an explicit biological mechanism of genetic change.

Breakpoint—Sankoff et al. (1996) defined the breakpoint distance based on the minimum number of breaks required to edit one chromosome into another. The distance refers to the number of adjacent pairs of loci present on one chromosome but not another (Fig. 10.24). The breakpoint distance can be calculated in linear time, but does not include orientation information.

Inversion—Unlike breakpoint distances, inversion distances can make use of the orientation, or sign, of loci. An inversion is a flipping of a segment of a chromosome (Fig. 10.25). Without orientation information, the calculation of inversion distance is NP–hard. If gene orientation information is present, however, the distance can be calculated in linear time (Hannenhalli and Pezvner, 1995).

Tandem Duplication Random Loss (TDRL) Boore (2000) attributed the vast majority of metazoan mitochondrial evolution to this TDRL mode. In this model, a segment of a chromosome (or even entire genome) is duplicated, and one copy of each duplicated locus is deleted at random (Fig. 10.26). The distance between two genomes can be determined in $O(n \log n)$ time (Chaudhuri et al., 2006). The method has great appeal for the ordering of loci, but does not explain/take account of their differences in orientation.

$$L_1 L_2 L_3 L_4 L_5 L_6 \rightarrow (L_1 L_2)(L_2 L_3)(L_3 L_4)(L_4 L_5)(L_5 L_6)$$

$$L_1 \bar{L}_3 \bar{L}_2 \bar{L}_6 L_4 L_5 \rightarrow (L_1 \bar{L}_3)(\bar{L}_3 \bar{L}_2)(\bar{L}_2 \bar{L}_6)(\bar{L}_6 L_4)(L_4 L_5)$$

distance = 3

Figure 10.24: Breakpoint distance between two chromosomes. The bars denote inverse orientation. There are three locus adjacencies in the upper chromosome not found in the lower. Note that neither orientation nor within-pair order are relevant.

$$L_1 L_2 L_3 L_4 L_5 L_6 \rightarrow L_1 (\bar{L}_3 \bar{L}_2) L_4 L_5 L_6 \rightarrow$$

$$L_1 \bar{L}_3 \bar{L}_2 (L_4 L_5 L_6) \rightarrow L_1 \bar{L}_3 \bar{L}_2 \bar{L}_6 (L_5 L_4) \rightarrow L_1 \bar{L}_3 \bar{L}_2 \bar{L}_6 L_4 L_5$$

distance = 2

Figure 10.25: Inversion distance between two chromosomes. The bars denote inverse orientation. Three inversions are required to edit the top left sequence into the bottom right ($L_2 L_3$ to $\bar{L}_3 \bar{L}_2$, $L_4 L_5 L_6$ to $\bar{L}_6 \bar{L}_5 \bar{L}_4$ and $\bar{L}_5 \bar{L}_4$ to $L_4 L_5$).

$$L_1 L_2 L_3 L_4 L_5 L_6 \rightarrow L_1 \cancel{L_2} L_3 \cancel{L_4} \cancel{L_5} \cancel{L_6} \cancel{L_1} L_2 \cancel{L_3} L_4 L_5 L_6$$

$$\rightarrow L_1 L_3 \cancel{L_2} L_4 L_5 L_6 \rightarrow L_1 L_3 L_2 \cancel{L_4} \cancel{L_5} L_6 L_4 L_5 \cancel{L_6}$$

$$\rightarrow L_1 L_3 L_2 L_6 L_4 L_5$$

Figure 10.26: Tandem duplication random loss transformation between chromosomes. There are two rounds of duplication and loss to edit the upper left sequence into the lower right.

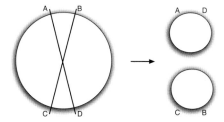

Figure 10.27: Double-Cut-Join rearrangement model of Yancopoulos et al. (2005). The two cuts on the left chromosome yield the two rearranged chromosomes on the right.

Double-Cut-Join (DCJ)—DCJ was proposed by Yancopoulos et al. (2005) as a rearrangement mechanism (Fig. 10.27). In this operation, a chromosome is cut in two places and rejoined, bringing the loci next to the cuts into adjacency. The distance between two chromosomes can be calculated in linear time (Yancopoulos et al., 2005). DCJ mediated rearrangements can be very flexible, allowing for a large diversity of moves, inversions, duplications and losses to occur.

10.13.3 Median Evaluation

For the distances described above (with the exception of unsigned inversions), there are linear or nearly linear algorithms for their calculation. The median optimization, however, for each of them is known (or in the case of TDRL, thought) to be NP–hard. Many heuristic tree optimization algorithms optimize vertices as medians of their three adjacent vertices. This makes the tree optimizations intractable even when based on linear time distances. Implementations such as GRAPPA (Bader et al., 2002) have implemented a variety of median solvers for this purpose.

10.13.4 Combination of Methods

The methods above optimize locus-level rearrangements alone. In order to include the nucleotide sequence information as well, combination methods are required to include the broadest collection of available data (Darling et al., 2004).

Wheeler (2007b) coupled genomic rearrangement (via the use of GRAPPA) with sequence optimization (DO and FS—see above) to include the information in annotated sequences. This annotation was not required for locus homology (= identity), but to mark the potential break points in the rearrangement analysis. LeSy et al. (2006, 2007) removed the annotation requirement, resulting in an entirely dynamic approach to both nucleotide and locus homology (Fig. 10.28).

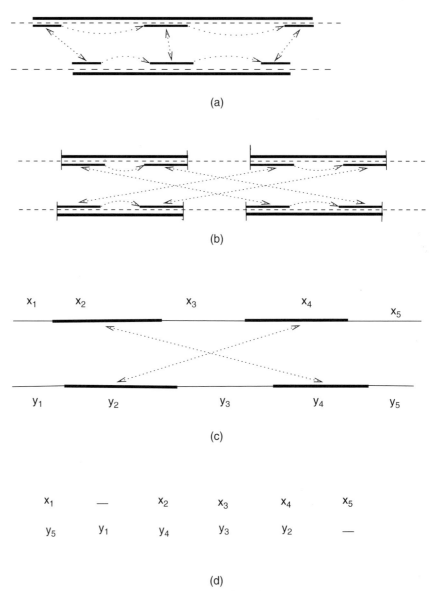

Figure 10.28: The method of LeSy et al. (2007) to identify homologous regions and loci dynamically: (a) non-rearranged seeds constitute a block, (b) consecutive blocks are connected into a large block, (c) blocks are used as anchors to divide genomes into loci, (d) loci are aligned allowing order rearrangements.

10.14 Horizontal Gene Transfer, Hybridization, and Phylogenetic Networks

There are two sorts of phylogenetic networks that occur in discussions and analysis of reticulate scenarios. The first is motivated by ideas of horizontal gene transfer (HGT, Fig. 10.29) and involves the addition of edges to trees such that they accommodate transfer events for specific sets of characters. The second, hybridization networks, produces trees possessing vertices with in-degree 2 and out-degree 1, motivated by the hybridization of lineages (= vertices) (Fig. 10.30).

When dealing with networks as explanations of HGT events, directed edges are added to binary trees to allow for the individual patterns of sets of characters ($N = (V, E)$). The resolution of the alternate binary trees implicit in the network account for the diversity of postulated events. In order to arrive at the optimality (cost) of such a network, the sum of cost of the most parsimonious trees ($T_i \in N$) for each set of characters ($b_i \in B$) is calculated (Eq. 10.9).

$$N = (V, E) \tag{10.9}$$

$$N_{cost} = \sum_{b_i \in B} \left(\min_{T \in T(N)} T_{cost}(b_i) \right) \tag{10.10}$$

The construction of hybrid networks involves the calculation of the *hybrid number*. This is determined by the SPR (Subtree-Pruning and Regrafting) distance between component trees. This calculation is NP–hard even for a pair of trees (Bordewich and Semple, 2005), hence many operations on this class of phylogenetic networks have exponential time complexity.

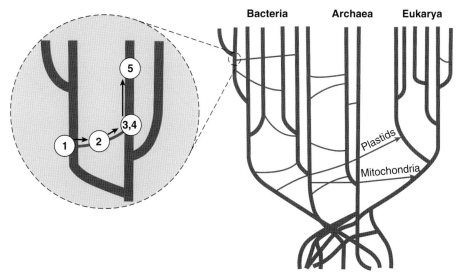

Figure 10.29: Depiction of potential horizontal gene transfer events (Smets and Barkay, 2005).

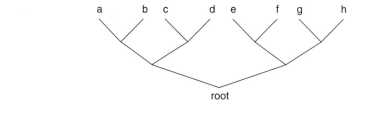

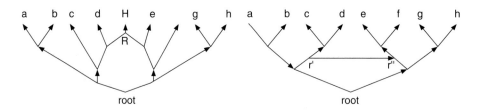

Figure 10.30: Phylogenetic networks. Underlying tree (top), modified to include the hybrid node R (bottom left), and to accommodate a horizontal event along edge (r', r'') (bottom right).

10.15   Exercises

1. Consider Figure 10.31. Give the down-pass, up-pass optimizations (showing rules) and tree cost assuming the character is additive.

2. Consider Figure 10.31. Give the down-pass, up-pass optimizations (showing rules) and tree cost assuming the character is non-additive.

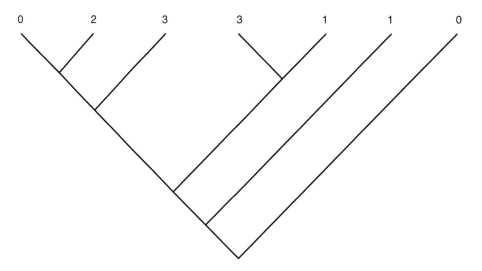

Figure 10.31: Cladogram and character states for exercises 1 and 2.

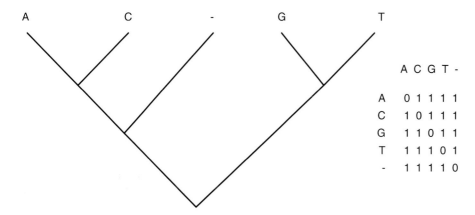

Figure 10.32: Cladogram, character states, and cost matrix for exercise 3.

3. Consider Figure 10.32. Give the down-pass, up-pass (final states) optimizations and tree cost assuming the character is optimized according to the cost matrix in the figure using dynamic programming.

4. For the above three examples, collapse branches according to rules 1 and 3 of Coddington and Scharff (1994). Do you think either (or another) is more reasonable? For all situations?

5. Consider Figure 10.33. Give the down-pass optimizations and tree cost assuming the character is optimized according to the cost matrix in the figure using direct optimization. Use IUPAC codes to represent nucleotide ambiguity.

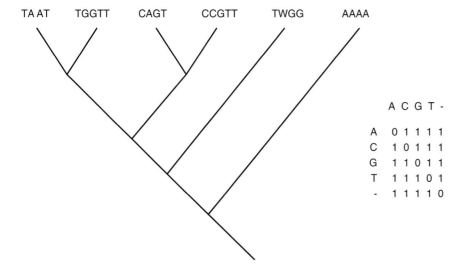

Figure 10.33: Cladogram, sequence states, and cost matrix for exercise 5.

$$12345$$

$$13\bar{5}\bar{4}2$$

Figure 10.34: Chromosomes with number loci, $\bar{x}$ signifies complement of x.

6. Consider Figure 10.34. If the numbers represent loci (and bar above complement) on a chromosome, what is the breakpoint distance between these two chromosomes?

7. Briefly discuss the differences between dynamic and static homology concepts with reference to a morphological example.

Chapter 11

Optimality Criteria–Likelihood

The previous two chapters discussed methods to determine the cost of a tree based on overall distance and the minimization of weighted transformations. We discuss here the determination of tree cost using stochastic models of character change optimizing the probability of the observed data on T given some set of parameters. This probability is proportional to the likelihood function of Section 6.1.7 and is referred to as the *maximum likelihood* (ML) criterion.

As with parsimony, ML methods assign median (ancestral) states (either in an optimal or average context) such that the overall likelihood of the tree is maximized. Unlike minimization-based parsimony, ML methods require explicit models of character evolution (as opposed to edit cost regimes) and edge parameters (branch lengths; parsimony requires none) to determine tree optimality.

The presentation here will also divide characters into static and dynamic types since they require different analytical techniques.

11.1 Motivation

One might explore alternate optimization criteria for their own sake. ML, however, was proposed in the context of purported problems with parsimony analysis. Although Camin and Sokal (1965) and Farris (1973a) had discussed ML methods, Felsenstein (1973) was the first to identify concerns with parsimony and advocate ML as a solution. Much of the discussion centering on the relative merits of parsimony and likelihood in systematics is based on the simple scenario described by Felsenstein (1978).

Joseph Felsenstein

11.1.1 Felsenstein's Example

Felsenstein posited a four-taxon example (Fig. 11.1) with a simple model of change in binary characters to make his point. In this scenario, there are taxa

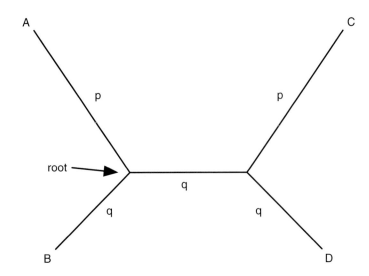

Figure 11.1: Felsenstein (1978) scenario for the statistical inconsistency of parsimony. Probability calculations begin at the root vertex.

A, B, C, and D related by a tree. A and B are on one side of a central split and C and D the other. All characters are posited to have state 0 at the arbitrarily labelled root position. The probabilities of change on each branch are either p or q as labelled. The probabilities of character change are symmetrical so that for all characters $pr(0 \rightarrow 1) = pr(1 \rightarrow 0)$.

Felsenstein was concerned with the issue of *statistical consistency*; in this context, consistency refers to the conditions under which characters would recover the model tree ($AB|CD$) as opposed to the alternatives ($AC|BD$ or $AD|BC$). There are six character distributions relevant to this problem: two for each of the three alternate splits (Eq. 11.1), where the number of each characters supporting a split (n_{ABCD}) are:

$$AB|CD : n_{1100} + n_{0011} \tag{11.1}$$
$$AC|BD : n_{1010} + n_{0101}$$
$$AD|BC : n_{1001} + n_{0110}$$

Each of these conditions has an associated probability (starting from the root) based on p and q (Eq. 11.2):

$$
\begin{aligned}
pr_{1100} &= pq\left[(1-q)^2(1-p) + q^2 p\right] \tag{11.2}\\
pr_{0011} &= (1-q)(1-p)\left[q(1-q)(1-p) + (1-q)pq\right]\\[6pt]
pr_{1010} &= p(1-q)\left[q^2(1-p) + (1-q)^2 p\right]\\
pr_{0101} &= (1-p)q\left[q(1-q)p + (1-q)q(1-p)\right]\\[6pt]
pr_{1001} &= p(1-q)\left[q(1-q)p + (1-q)q(1-p)\right]\\
pr_{0110} &= (1-p)q\left[q^2(1-p) + (1-q)^2 p\right]
\end{aligned}
$$

In order for the parsimonious result to return the model tree, the probability of those characters supporting the tree must be greater than that for the two alternatives (Eq. 11.3).

$$pr_{1100} + pr_{0011} \geq pr_{1010} + pr_{0101}, \; pr_{1001} + pr_{0110} \qquad (11.3)$$

If $q \leq \frac{1}{2}$ (which we assume), then $pr_{1010} + pr_{0101} \geq pr_{1001} + pr_{0110}$. Hence, the condition we require is that $pr_{1100} + pr_{0011} \geq pr_{1010} + pr_{0101}$. This will be achieved when the probability of two parallel changes in p exceeds that of a single change in q (Eq. 11.4).

$$p^2 \leq q(1-q) \qquad (11.4)$$

The key relationship is between p and q. As long as p grows with respect to q, parsimony will be increasingly unlikely to return the model tree (Fig. 11.2)[1].

Criticisms and qualifications of this result are argued in discussions of the relative merits of optimality criteria and are discussed in Chapter 13.

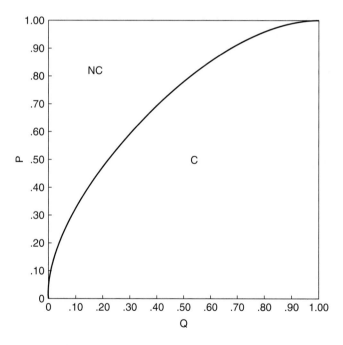

Figure 11.2: The "Felsenstein Zone" (NC) of statistical inconsistency of parsimony (Felsenstein, 1978).

---

[1]This effect is removed if p/q is constant and p and q become adequately small (Felsenstein, 1973) or the number of states increases sufficiently (Steel and Penny, 2000).

11.2   Maximum Likelihood and Trees

From the discussion of Section 6.1.7, the likelihood of a hypothesis (in this case a tree T) given data D, is proportional to the probability of the data given the tree (and some model; Eq. 11.5, Edwards, 1972).

$$l(T|D) \propto pr(D|T) \tag{11.5}$$

A systematic ML method selects T such that $pr(D|T)$ is maximized. This statement includes the requirement of knowledge of a broad variety of quantities needed to determine the likelihood. These include transformation models, edge distribution (branch lengths), and other parameters bundled together under the term "nuisance parameters."

11.2.1   Nuisance Parameters

Nuisance parameters are all those aspects required to calculate $pr(D|T)$ other than the data and tree topology. The three most important and commonly specified nuisance parameters are 1) transformation model (probabilities of change between character states), 2) edge parameters (time and rate of change along branches), and 3) distribution of rates of change among characters. These parameters can be denoted collectively by θ, and are estimated from observed data (as with edge parameters), or chosen to maximize the likelihood of a tree or trees. An important assumption for the analysis of character data is that they are *independent and identically distributed* (i.i.d.). This allows the joint likelihood of several characters to be calculated as the product of their individual values. Certainly, for many character types, this is not reasonable (*e.g.* stem and loop sequence characters in rRNA). However, distributional models can account for this to a large extent (although dynamic character types would be an exception).

   If we have knowledge of the distribution of the nuisance parameters $\Phi(\theta|T)$, we can integrate out θ (within parameter space Θ) to determine $p(D|T)$ (Eq. 11.6).

$$p(D|T) = \int_{\theta \in \Theta} p(D|T, \theta) d\Phi(\theta|T) \tag{11.6}$$

That T which maximizes $p(D|T)$ in this way is referred to as the *maximum integrated likelihood* (MIL) (Steel and Penny, 2000). The MIL is also the MAP Bayesian estimate (Chapter 12) if the distribution of tree priors is uniform (flat).

   When discussing stochastic model-based systematic methods, it can be useful to determine the probability that a given method, M, will return the "true" tree given a tree T and set of model parameters θ, $\rho(M, T, \theta)$. If we have $\Phi(\theta|T)$ and a prior distribution of trees, $p(T)$, the nuisance parameters and tree can be integrated out, identifying M with the highest expectation of success (Eq. 11.7).

$$\rho(M) = \sum_T p(T) \int_{\theta \in \Theta} \rho(M, T, \theta) d\Phi(\theta|T) \tag{11.7}$$

Székely and Steel (1999) showed that $\rho(M)$ is maximized for the method that returns T with maximum $p(T)pr(D|T)$. This is the Bayesian *maximum a posteriori* or MAP tree. As mentioned above, this is identical to the MIL tree when all prior probabilities of trees are equal. The use of non-uniform tree priors (such as empirical or Yule) breaks this identity.

11.3 Types of Likelihood

As mentioned above, θ can have many complex components, and we are unlikely to have much knowledge of their distribution. One approach to circumvent this problem is to choose θ such that $p(D|T, \theta)$ is maximized. This is referred to as *maximum relative likelihood* (MRL). In general, this is the methodology used in empirical analyses. Problems may arise when $p(D|T, \theta) > p(D|T', \theta')$ for a low probability θ (if we were to have $\Phi(\theta|T)$) while for a set of high probability θ, $p(D|T', \theta') > p(D|T, \theta)$. Steel and Penny (2000) cite such an example in a four-taxon case where parsimony outperforms MRL. MRL operates in absence of $p(T)$ and $\Phi(\theta|T)$, allowing likelihood analysis of systematic data. There are, however, further distinctions among MRL methods.

11.3.1 Flavors of Maximum Relative Likelihood

There are three variants in the manner in which non-leaf character states are determined. The most usual method is to sum over all possible vertex state assignments weighted by their probabilities. In the nomenclature of Barry and Hartigan (1987), this is referred to as *maximum average likelihood* (MAL). An alternative would be to assign specific vertex states (as well as other parameters) such that the overall likelihood of the tree is maximized. Barry and Hartigan (1987) suggested this method, naming it *most parsimonious likelihood* (MPL, sometimes referred to as *ancestral maximum likelihood*). This would appear to be convergent with parsimony, but the edge probabilities are the same over all characters hence MPL will not (in general) choose the same tree as parsimony.

John Hartigan

A third variant was proposed by Farris (1973a) and termed *evolutionary path likelihood* (EPL). In this form, the entire sequence of intermediate character states between vertices are specified such that the overall tree likelihood is maximized. Interestingly, the tree which maximizes this form of likelihood is precisely the most parsimonious tree. This result holds for a broad and robust set of assumptions (there is no requirement of low or homogeneous rates of character change for example). This would conflict with Felsenstein's assertion of ML methods being consistent and Farris' result that MP is an ML method. This seeming paradox is resolved when it is realized that the forms of likelihood discussed by Farris and Felsenstein (and the separate analogous MP = ML results of Goldman, 1990 and Tuffley and Steel, 1997) differ (Fig. 11.3). For the remainder of this discussion, when we talk of ML methods, we will be referring to MAL.

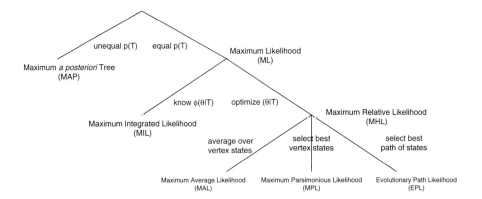

Figure 11.3: A classification of likelihood methods employed in systematics.

11.4   Static-Homology Characters

11.4.1   Models

Character Transformation

We can create a general model for a character of n states, with instantaneous transition (rate) parameters between states i and j, R_{ij}, and a vector of state frequencies Π (Eq. 11.8; Yang, 1994a).

$$R = \begin{bmatrix} R_{00} & \dots & R_{0n} \\ \cdot & \dots & \cdot \\ \cdot & \dots & \cdot \\ \cdot & \dots & \cdot \\ R_{n0} & \dots & R_{nn} \end{bmatrix} \Pi = \begin{bmatrix} \pi_0 \\ \cdot \\ \cdot \\ \cdot \\ \pi_n \end{bmatrix} \tag{11.8}$$

In general, we require several symmetry conditions of R (Eq. 11.9).

$$\forall i \; R_{ii} = 0 \tag{11.9}$$
$$\forall i, j \; R_{ij} = R_{ji}$$
$$\sum_{i=1}^{n} \sum_{j=1}^{n} \pi_i \cdot \pi_j \cdot R_{ij} = 1$$

The combination of these two matrices yields the Q, or rate matrix, of Tavaré (1986) (Eq. 11.10).

$$Q_{ij} = \begin{cases} R_{i,j} \cdot \pi_j & i \neq j \\ -\sum_{m=1}^{n} R_{i,m} \cdot \pi_m & i = j \end{cases} \tag{11.10}$$

The probability of change (P) between states i and j in time t can be calculated from elementary linear algebra (Eq. 11.11; Sect. 6.2):

$$P_{i,j}(t) = \sum_{m=1}^{n} e^{\lambda_m t} \cdot U_{m,i} \cdot U_{j,m}^{-1} \tag{11.11}$$

with λ_m, the eigenvalues of Q, U the associated matrix of eigenvectors and U^{-1}, its inverse (Strang, 2006). The time-reversible constraint of the matrix allows efficient computation of tree likelihoods.

This formulation is the most general (if symmetrical) description of a Markov process for n character states. This model has, at most, $n - 1$ independent frequency parameters (Π; one for each state, but the total must sum to 1) and $\binom{n}{2} - 1$ independent rate parameters (R) due to the constraints above (Eq. 11.9).

Special Cases

All character transformation models in use today, from the simple binary model of Felsenstein (1973), through the four state homogeneous Jukes and Cantor (1969) to General-Time-Reversible models for four (Lanave et al., 1984; Tavaré, 1986) and five states (McGuire et al., 2001; Wheeler, 2006), are simplifications of the most general process through symmetry requirements (*e.g.* transversions equal). All of the named models other than GTR (*e.g.* JC69) are special cases where analytical solutions are known (as opposed to computationally determining eigenvalues and applying Eq. 11.11). The hierarchy of simplifications for four states is illustrated in Swofford et al. (1996) (Fig. 11.4).

11.4.2   Rate Variation

In addition to models of character transformation, there are also distributional models of character change rates. These are most frequently used in the analysis of molecular sequence data where aligned nucleotide characters are analyzed as

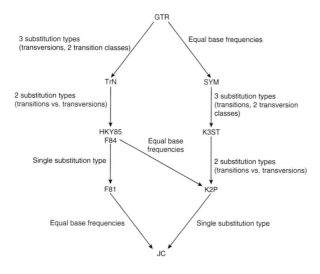

Figure 11.4: Swofford et al. (1996) relationships among DNA substitution likelihood models from the least parameterized JC69 to the most, GTR.

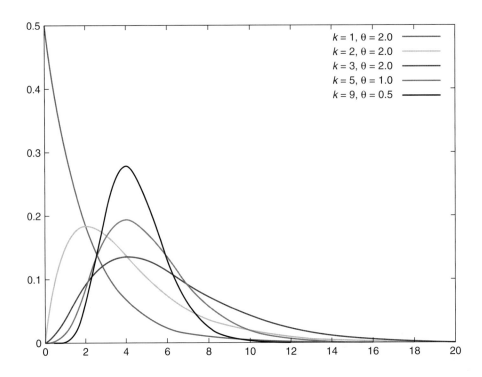

Figure 11.5: The gamma distribution with shape parameter $\alpha = k$, $\beta = \theta$.

a block. Positions vary in their observed levels of variation (hence, evolutionary rates), and this is accommodated by adding variation to the global rates of change used to calculate tree likelihoods.

The two most common are the fraction of invariant sites (Hasegawa et al., 1985) and discrete-gamma distribution (Yang, 1994a). The notion behind the use of an invariant sites parameter (usually referred to by I) is that one frequently observes many invariant positions with sequence data and accounting for this class of positions with a global rate is undesirable. Hence, a parameter is added to account for the fraction of sites available for substitution.

The gamma distribution (used in its computable discrete form), adds additional classes of positional rates based on a shape parameter α. The distribution (Eq. 11.12, Fig. 11.5) has a mean of α/β and variance of α/β^2, but we usually set $\beta = \alpha$ for a mean of 1.

$$g(x; \alpha, \beta) = \frac{\beta^\alpha x^{\alpha-1} e^{-\beta x}}{\Gamma(\alpha)} \tag{11.12}$$

The user specifies a number of rate classes (often in concert with invariant sites) and estimates α such that the tree likelihood is maximized. It is important to note that all rate classes are applied to each position, as opposed to a single class to a given position. For n taxa, m characters (*e.g.* aligned nucleotide sites), s states, and r rate classes, the overall memory consumption will be $O(nmsr)$.

11.4.3 Calculating $p(D|T, \theta)$

For a single character (x) on a tree, the likelihood of internal vertex i (L_i) with descendant vertices j and k would be the sum of the probability between x_i and each state in each descendant (given the edge parameter t; Fig. 11.6) multiplied by its respective likelihood and summed over all states. The character likelihoods are multiplied over the entire data set to determine the tree likelihood (Eq. 11.13).

$$L_i(x) = \sum_i^{states} \left[\left(\sum_{x_j} p_{x_i,x_j}(t_j) L_j(x_j) \right) \times \left(\sum_{x_k} p_{x_i,x_k}(t_k) L_k(x_k) \right) \right] \quad (11.13)$$

When edge weights are not known (as in nearly all real data situations), they must be estimated. This can be done in several ways, but all rely on calculation of the marginal likelihood (holding all other parameters constant) of a given edge assuming a variety of weights (t parameter) and choosing the optimal value (Fig. 11.7). Often Brent's Method (Brent, 1973) or Newton–Raphson (Ypma, 1995) is used to estimate the edge parameters.

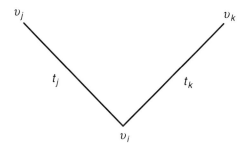

Figure 11.6: Labeled subtree for likelihood calculations.

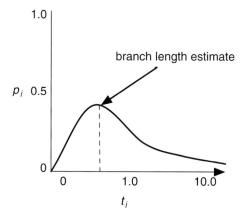

Figure 11.7: Estimate of edge weight parameter t by maximizing the probability of transformation along the edge, p_i.

The determination of the MAL of a tree is a heuristic procedure due to the large number of parameter estimations involved. As with parsimony optimization (Chapter 10), the tree is traversed setting median states recursively. This recursion is initialized with the likelihood of leaf states at 1 (no need to sum to one for likelihood) and all other leaf states 0. The likelihood is calculated via a post-order tree traversal from the tips to the root multiplied by the prior probabilities of the states themselves (Eq. 11.14).

$$L_T(x) = \prod_{i=1}^{states} \pi_i \prod_{\forall u,v \in E} L_{u,v} \qquad (11.14)$$

Given that these values can be quite small, it is often convenient to speak of log or $-$ log likelihood values[2]. The following example assumes that the edge parameters are known. If this is not so (which is usually the case), such a single post-order traversal will not be sufficient to determine the tree likelihood. An iterative edge refinement procedure will be required to optimize the edge parameters (Felsenstein, 1981).

An Example

Consider a single nucleotide character analyzed under the JC69 model (Fig. 11.9). If we fix all the edge probabilities, $\mu t = 0.1$, we can calculate the likelihood of the topology given the analytical probabilities in Equation 11.15.

$$P_{ij}(t) = \begin{cases} \frac{1}{4} + \frac{3}{4}e^{-\mu t} & i = j \\ \\ \frac{1}{4} - \frac{1}{4}e^{-\mu t} & i \neq j \end{cases} \qquad (11.15)$$

Hence, the edge probabilities are given in Equation 11.16.

$$P_{ij}(t) = \begin{cases} 0.929 & i = j \\ \\ 0.0238 & i \neq j \end{cases} \qquad (11.16)$$

A subtree example with leaf states (A and C) and edge parameters 0.1 is shown in Figure 11.8.

The overall likelihood for the tree in Figure 11.9 is 1.76×10^{-6} or, in familiar $-$log (base e) units, 13.25.

11.4.4   Links Between Likelihood and Parsimony

Typical likelihood analyses employ several homogeneity conditions. Usually the same edge parameter is applied to all characters (although it may vary over

---

[2]The finite precision of computers can cause problems for likelihood calculations (floating point error) due to the large number of operations required when evaluating trees. Alternate implementations of the same algorithm may well generate likelihoods that differ non-trivially. Extreme care must be taken to avoid this problem.

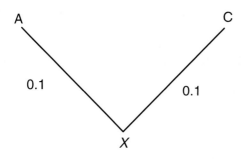

$$L(x = A) = [0.929 \cdot 1.0] \times [0.0238 \cdot 1.0] = 0.0221$$

$$L(x = C) = [0.0238 \cdot 1.0] \times [0.929 \cdot 1.0] = 0.0221$$

$$L(x = G) = [0.0238 \cdot 1.0] \times [0.0238 \cdot 1.0] = 0.000566$$

$$L(x = T) = [0.0238 \cdot 1.0] \times [0.0238 \cdot 1.0] = 0.000566$$

$$\text{Total } L(x) = 0.0453$$

Figure 11.8: Labeled subtree with likelihood calculations.

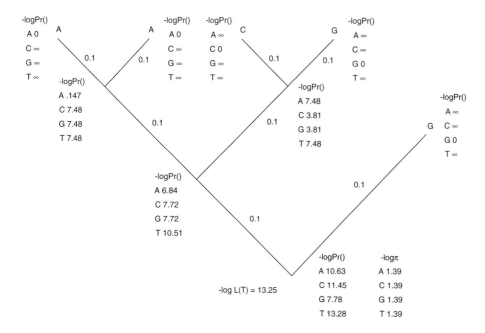

Figure 11.9: An example likelihood calculation under JC69 model with all edge parameters set to 0.1.

Nicholas Goldman

Chistopher Tuffley

edges), and the same model as well. Under these conditions, MAL will frequently lead to results at variance with parsimony. As mentioned earlier, Farris (1973a) employed a simple model to show that parsimony and EPL would choose the same optimal tree. Such connections are not limited to this scenario.

Goldman (1990) discussed a number of scenarios involving likelihood, parsimony, and compatibility. Goldman showed that when edge weights are constant over the tree, likelihood and parsimony will converge. More recently, Tuffley and Steel (1997) discussed the No-Common-Mechanism (NCM) model, where each character has a potentially unique rate that may vary among edges as well (hence the name). The rate for each character on each edge is optimized (either zero, or infinite) to maximize the likelihood. Under a Neyman (1971) type model with r states, the overall likelihood of the tree can be determined as a function of the number and distribution of parsimony changes on the tree (Eq. 11.17), with r_i, the number of states exhibited by character i, χ_i, the parsimonious vertex states assignments for character i, and $-l(\chi_i, T)$, the parsimony length of assignment χ for character i on tree T.

$$L_T(X) = \prod_{i=1}^{k_{characters}} r_i^{-l(\chi_i, T)-1} \tag{11.17}$$

For the tree and leaf states of Figure 11.9, the likelihood would be $3^{-(2+1)} = 0.037$. It is often said that this model leads to equivalent results between parsimony and likelihood, but this will only occur when the number of states of each character (r_i) is a constant over the data set. In this way, NCM can be viewed as a likelihood-based character weighting scheme in parsimony analyses.

11.4.5   A Note on Missing Data

Missing data are not, in principle, a problem for likelihood analyses. Leaf state vectors can be set to 1.0 for each of the observed (in the case of polymorphism) or implied (all states = 1.0) states in the case of entirely missing observations. Implementations, however, may differ in the treatment of these unknown observations. Currently, implementations treat missing data in this manner. Obviously, this can have an effect on analyses. This issue can become all the more pernicious when coupled with the practice of treating indels or "gap" characters as missing values (as opposed to a 5th state). Though clearly suboptimal (and unnecessary, as shown later), such a treatment of indels is common and problematic.

11.5   Dynamic-Homology Characters

As with parsimony, maximum likelihood can be applied to the analysis of dynamic-homology characters. With sequence (nucleotide and amino-acid) and higher order characters (*e.g.* gene rearrangement), two general approaches have been taken in the construction of stochastic models. The first uses a simple extension of 4-state nucleotide or 20-state amino-acid models to include "gaps" as a 5th or 21st state (Wheeler, 2006). These models treat indels as atomic events, emphasize simplicity and make little attempt to model reality *per se.* The second approach makes an explicit attempt to model the process of sequence change including indel events (Thorne et al., 1991, 1992), resulting in more complex scenarios.

In general, models describing the process of gene rearrangement are not attempts to describe the mechanisms of genomic change as much as descriptive statements of the frequency and patterns of change (*e.g.* Larget et al., 2004).

11.5.1   Sequence Characters

In order to perform a dynamic homology analysis (Tree Alignment Problem; Chapter 10) of multiple leaf sequences related by a tree, several components are required. First, a model must be specified allowing both element substitution and insertion–deletion (indel). Second, a procedure needs to be identified to calculate the likelihood "distance" between any pair of sequences. And third, a method of creating sequence medians (vertex or HTU sequences) must be described.

Models

$n + 1$ *State Models*—A simple expansion of sequence substitution models to include an extra state for "gaps" representing indels (such as the r-state model of Neyman, 1971) has been used by McGuire et al. (2001) in their Bayesian analysis of pre-aligned sequences and the ML Direct Optimization (ML–DO) of Wheeler (2006).

In the symmetrical ($r_{ij} = r_{ji}$, R of Tavaré, 1986) general 5-state case there are five state frequencies to be specified (A, C, G, T, -), although they must sum to 1, and 10 transition rates among the states (Fig. 11.10). As with the GTR model of sequence substitution above, there are a broad variety of special case models that can be constructed by enforcing various additional symmetry conditions (such as JC69+Gaps, Eq. 11.18; Wheeler, 2006).

$$
P_{ij}(t) = \begin{cases} \frac{1}{5} + \frac{4}{5}e^{-\mu t} & i = j \\[2mm] \frac{1}{5} - \frac{1}{5}e^{-\mu t} & i \neq j \end{cases} \tag{11.18}
$$

Considering the example alignment of 11.19 under the model in Equation 11.18 with an edge weight (branch length) of 0.1 (μt), $p(I, II) = (0.01903)^3(0.9239)^2 = 5.882 \times 10^{-6}$.

$$\text{Sequence I AC-GT} \tag{11.19}$$

$$\text{Sequence II AGC-T}$$

	A	C	G	T	-
A	$-(\pi_C\alpha+\pi_G\beta+\pi_T\gamma+\pi_-\delta)$	$\pi_C\alpha$	$\pi_G\beta$	$\pi_T\gamma$	$\pi_-\delta$
C	$\pi_A\alpha$	$-(\pi_A\alpha+\pi_G\varepsilon+\pi_T\zeta+\pi_-\eta)$	$\pi_G\varepsilon$	$\pi_T\zeta$	$\pi_-\eta$
G	$\pi_A\beta$	$\pi_C\varepsilon$	$-(\pi_A\beta+\pi_C\varepsilon+\pi_T\theta+\pi_-\kappa)$	$\pi_T\theta$	$\pi_-\kappa$
T	$\pi_A\gamma$	$\pi_C\zeta$	$\pi_G\theta$	$-(\pi_A\gamma+\pi_C\zeta+\pi_G\theta+\pi_-\nu)$	$\pi_-\nu$
-	$\pi_A\delta$	$\pi_C\eta$	$\pi_G\kappa$	$\pi_T\nu$	$-(\pi_A\delta+\pi_C\eta+\pi_G\kappa+\pi_T\nu)$

Figure 11.10: A general, symmetrical, 5-state model (states A, C, G, T, '-').

These models have the virtue of simplicity and ease of calculation, hence can be applied to real data sets with multiple loci and empirically interesting (>100) numbers of taxa (Whiting et al., 2006).

Birth–Death Model—The Thorne et al. (1991) and Thorne et al. (1992) models (TKF91 and TKF92), treat the insertion–deletion process in an alternate fashion. There are two components to the calculation of the probability of transforming one sequence into another: the probability of an alignment (α as in 11.19) given a set of insertions, deletions, and matches and model $[p(\alpha|\alpha',\theta)]$; and the probability of a specific pattern of indels and matches given a model $[p(\alpha'|\theta)]$. The method couples a birth–death process (parameters λ—insertion or birth rate; μ—deletion or death rate) with standard four-nucleotide substitution models.

Both TKF91 and TKF92 model the indel process in the same way, transforming one sequence into another (the model is symmetrical). There are three sorts of events. The first is an insertion (not leading) in the first sequence to yield the second (p). The second transformation type is a deletion (p'), and the third, a leading insertion, takes place before the left-most residue (p''). The probabilities of these structural events are as in Equation 11.20, with λ birth rate (insertion), μ death rate (deletion), $n > 0$ indel size, and time t.

$$
\begin{aligned}
p_n(t) &= e^{-\mu t}\left[1 - \lambda\beta(t)\right]\left[\lambda\beta(t)\right]^{n-1} \\
p'_n(t) &= \left[1 - e^{-\mu t} - \mu\beta(t)\right]\left[1 - \lambda\beta(t)\right]\left[\lambda\beta(t)\right]^{n-1} \\
p''_n(t) &= \left[1 - \lambda\beta(t)\right]\left[\lambda\beta(t)\right]^{n-1}
\end{aligned}
\tag{11.20}
$$

with

$$
\beta(t) = \frac{1 - e^{(\lambda-\mu)t}}{\mu - \lambda e^{(\lambda-\mu)t}}
$$

The substitution process follows standard models with state frequencies determining the probability of inserting a specific sequence.

11.5.2   Calculating ML Pairwise Alignment

Both the above models can be optimized for two sequences by versions of the familiar dynamic programming procedure used for pairwise sequence alignment (Sect. 8.4). Here, we discuss the algorithm for the $n + 1$ state model. The recursions are more complex for TKF92, but they follow the same basic outline (see Thorne et al., 1992 for specifics).

Dynamic Programming

In order to calculate the probability of transforming one sequence into another (or a pairwise alignment; as with parsimony the cost is identical for two sequences), three elements are required: the sequences, the transformational model, and a time parameter to mark the differentiation between the sequences $[p(I, II|\theta, \tau)]$. Dynamic programming will optimize the likelihood for a given t, but as with edge weight/branch length optimization, the procedure must be repeated, varying or estimating t until the likelihood is optimized (Eq. 11.21).

$$p(I, II|\theta) = \max_t p(I, II|\theta) \tag{11.21}$$

Since t is chosen to maximize the pairwise probability, the method will yield an MRL.

It is often convenient to work with the negative logarithm of likelihood and probability values as opposed to their absolute values and, in this case, it allows an elegant modification of the Needleman and Wunsch (1970) algorithm (Alg. 8.1). Based on model and time, the conditional probability of an indel or element match can be calculated *a priori*. In the scenario above (JC69+Gaps with $t = 0.1$), the probability of an indel is $\frac{1}{5} - \frac{1}{5}e^{-0.1} = 0.01903$, an element mismatch (substitution) is the same $\frac{1}{5} - \frac{1}{5}e^{-0.1} = 0.01903$, while an element match $\frac{1}{5} + \frac{4}{5}e^{-0.1} = 0.9239$. Using the logarithms of these values, the multiplicative probabilities of a scenario can be optimized as additive sums $[\log(p(x_i) \cdot p(x_j)) \rightarrow \log p(x_i) + \log p(x_j)]$ by treating them as match, mismatch, and indel costs.

Although the log transform probabilities can be used as edit costs (*i.e.* $cost[i][j] = \log p(I_i, II_j|\theta, t)$), the core recursion requires a modification. The probability of inter-transforming (or aligning) two sequences is the sum of the probabilities of all potential transformation (or alignment) scenarios between the two. As we know (Eq. 8.6; Slowinski, 1998), there are a large number of these to calculate. The Needleman–Wunsch algorithm can accomplish this when the central alignment recursion is changed to a sum as opposed to the minimum of three paths. This sum is taken among the probabilities (*not* log probabilities) of the three options (element insertion, deletion, and match) at each cell (Eq. 11.22) ($cost[i][j]$ is the $-\log$ transformed likelihood).

$$
\begin{aligned}
cost[i][j] = \ & \log(e^{-(cost[i-1][j-1]+\sigma_{i,j})} \\
& + e^{-(cost[i-1][j]+\sigma_{indel})} \\
& + e^{-(cost[i][j-1]+\sigma_{indel})})
\end{aligned} \tag{11.22}
$$

For sequences of lengths n and m, $p(I, II|\theta, t) = e^{-cost[n][m]}$. The traceback diagonal marks the maximum likelihood path as before. The complete matrix (in $\log_e$ units) is shown in Figure 11.11 resulting in a $p(I, II|\theta = JC69+Gaps, t = 0.1) = 0.0007849$. If one were to calculate the probability directly from the four aligned positions the value would be $0.01903^2 \cdot 0.8535^2 = 0.0002638$, considerably lower than that yielded by the algorithm. This is because the specific

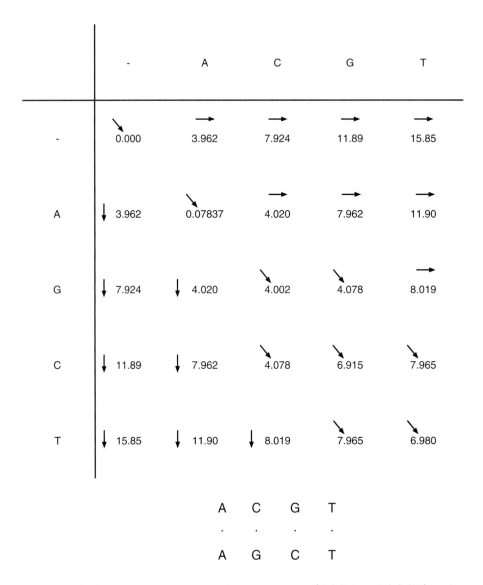

Figure 11.11: Likelihood alignment of two sequences (ACGT and AGCT) under the JC69+Gaps (5-state Neyman) model with a time parameter (μt) of 0.1 ($\log_e$ units).

alignment produced is only one of many alignment scenarios that contribute to the total probability of transformation between the sequences. This particular alignment has the highest probability of all possible alignments, hence is termed the *dominant* likelihood alignment (in the terminology of Thorne et al., 1991). We can search directly for this by choosing the maximum probability choice (insertion, deletion, or element match) in Eq. 11.22 as opposed to the sum (Fig. 11.12). The probability produced in this way jibes precisely with that

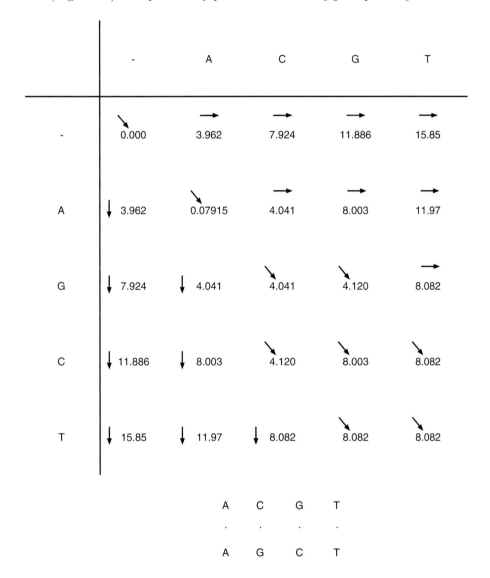

Figure 11.12: Dominant likelihood alignment of two sequences (ACGT and AGCT) under the JC69+Gaps (5-state Neyman) model with a time parameter (μt) of 0.1 ($\log_e$ units).

expected ($e^{-8.082} = 0.0003091$). In this case, both procedures yielded the same alignment, but this need not be the case in general.

The distinction between *dominant* and *total* likelihood is an important one. A single alignment may be "best" in a likelihood context, but may contain a very small fraction of the total likelihood (in this case 33%). In the context of likelihood forms discussed above, the dominant likelihood is akin to an MPL object, and the total likelihood score MAL. When sequence change is analyzed on a tree, these distinctions have downstream ramifications in the identification of ML trees, and character change maps on those trees.

11.5.3   ML Multiple Alignment

As with parsimony, there are relatively direct extensions of pairwise alignment to multiple sequence alignment (MSA). The approach of Wheeler (2006) was to create an implied alignment (Wheeler, 2003a) using the maximum likelihood form of Direct Optimization (Wheeler, 1996). In this ML–TAP approach, medians (and tree topologies) are chosen to optimize likelihood under a variety of models from a 5-state Neyman scenario to an enhanced GTR+Gaps model. The relative performance of parsimony and ML implied alignments was tested by Whiting et al. (2006), showing (comfortingly) that ML MSA were superior for ML (by 10% log likelihood units) while those based on parsimony were superior for parsimony analysis (by 30%; manual alignments were distant finishers; Table 11.1).

MSA methods based on the TKY92 model (Thorne et al., 1992), such as Fleissner et al. (2005) and Redelings and Suchard (2005), make use of Bayesian Hidden Markov Models and are discussed briefly above and in Chapter 12 in more detail.

11.5.4   Maximum Likelihood Tree Alignment Problem

Although it is as yet unstudied, given the NP–hard nature of the parsimony version of the TAP, the ML variant is likely to be extremely challenging if not NP–hard itself. As with parsimony heuristics to the TAP, we can generate several heuristic ML–TAP procedures. Unfortunately, almost nothing is known about the quality of these solutions (boundedness).

	ClustalX	Manual	DO–MP	DO–ML
Mixed Model Likelihood	61,489.630	55,329.945	51,611.928	50,496.073
Single Model Likelihood	61,548.268	55,858.397	51,554.225	51,014.655
Parsimony Tree Length	15,154	20,341	11,483	11,702

Table 11.1: Performance of ClustalX (Higgins and Sharp, 1988), Manual, DO–MP, and DO–ML multiple sequence alignment (Whiting et al., 2006). DO implied alignment runs were created using Wheeler et al. (2005) and ML scores by Huelsenbeck and Ronquist (2003).

Medians and Edges

As with parsimony heuristics to the TAP, the identification of median sequences is crucial to the quality of the solution. ML–TAP has the added factor of edge or branch time. As mentioned above for the two sequence case, the probability of the alignment is dependent on the time parameter that is identified numerically through repeated (or estimated) likelihood optimizations.

Sections 10.9.2, 10.9.3, and 10.9.4 identified sequence medians in ways that are directly applicable to ML. The algorithm for determining sequence medians using Direct Optimization (Alg. 10.7; DO, Wheeler, 1996) can be applied largely without modification. Two issues merit attention. The first is the use of dominant or total likelihood for tree likelihood values and medians. Total likelihood will reflect more of the probability of alternate medians, but unless these medians are of optimal (in this case highest probability) cost, they will not be reflected in the median sequences. Dominant likelihood calculations maintain a more consistent approach in that the tree likelihoods are directly traceable to these specific sequences. When the total likelihood is used, this connection can be lost (Wheeler, 2006). The second issue centers around the median determination and time parameter. The time parameter interacts with the median identification process, not only to determine the probability of an ancestor–descendant transformation, but the ancestral sequences (= medians) themselves. When edge times are estimated to optimize likelihood scores, the medians themselves are likely to change, creating additional time complexity in the process. This is especially prominent when using iterative improvement methods (Sankoff and Cedergren, 1983; Wheeler, 2003b, 2006). With iterative improvement, there are three edges incident on a vertex which require simultaneous optimization in addition to the 3-dimensional median calculation.

Lifted, Fixed-States, and Search-Based (Sect. 10.9.3) procedures deal with a fixed pool of medians, hence that component of time complexity is reduced. Edge iteration is still an issue in two ways. First, the pairwise probability of transformation between states is time dependent. This can be either held constant over all state pairs, or be optimized (in a fashion akin to Tuffley and Steel, 1997) uniquely for each sequence pair. Secondly, edge times can be applied while a tree is optimized (using a single time) or using optimized times from the pairwise sequence comparisons.

The issue of dominant and total likelihood also enters in this class of heuristics through the summing (as in average likelihood) over all potential sequence medians, or the identification of the most likely medians (MPL) and determining tree likelihood on that basis.

Wheeler (2006) discussed the above ML–TAP heuristics in the context of a 5-state model, although they could be applied to other models. Fleissner et al. (2005) developed a heuristic ML–TAP procedure specifically for the TKF92 model. In their approach, simulated annealing (Sect. 14.7) is used in two ways alternately. The first is to optimize the analytical parameters (substitution model parameters, indel birth and death values, fragment length), and the second to optimize the alignment patterns of indels (h, α' of Thorne et al., 1991) and tree topology. The method initializes with a Neighbor-Joining (Saitou and Nei, 1987) tree and performs NNI (Sect. 14.3.2) to break out of local topological minima. The parameter and topology/indel pattern optimizations proceed alternately until improvements are no longer found. Due to the complexity of the TKF92 model and the simulated annealing approach, the method does not scale well and can only be used on a handful (<20) of sequences of moderate (< 500bp) length.

11.5.5 Genomic Rearrangement

As with all stochastic procedures, the root of likelihood-based reconstruction of genomic rearrangement data is the model. Currently, models are descriptive, that is, distributions of gene rearrangements are chosen and fit to empirical patterns, not based on any first principles analysis of the biological mechanism of inversion or transposition (an exception exists in the Birth–Death model of gene family evolution of Zhang and Gu, 2004).

$$p(k, \lambda) = \frac{\lambda^k e^{-\lambda}}{k!} \tag{11.23}$$

The basic descriptive model was set out by Nadeau and Taylor (1984), grounded in the empirical observation of the distribution of chromosomal rearrangements between humans and mice (Fig. 11.13). They posited a Poisson distribution (Eq. 11.23) of rearrangement events (k) on the genome and along a tree edge at average rate λ. This was expanded by Wang and Warnow (2001, 2005) to create corrected distances for use in distance-based phylogenetic analysis (Chapter 9).

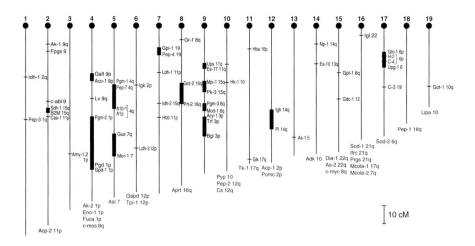

Figure 11.13: Mouse—Human rearrangements as illustrated by Nadeau and Taylor (1984).

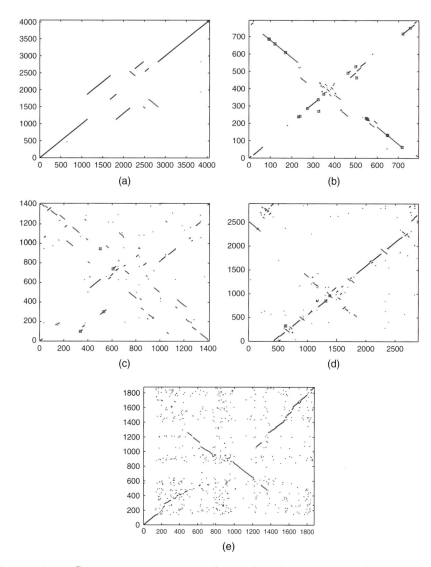

Figure 11.14: Genomic rearrangement locus dot-plot scenarios of Dalevi and Eriksen (2008): (a) = "Whirl," (b) = "X-model," (c) = "Fat X-model," (d) = "Zipper," and (e) = "Cloud." See Plate 11.14 for the color Figure.

Empirical Models

Dalevi and Eriksen (2008) presented a series of corrected distance estimates for five rearrangement scenarios named according to patterns on pairwise dot-plots (Fig. 11.14).

(a) "Whirl"—caused by an overrepresentation of uniformly distributed reversals across the genomes.

(b) "X-model"—due to a preponderance of reversals symmetrically distributed around the origins and terminations of replication.

(c) "Fat X-model"—explained by symmetrically distributed reversals with enhanced variation in their position with respect to the origins and terminations of replication.

(d) "Zipper"—thought to result from a large amount of short reversals (up to 5% of the genome) distributed uniformly over the genome.

(e) "Cloud"—as rearrangements accrue, the gene order becomes randomized loosing the previous patterns into a "cloud."

In general, these descriptions of rearrangement patterns are not used to reconstruct trees directly, but to estimate overall dissimilarity for distance analysis.

11.5.6 Phylogenetic Networks

As with parsimony (Sect. 10.14), horizontal gene transfer and hybridization can be explained by networks and in an analogous fashion (Strimmer and Moulton, 2000). Jin et al. (2006) proposed no biological model of horizontal gene transfer or hybridization, but two methods to calculate the likelihood of the network. In the same way that the parsimony score of a network is calculated by summing the minimum tree costs consistent with the network (Eq. 10.9) for each block of characters, likelihoods can be multiplied over the best likelihood tree for each character block. A second option would be to sum the likelihoods of all tree scenarios consistent with the network (Fig. 11.15; Eq. 11.24).

$$N = (V, E) \tag{11.24}$$

$$L_N^{all}(S|N, \theta) = \sum_{T \in N} (p(T) \cdot L(S|T, \theta))$$

$$L_N^{best}(S|N, \theta) = \max_{T \in N} (p(T) \cdot L(S|T, \theta))$$

It is unclear which, if either, procedure is appropriate. The first method assumes all blocks are independent. This may or may not be reasonable given that the recognition of blocks is dependent on their relative positions and behavior. The second model has the advantage of including alternate scenarios, weighted by their likelihoods, but allows for multiple histories for all blocks.

11.6 Hypothesis Testing

11.6.1 Likelihood Ratios

Often, it is desirable to know whether a difference between two likelihood values is "significant." As odd as such a concept may seem within the rationale of

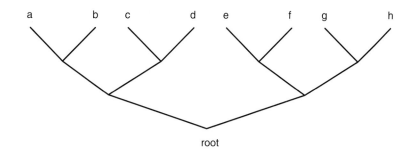

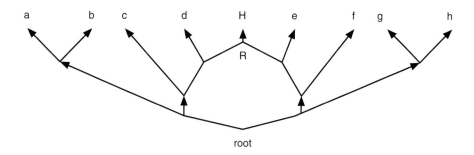

Figure 11.15: Phylogenetic tree (above) and network (below).

likelihood, given a few simple assumptions, such statements can be made (DeG-root and Schervish, 2006). The basic idea is that if the likelihood functions are well-behaved, twice the difference in the log of the ML value is distributed as χ^2 (Eq. 11.25). When the hypotheses to be compared are simple estimates of parameters (such as a branch length or comparison of two trees), this distribution will have one degree of freedom.

$$2\Delta l_{T,T'} = 2\log(l_T/l_{T'}) = 2(\log l_T - \log l_{T'}) \qquad (11.25)$$

Likelihood ratio tests are used to determine whether edge weights (time parameter) are greater than 0 and should be collapsed, or whether one of two competing and nearly optimal hypotheses is superior. The confidence value is (given the single degree of freedom) 1.9207 log likelihood units, or a likelihood ratio of 6.826. If two tree likelihoods are differ by at least this value, their difference is statistically significant (Felsenstein, 2004).

Branch Collapsing

The likelihood ratio can also be used to test if an edge probability (ML branch length) is significantly greater than zero. The likelihood of a tree with an edge constrained to have $\mu t = 0$ can be determined and compared to the likelihood of the tree optimized for the time parameter of that edge. As above (Eq. 11.25), the likelihood ratio can be tested via χ-squared with a single degree of freedom.

11.6.2   Parameters and Fit

As with all statistical fitting operations, increasing the number of parameters will increase the fit and decrease the error. In general, if the addition of a parameter results in a large increase in fit (here likelihood), we accept that parameter. The problem comes as more and more parameters are added and the increases in quality of solution (in terms of error) are less dramatic, leading to overparameterization and loss of predictivity (Fig. 11.16). An analysis of molecular sequence data using the JC69 model is based on zero parameters (everything is equal and nothing specified)[3]. The same data modeled using GTR would no doubt yield a better likelihood score using its eight parameters (five rate and three frequency). This might be further improved with invariant sites and discrete-gamma rate parameters. When should this stop? How can overparameterization be avoided?

There are two commonly used statistics to decide this. The first is the ratio of the likelihoods of solutions with different parameterization—the likelihood ratio test above. In the case of testing models, Equation 11.25 is distributed as $\chi^2_{p'-p}$ where p and p' are the number of parameters in the models to be compared.

Large sequence data sets nearly invariably choose the most complex models $(\text{GTR}+I+\Gamma)$[4] under this criterion, motivating the use of the alternate Akaike Information Criterion (AIC; Akaike, 1974). In the AIC, the test statistic is calculated as $-2\log l_T + 2p$ where p is the number of parameters used in the likelihood calculation for a tree T, l_T. An extra parameter is favored if it improves the likelihood by one log unit.

A third criterion, Bayesian Information Criterion (BIC), penalizes extra parameters more harshly with a term that depends on the data size n (Schwaz, 1978).

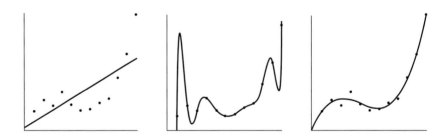

Figure 11.16: Data with various polynomial curves fitted to them.

---

[3]Even for JC69 there are other parameters in an analysis—one for each edge of the tree for example—but this is constant among models analyzing the same data set, hence plays no role in the marginal complexity of one model over another.

[4]The use of invariant sites simultaneously with Γ classes is problematic, since the parameters are not independent.

Model	*l*	K	AIC$_c$	Δ AIC$_c$	*w*	Cum(*w*)
TN93+I+Γ	5441.4600	78	11045.5888	0.0000	0.5221	0.5221
TIM+I+Γ	5441.3765	79	11047.5965	2.0077	0.1913	0.7134
HKY85+I+Γ	5443.6729	77	11047.8422	2.2534	0.1692	0.8826
K81uf+I+Γ	5443.5566	78	11049.7821	4.1934	0.0641	0.9468
GTR+I+Γ	5440.9150	81	11051.0301	5.4413	0.0344	0.9811
TVM+I+Γ	5442.7393	80	11052.4991	6.9103	0.0165	0.9976
TN93+Γ	5448.6792	77	11057.8549	12.2661	0.0011	0.9988
HKY85+Γ	5450.5068	76	11059.3402	13.7514	0.0005	0.9993
TIM+Γ	5448.6577	78	11059.9843	14.3955	0.0004	0.9997
K81uf+Γ	5450.4883	77	11061.4730	15.8843	0.0002	0.9999
GTR+Γ	5448.0298	80	11063.0802	17.4914	0.0001	1.0000
TVM+Γ	5449.6685	79	11064.1804	18.5917	0.0000	1.0000
TN93+I	5470.7568	77	11102.0102	56.4214	0.0000	1.0000
TIM+I	5470.7417	78	11104.1522	58.5635	0.0000	1.0000
GTR+I	5470.3452	80	11107.7110	62.1223	0.0000	1.0000
HKY85+I	5476.8496	76	11112.0257	66.4370	0.0000	1.0000
K81uf+I	5476.8208	77	11114.1381	68.5493	0.0000	1.0000
TVM+I	5476.1650	79	11117.1736	71.5849	0.0000	1.0000
F81+I+Γ	5769.1118	76	11696.5501	650.9614	0.0000	1.0000
F81+Γ	5782.0566	75	11720.2721	674.6834	0.0000	1.0000
F81+I	5807.4927	75	11771.1442	725.5554	0.0000	1.0000
GTR	5805.0576	79	11774.9588	729.3700	0.0000	1.0000
TVM	5808.4727	78	11779.6141	734.0254	0.0000	1.0000
TIM	5810.4102	77	11781.3168	735.7280	0.0000	1.0000
TN93	5813.4780	76	11785.2825	739.6938	0.0000	1.0000
K81uf	5813.5190	76	11785.3646	739.7758	0.0000	1.0000
HKY85	5816.5894	75	11789.3375	743.7488	0.0000	1.0000
SYM+I+Γ	5861.0859	78	11884.8407	839.2520	0.0000	1.0000
TVMef+I+Γ	5867.6128	77	11895.7221	850.1333	0.0000	1.0000
SYM+Γ	5876.7803	77	11914.0570	868.4683	0.0000	1.0000
TVMef+Γ	5884.4272	76	11927.1810	881.5922	0.0000	1.0000
TIMef+I+Γ	5885.0684	76	11928.4632	882.8745	0.0000	1.0000
K81+I+Γ	5893.7642	75	11943.6872	898.0984	0.0000	1.0000
TN93ef+I+Γ	5897.7529	75	11951.6647	906.0759	0.0000	1.0000
TIMef+Γ	5899.2588	75	11954.6764	909.0877	0.0000	1.0000
K80+I+Γ	5906.2329	74	11966.4593	920.8706	0.0000	1.0000
K81+Γ	5908.7876	74	11971.5687	925.9800	0.0000	1.0000
TN93ef+Γ	5911.5659	74	11977.1254	931.5366	0.0000	1.0000
SYM+Γ	5908.7021	77	11977.9008	932.3120	0.0000	1.0000
TVMef+I	5917.6128	76	11993.5521	947.9633	0.0000	1.0000
K80+Γ	5920.9038	73	11993.6382	948.0494	0.0000	1.0000
TIMef+I	5928.9629	75	12014.0846	968.4959	0.0000	1.0000
K81+I	5938.0137	74	12030.0209	984.4321	0.0000	1.0000
TN93ef+I	5940.7383	74	12035.4701	989.8813	0.0000	1.0000
K80+I	5949.5186	73	12050.8677	1005.2789	0.0000	1.0000
F81	6088.2227	74	12330.4388	1284.8501	0.0000	1.0000
JC69+I+Γ	6101.2656	73	12354.3618	1308.7730	0.0000	1.0000
JC69+Γ	6114.8408	72	12379.3515	1333.7628	0.0000	1.0000
JC69+I	6142.1719	72	12434.0137	1388.4249	0.0000	1.0000
SYM	6170.8916	76	12500.1097	1454.5209	0.0000	1.0000
TVMef	6190.3394	75	12536.8375	1491.2488	0.0000	1.0000
TIMef	6194.5806	74	12543.1547	1497.5659	0.0000	1.0000
TN93ef	6210.6353	73	12573.1011	1527.5123	0.0000	1.0000
K81	6214.1152	73	12580.0610	1534.4723	0.0000	1.0000
K80	6230.2100	72	12610.0898	1564.5011	0.0000	1.0000
JC69	6411.5161	71	12970.5438	1924.9551	0.0000	1.0000

Figure 11.17: Model test (Posada and Buckley, 2004) based on mitochondrial data of Sota and Vogler (2001). *l* is the log likelihood, *K* the number of parameters, AIC_l the Akaike Information Criterion, $DeltaAIC_l$ the difference in AIC_l with the next "best," *w* the Akaike weights, and Cum(*w*) the cumulative Akaike weights.

$$\text{BIC} = -2\log l_T + p\log n \qquad (11.26)$$

These tests are implemented in Posada and Crandall (1998) and well summarized in Posada and Buckley (2004). An example of a test of a broad variety of models in an empirical context is given by Posada and Buckley (2004) in their reanalysis of Sota and Vogler (2001) (Fig. 11.17).

11.7   Exercises

1. What is the probability of transformation between the aligned sequences ACGT and AGCT under the JC69 model with time parameters $\mu t = \{0.1, 0.2, 0.5, 1.0\}$?

2. What is the probability of transformation between the aligned sequences $\frac{ACGT}{AGCT}$, $\frac{ACG-T}{A-GCT}$, and $\frac{A-CGT}{AGC-T}$ under a 5-state Neyman model with time parameters $\mu t = \{0.1, 0.2, 0.5, 1.0\}$?

3. What were the maximum likelihood estimators of the time parameter in the previous two exercises? If the four given time parameter values were the only ones possible, what would the integrated likelihoods be? What fraction of the integrated likelihoods were the maximum values?

4. Using a Neyman model for binary characters, what would the likelihoods be for the two cladograms in Fig. 11.18 where all time parameters (μt) were 0.1? 0.2?

5. Under the No-Common-Mechanism model, what are the likelihoods for the cladograms in exercise 4?

6. Two systematists argue the question "ML using No-Common-Mechanism and parsimony will yield the same tree for this data set," one taking the affirmative and one the negative, who is correct? Why?

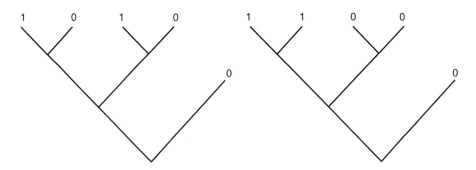

Figure 11.18: Example cladograms.

7. Using a Neyman model for sequence characters, and sequences ACGT and AGCT with time parameter $\mu t = 0.2$, determine the maximum likelihood alignment of the two sequences. What are the values of the "total" and "dominant" likelihoods? What fraction of the total likelihood is the dominant? Give an example of a non-dominant likelihood alignment (and its likelihood) included in the total likelihood calculation.

Chapter 12

Optimality Criteria–Posterior Probability

Thomas Bayes
(1702–1761)

Bayesian argumentation and methods have a long history in systematics. Edwards (1970) discussed tree priors in order to construct Bayesian estimators of phylogeny. Farris (1973a) based his likelihood analysis of parsimony on an initial Bayesian framework, and his (1977) criticism of Dollo's law was also based in Bayesian formalism. Felsenstein (2004) cites work by Gomberg in 1968 on Bayesian analysis in phylogeny. Harper (1979) contrasted his Bayesian approach to phylogenetic systematics with the reasoning of Popper. Smouse and Li (1987) assumed flat priors (for three competing topologies), justifying a likelihood analysis of primate mtDNA restriction patterns. Wheeler (1991) used explicit Bayesian methods with empirical topology priors (and loss function) to justify data combination (total evidence *sensu* Kluge, 1989).

Although these methods sought to maximize posterior probability, they lacked numerical techniques to integrate over nuisance parameters (Sect. 11.2.1) and non-uniform prior distributions of tree topologies (Wheeler, 1991 excepted) allowing a fully Bayesian analysis. The introduction of Monte Carlo Markov Chains (Metropolis and Ulam, 1949; Metropolis et al., 1953; Hastings, 1970) by Yang and Rannala (1997) made Bayesian phylogenetics tractable and its recent popularity is due to this advance.

12.1 Bayes in Systematics

The *sine qua non* of Bayesian analysis (Sect. 6.1.5) is the prior distribution of the parameter(s) to be estimated. In the case of systematics, the prior of paramount

Systematics: A Course of Lectures, First Edition. Ward C. Wheeler.
© 2012 Ward C. Wheeler. Published 2012 by Blackwell Publishing Ltd.

interest is tree topology. By Bayes' (1763) theorem, the phylogeny problem is recast as finding the topology ($T \in \tau$) with maximum posterior probability (Eq. 12.1) given the data (D) and nuisance parameters

$$p(T|D, \theta, \mathbf{t}_T) = \frac{p(T) \cdot p(D|T, \theta, \mathbf{t}_T)}{\sum_{T_j \in \tau} pT_j \cdot p(D|T_j, \theta, \mathbf{t}_T)} \qquad (12.1)$$

(with data D, tree T, θ all aspects of the substitution model, and $\mathbf{t}_T$ the edge weights for T). Integration over the nuisance parameters (or a portion of them such as edge parameters) can be performed to remove that element of conditionality (Eq. 12.2).

$$p(T|D) = \frac{\int \int p(\theta)p(T|\theta)p(\mathbf{t}_T|\theta, T)p(D|\theta, T, \mathbf{t}_T)d\mathbf{t}_T d\theta}{\sum_{T_j \in \tau} \int \int p(\theta)p(T_j|\theta)p(\mathbf{t}_{T_j}|\theta, T_j)p(D|\theta, T_j, \mathbf{t}_{T_j})d\mathbf{t}_{T_j}d\theta} \qquad (12.2)$$

This is often interpreted as the probability that the topology is "true" given the data, model, and edge probabilities. This view is not necessary and posterior probability can be viewed as an optimality criterion in the same vein as parsimony and likelihood. The denominator, marginal probability of the data D, requires summation over all trees for all weighted combinations of edge and model parameters. Clearly, this will be nearly impossible for non-trivial data sets. Luckily enough, it is a constant for a given analysis, hence it is sufficient to maximize the numerator. MCMC algorithms (below) also make use of this, determining acceptance ratios based on this numerator alone.

The tree with maximum *a posteriori* probability has been acronymed the MAP tree (Rannala and Yang, 1996) and is the Bayesian optimality criterion for trees. This is distinguished from other trees produced by Bayesian analysis such as 95% credibility (that set of trees with $\geq 95\%$ posterior probability) or based on the posterior probabilities of clades summed over multiple trees (Mau et al., 1999). These have been termed "Topology–Bayes" (= MAP) and "Clade–Bayes" trees (Wheeler and Pickett, 2008) to make the distinction emphatic.

12.2 Priors

Ideally, distribution information would be available for all parameters in an analysis, but this is rarely, if ever, the case. The three prior distributions of greatest concern are those of the tree topologies, evolutionary model and edge weights (branch lengths).

12.2.1 Trees

Distributions of trees can have multiple components, two of these are topology and edge weights. These are usually treated separately, but birth–death models can yield distributions of both. In these models, lineages split yielding new lineages (birth) and go extinct (death), reducing the number (although the original Yule (1925) process modeled only birth). Since these occur with specified rates, not only are tree shapes determined, but the weights (= time lengths) of all the edges (Raup et al., 1973). Rannala and Yang (1996) used such a model in their initial work. Most frequently, tree priors are modeled as topology distributions with edge weights treated separately (see below).

There are three sorts of prior distributions we can place on trees: uniform, non-uniform, and empirical.

Uniform—In a uniform distribution of trees, all trees have equal probability. If we have rooted trees with four leaves, there are 15 possible topologies, and each has a probability of $\frac{1}{15}$. Since all topologies have, *a priori*, the same probability, this distribution is often referred to as a "flat" or ignorance distribution. Additionally, this distribution makes no mechanistic assumptions as to origination (speciation) or loss (extinction) of leaf taxa. Hence, the uniform distribution is thought to mirror our expectations, in the absence of data to describe ignorance in an appropriate manner. From the perspective of phylogenetic reconstruction, this would seem to be the most reasonable prior on trees, given that all trees are, in principle, equally possible. However, simulation studies often employ a Yule-type process to model lineage diversification. Analysis of trees generated in this way may be more reasonably treated with Yule-based priors.

Non-uniform—A second type of tree distribution is non-uniform in that not all trees possess the same *a priori* probability, but are thought by some (*e.g.* Rannala and Yang, 1996) to be appropriate ignorance priors. The most common example of this is the Yule (1925) distribution. As opposed to the uniform distribution, which is created by adding leaves with uniform probability in turn to *all* edges, the Yule distribution is created by adding leaves only to *pendant* edges (those incident on leaves) with uniform probability. This leads to quite different distributions on tree shapes (Fig. 12.1) with the balanced tree ((1,2)(3,4)) having probability $\frac{2}{6}$ for Yule and $\frac{3}{15}$ for uniform. The motivation behind the Yule process is a model of diversification. Only pendant lineages can split to create new leaves sister to existing ones. New leaves are never added to internal edges (since they become extinct upon splitting). This model has been augmented (Hey model) by birth–death parameters for lineages (Raup et al., 1973; Hey, 1992; Moores et al., 2007), which does not affect the distribution of trees, but can place a distribution on edge weights (Fig. 12.2).

Empirical—These distributions are based on prior information. Although a staple of applied Bayesian statistics (Martiz and Lwin, 1989), empirical methods have not found much traction in systematics. In principle, data combination could lead to a form of empirical priors if there are pre-existing data bearing on a problem (Wheeler, 1991). Other than exemplar cases (*e.g.* Wheeler and Pickett, 2008), this approach has been little used.

12.2.2   Nuisance Parameters

Although a terrible term (even topology can be a "nuisance" term in some situations), distributions on nuisance parameters are required to determine the posterior probability for a tree. There are two standard components, the edge

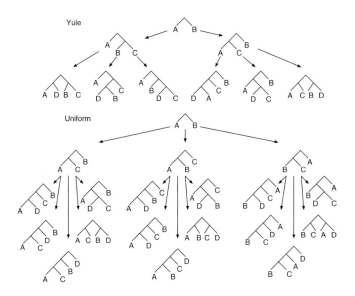

Figure 12.1: A Yule (upper) and uniform (lower) distribution of trees with four leaf taxa.

weights (t) and the substitution model (θ, although all nuisance parameters are sometimes referred to by θ).

Edges/Branches—Edge weight distributions are strange beasts. In the first place, which edges do they refer to? The edges of a particular tree? Those of all trees for a particular data set? Those of all data sets with the same leaf set size (since one would expect edges to be half as long if the leaf set were doubled)? All problems of all sizes? Are these edge weights determined by evolutionary change alone, or does sampling play a role? These factors make general statements about edges problematic.

Edge priors have been dealt with in several ways. Initially, edge transformation probabilities were constrained to be clocklike (Rannala and Yang, 1996), but this undesirable restriction has been removed and a variety of models are now available. Two general distributions of edge weights are in general use: uniform (Sect. 6.1.3) and exponential (Sect. 6.1.3) (Huelsenbeck and Ronquist, 2003), although other, more complex priors have been proposed (Kishino et al., 2001; Thorne et al., 1998). In the uniform distribution, the probabilities of edge weights (up to a constant value) are equal, whereas the exponential distribution (Eq. 12.3, Fig. 12.3) has a greater probability for shorter branches than for longer (see Chapter 6).

$$\forall x \geq 0; p(x) = \lambda e^{-\lambda x} \tag{12.3}$$

In general, uniform edge parameter distributions will have a higher probability of longer branches and higher clade posterior probabilities (Yang, 2006), a property thought undesirable by some (Huelsenbeck and Ronquist, 2003). The use of the exponential distribution is somewhat arbitrary other than for modeling the

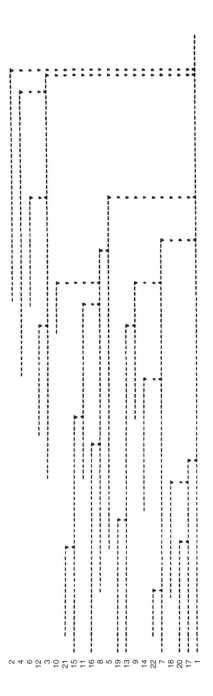

Figure 12.2: Stochastic origination and extinction from Raup et al. (1973) showing 22 lineages evolving over 115 time intervals from bottom to top.

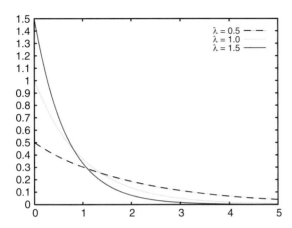

Figure 12.3: The exponential distribution.

wait time between Poisson-distributed lineage splitting events. Of course, this justification would assume we know *all* splitting events, which we cannot know due to extinction of unobserved lineages.

In my opinion, this reflects an unreasonable interpretation of the significance of branch lengths as statements of evolutionary process as opposed to one where taxonomic sampling plays a major role. Given that samples of extant taxa are nearly always incomplete, and knowledge of extinct taxa extremely limited if not entirely absent, the uniform edge weight distribution seems at least equally plausible as a statement of prior information. Edge weights are as much statements of sampling as of evolutionary process.

A further complication comes in the use of a single parameter for all edges. Some will be larger than others, and almost all (in the course of an analysis) will be "wrong." For this reason, Yang and Rannala (2005) suggested an empirical Bayes' method where edge parameters are likelihood estimates determined uniquely for each edge on each tree with the data at hand. Given that these data are not "prior" in any way, strict Bayesians may howl, but this process (though potentially time-consuming) seems a powerful and reasonable means to establish appropriate edge parameters.

No matter which distributions of edge weights are used, the objective is the same, to determine the posterior probability of the tree conditioned solely on the substitution model.

Substitution Models—Ideally, we would employ a distribution of substitution models in order to integrate out this nuisance parameter and determine $p(T|D)$ directly. This cannot (at present) be done completely (although Hidden Markov Models—HMM—approach this "model of models"). Typically, the General-Time-Reversible (GTR) model (Lanave et al., 1984; Tavaré, 1986) is used whereby the potential parameter values encompass the variety of simpler models with specific symmetry conditions (*e.g.* JC69). A distribution on the parameters of this model should contain a variety of other models. Typically, a Dirichlet distribution (Sect. 6.1.3) is used to accomplish this. At least for those models reducible from GTR, this is a reasonable procedure.

12.3   Techniques

Since, in general, we are not concerned with the actual value of $p(T|D)$, and the denominator of Equation 12.1 is a constant for any particular data set, we can restrict ourselves to maximizing $p(T) \cdot p(D|T)$. This reduces to maximizing the product of the tree prior and the tree likelihood with the nuisance parameters integrated out (MIL). In general, analyses use a flat (= uninformative) prior for tree topologies, hence the Bayesian analysis boils down to a search for trees with high MIL (Chapter 11). This could be accomplished via a variety of standard tree-search techniques (Chapter 14) such as exhaustive tree enumeration (Rannala and Yang, 1996) or heuristic branch-swapping (Wheeler and Pickett, 2008). These tried and true methods should yield useful MAP trees—if supplied with integrated likelihoods for the trees themselves. Without this information, however, these techniques are less directly applicable, and other approaches are more commonly employed.

Bruce Rannala

12.3.1   Markov Chain Monte Carlo

Typically, the problem of determining the posterior probability of a tree requires knowledge of the tree prior, the distribution of model parameters and of edge weights. Even if the priors are not an issue (*e.g.* flat, as above), there are many parameters to integrate away. For example, a GTR-model analysis for 100 taxa has at least 205 parameters. Numerical integration in such high dimensionality will be extremely time consuming.

Markov Chain Monte Carlo (MCMC) techniques based on the Metropolis–Hastings algorithm of simulated annealing offer an alternate means of determining the relative posterior probabilities (based on the numerator of Eq. 12.2) of competing trees. This technique (Sect. 14.7) has general use in complex optimization problems and has been adapted for use in tree-searching (Goloboff, 1999a, as "Tree-Drifting").

Ziheng Yang

12.3.2   Metropolis–Hastings Algorithm

As discussed in Chapter 14, the Metropolis–Hastings algorithm (Metropolis et al., 1953; Hastings, 1970) has been used to allow a search to escape local optima by accepting suboptimal solutions in the hope of later finding more global solutions. In the context of Bayesian estimation in systematics, the objective is a stable population of trees where trees are represented in proportion to their posterior probability. The Metropolis–Hastings transition probabilities between pairs of trees are determined by the ratio of their posteriors (Eq. 12.2). Since they share the same denominator, this ratio is based solely on the numerator of Equation 12.2. This saves us from dependency on the difficult to calculate marginal data probability.

This is accomplished via a series of proposed changes to the current tree (including the edge parameter vector and substitution model). If we define the

$$\pi(T) = p(\theta)p(T|\theta)p(\mathbf{t}_T|\theta, T)p(D|\theta, T, \mathbf{t}_T) \tag{12.4}$$

acceptance ratio based on the relative quality of solutions (Eqs. 12.5 and 12.6) after a change from T to T^*.

$$\alpha = \min\left(1, \frac{\pi(T^*)}{\pi(T)}\right) \tag{12.5}$$

The probability of accepting a solution (a tree in this case) T^* is 1 if T^* is "better" (higher posterior probability) than T. A worse solution will be accepted with probability $f(\alpha) > 0$ depending on the "temperature" of the system. If the temperature is relatively high, suboptimal solutions are more likely to be accepted. If the system is too hot, any solution will be accepted, and the procedure becomes a random walk about the solution space. If the temperature is too low, only better solutions are accepted with the familiar problems of local optima. The sequence of states forms a Markov "chain" that will never (on its own) terminate. The method should converge on the posterior probability as long as two conditions are met: 1) the transition system allows all states to transition into all other states, and 2) the chain is aperiodic. These conditions are easily met.

As an example, consider a coin. A sequence of tosses (yielding three heads and seven tails) constitutes D. The coin may be fair (state 0; prior probability 0.5), biased towards tails (state 1; prior probability 0.25), or biased towards heads (state 2; prior probability 0.25). These are the three "states" whose posterior probabilities we want to estimate (Fig. 12.4). The transition probabilities between the states will be:

$$\alpha_{0\rightarrow1} = \frac{p(\text{biased T}) \cdot r(D|\text{biased T})}{p(\text{fair}) \cdot r(D|\text{fair})} = \frac{0.25 \cdot 0.00209}{0.5 \cdot 0.000977} = 1.070 \rightarrow 1$$

$$\alpha_{1\rightarrow0} = 0.935$$

$$\alpha_{0\rightarrow2} = \frac{p(\text{biased H}) \cdot r(D|\text{biased H})}{p(\text{fair}) \cdot r(D|\text{fair})} = \frac{0.25 \cdot 0.0000257}{0.5 \cdot 0.000977} = 0.0132$$

$$\alpha_{2\rightarrow0} = 76.0 \rightarrow 1$$

$$\alpha_{1\rightarrow2} = \frac{p(\text{biased H}) \cdot r(D|\text{biased H})}{p(\text{biased T}) \cdot r(D|\text{biased T})} = \frac{0.25 \cdot 0.0000257}{0.25 \cdot 0.000209} = 0.123$$

$$\alpha_{2\rightarrow1} = 8.13 \rightarrow 1 \tag{12.6}$$

After some large number of probabilistic transitions, the relative number of visited states should settle down on the posterior probabilities of $p(\text{fair}) = 0.480$, $p(\text{biased T}) = 0.514$, and $p(\text{biased H}) = 0.00631$.

The Metropolis–Hastings algorithm for identifying the maximum posterior probability tree would follow the general procedure:

1. Choose an initial tree (T), edge weight vector ($\mathbf{t}$), substitution model (θ).

2. Propose a new tree (T^*).

      Propose new $\mathbf{t}$

      Propose new θ

3. Evaluate α for T and T^*.

    If $\alpha \geq 1$ accept T^*

    If $\alpha < 1$ accept T^* with probability depending on "temperature"

4. Save resulting tree and parameters each kth iteration.

5. Repeat steps 2 through 4 n times.

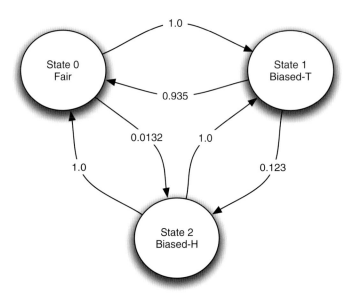

Figure 12.4: Metropolis–Hastings acceptance ratios for three states of a coin: fair, biased-Heads, and biased-Tails. Acceptance ratios of Equation 12.6.

A new tree may be proposed with new edge and model parameters, or nuisance parameters may undergo their own proposal process (step 2).

    There are three flavors of the Metropolis–Hastings algorithm in common use: 1) Single component, 2) Gibbs sampler, and 3) Metropolis-coupled MCMC or MC^3. MC^3 or Bayesian MCMC is the most frequently employed strategy.

12.3.3   Single Component

The single component Metropolis–Hastings procedure operates as above for a single parameter, but when there are multiple parameters to be estimated, cycles through them individually. Each parameter undergoes a proposal and transition process, while all others are held constant. The order of parameter evaluation may be deterministic, randomized, or probabilistic. This method can perform poorly if there are correlated parameters, where independent changes will find joint optimization with lower probability.

12.3.4 Gibbs Sampler

The Gibbs sampler uses the parameter conditional distribution with respect to all others to define the transition conditions (Gelman and Gelman, 1984). Each parameter is treated in turn as with the Single component above. The Gibbs sampler requires analytical information about these distributions that may be difficult to determine. This form of sampling was used by Jensen and Hein (2005) in their determination of sequence medians and multiple alignments under the TKF91 model (Thorne et al., 1991).

12.3.5 Bayesian MC^3

The MC^3 algorithm (Geyer, 1991) generalizes the MCMC procedure by running multiple (m) chains in parallel, each for n iterations. Chain π_0 is the *cold* chain (in terms of Metropolis–Hastings temperature), and the others are successively hotter (Fig. 12.5). Typically, the heating is incremental through the chains 1 to $m - 1$ (Eq. 12.7).

$$\pi_j(T) \propto \pi_0(T)^{\frac{1}{1+\lambda(j-2)}}, \quad \lambda > 0 \tag{12.7}$$

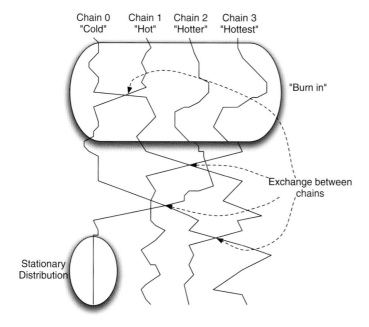

Figure 12.5: MC^3 process with four chains from "Cold" to "Hottest." The initial period of transitions far from stationarity is discarded as the "burn in" period. Only the results of the stationary phase "cold" chain are used for the calculation of posterior probabilities. Tree space traversal and chain mixing are represented by the jagged chain lines.

Only the cold chain converges to the estimated parameter—the others are for mixing between chains to improve solutions (Yang, 2006). These chains (i and j) exchange values periodically according to a second Metropolis–Hastings acceptance ratio (Eq. 12.8).

$$\alpha = \min\left(1, \frac{\pi_i(T_j)\pi_j(T_i)}{\pi_i(T_i)\pi_j(T_j)}\right) \tag{12.8}$$

The motivation is that the hot chains allow the movements between solution peaks, while the cold chains find the local minima in their neighborhood. Periodic mixing between chains allows solutions separated by large probability gaps to enter the cold chain. Such gaps may impair the ability of MC^3 to find valid stationary solutions (Yang, 2006). The end result of the MC^3 should be that each tree is represented in the stationary pool in proportion to its posterior probability in the cold chain.

Evaluating MC^3

Even with exchanges between m chains and after n iterations it can be difficult to determine whether the chain has converged on the distribution of the desired parameter reaching stationarity or is still changing (the "burn in" period).

There are several strategies for evaluating whether stationarity has been reached:

- Run the algorithm without data. The posterior distribution should converge on the (known) prior.

- Repeat the process multiple times. The results of independent runs should converge on the same estimates of posteriors.

- Examine the distribution of estimated parameters over the progress of the runs. All parameters should have settled down to their final estimates.

No matter what strategy is employed, great care should be taken to ensure that stationarity has been reached. The validity of the estimates depends on it.

12.3.6   Summary of Posterior

There are three commonly used methods of summarizing Bayesian analysis of tree topologies. The first is the straightforward (but somewhat uncommon) presentation of the MAP tree as one would present an ML or parsimony result (Rannala and Yang, 1996). The second is to present the minimal set of trees (or their consensus) with total posterior probability of 95% (or some other) level of Bayesian credibility (Rannala and Yang, 1996; Mau et al., 1999). By far the most common method is to present the majority consensus of the trees in the stationary pool. Each clade with > 50% posterior probability is presented on a consensus tree labeled by these clade posteriors (Fig. 12.7). This summary method was first presented by Mau et al. (1999) and is implemented in the program MrBayes (Huelsenbeck and Ronquist, 2003). It is important to note that the contributions of a given clade posterior probability will likely come from multiple, incompatible, tree topologies (Fig. 12.6). Combining such

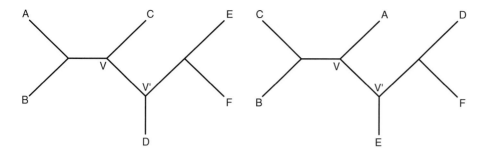

Figure 12.6: Two edges that are quite different ((V, V') on left and right), yet define identical splits (= clades if rooted on either side of the split) that would be summed to determine the posterior clade probability.

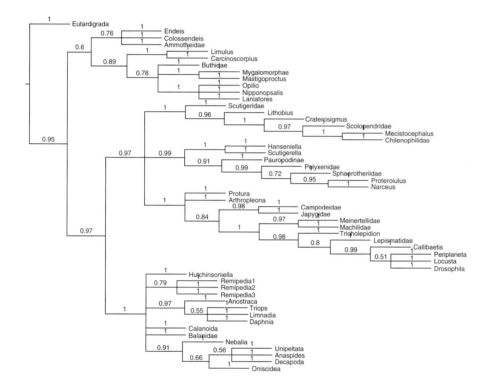

Figure 12.7: A Clade–Bayesian tree (*sensu* Wheeler and Pickett, 2008) of arthropod anatomical data (Wheeler and Pickett, 2008).

disparate objects is certainly questionable; are left and right $e_{V,V'}$ of Figure 12.6 equivalent? The posterior probabilities were determined for whole trees; the justification of their use on subtrees is unclear. Furthermore, any comparison of their lengths (time parameter) would seem to be invalid since the other nodes to which they are connected differ[1].

---

[1]This information could be gathered by resampling a fixed MAP tree (Yang, 2006).

Kurt Pickett
(1972–2011)

A tree constructed from the posterior probability of clades may differ from the MAP tree (Wheeler and Pickett, 2008) and have other less desirable properties (most prominently clade size bias—see below).

12.4   Topologies and Clades

Although a uniform tree prior distribution has no effect on the choice of MAP tree, the distribution of individual clades (by size) forced by this distribution is not uniform. Hence, the clade posteriors can give a strong prior bias (clade c of size m in a tree with n leaves; Eq. 12.9).

$$pr(c_m) = \frac{\left(\prod_{i=2}^{m} 2i - 3\right) \cdot \left(\prod_{i=m+1}^{n} 2i - 2m - 1\right)}{\prod_{i=2}^{n} 2i - 3} \tag{12.9}$$

Pickett and Randle (2005) first pointed this out in an empirical and experimental manner and Steel and Pickett (2006) later proved the impossibility of creating uniform clade priors (Fig. 12.8). The influence of these priors can be severe

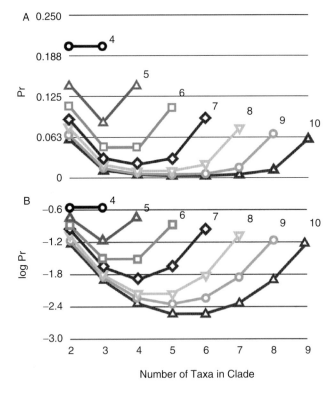

Figure 12.8: Pickett and Randle (2005) relationship between clade prior and clade size under a uniform topology prior.

(a factor of 10^{13} for groups of 2 and 25 in a tree of 50 taxa) and makes the interpretation of clade posteriors difficult even as a support measure. Goloboff and Pol (2005) presented an example where this effect results in a taxon with entirely missing values placed specifically (with 55% posterior clade probability) and other clades supported from 52 to 96% even though all placements of the all-missing taxon have equal posterior probability (Fig. 12.9). Yang (2006) based

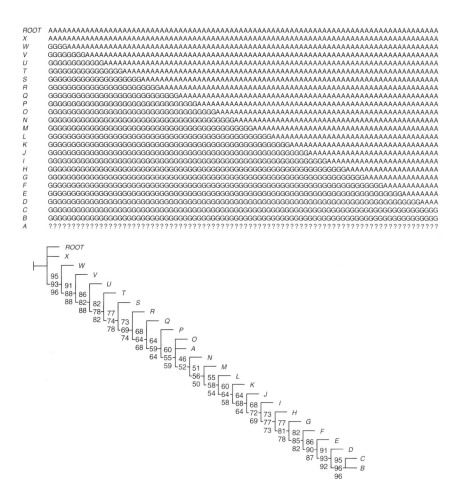

Figure 12.9: A completely uninformative taxon (A) placed near the center of a tree (posterior clade probabilities > 50% determined by MrBayes) of completely informative taxa (Goloboff and Pol, 2005). The numbers on the branches are posterior clade probabilities (determined by MrBayes; Huelsenbeck and Ronquist, 2003), the numbers above are the groups frequencies in the set of most parsimonious trees, and the numbers below bootstrap frequencies (determined by PAUP*; Swofford, 2002).

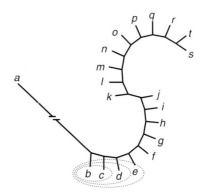

Figure 12.10: Example from Goloboff and Pol (2005) recast by Yang (2006) showing a distant taxon a from a series of easily placed taxa $b - s$. The edge parameter leading to a is 10 (basically randomized), whereas all the others are 0.2. For n taxa, the posterior probability of incorrect clade (b, c) will be $(2n - 7)/(2n - 5)$.

an example on Goloboff and Pol (2005) showing support for groups that are not present on the "true" tree approaching 1 (Fig. 12.10).

12.5   Optimality versus Support

Clearly, MAP is a well defined measure of optimality. Consistent analysis of systematic data can follow such a path as well as parsimony or likelihood. The posterior probabilities of clades cannot. There is no way for such summaries to participate in hypothesis testing, and they have clearly pathological behaviors in several simple cases. At best, clade posteriors offer a measure of the support of groups. However, even that utility awaits statistical justification (Yang, 2006).

12.6   Dynamic Homology

In moving from static to dynamic homology characters (in a stochastic framework), it is convenient to use mathematical descriptors that are more flexible and general than traditional Markov processes. In the cases of sequences, we can model indels with a simple $n + 1$ Markov model but, in general, this requires knowledge of the positions of these indels *a priori* such as in the case of the analysis of multiple sequence alignments. When the locations of sequence gaps are unknown, such simple models may be insufficient. Hidden Markov Models present a useful mechanism in such complex scenarios.

12.6.1 Hidden Markov Models

A standard Markov chain employs a single transition model (state) that is repeatedly applied, generating observations. Hidden Markov Models (HMM) add an extra layer of complexity in that the chain contains multiple *states* ($=$ Markov transition matrices) each of which *emits* an observation. The sequence of observations is the result not only of the randomized emission of an observation by a state, but also the randomized order of the states themselves. Normally, only the sequence of observations is known, hence the chain of states is "hidden." As with standard Markov processes, the generation of observations is "memoryless," depending only on the preceding state. When an HMM is run, there is first a sequence of states and second a sequence of emitted observations (Fig. 12.11). HMMs are broadly applicable and are used in diverse fields including speech recognition, ecology, and even financial market analysis.

The myriad relationships among the states (transitions) and between states and observations (emissions) can be summarized by a "trellis" diagram (Fig. 12.12).

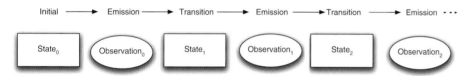

Figure 12.11: An HMM chain of events consisting of a start ($State_0$) and subsequent states ($State_i$), emissions from states ($Observation_i$), and transitions between states.

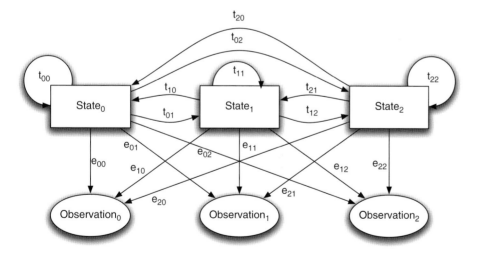

Figure 12.12: A trellis diagram of an HMM with three states and three possible emission states, showing transitions between states (t_{ij}) and emissions between states and observations (e_{ij}).

If we define a sequence of n states as $\pi = \pi_0, \pi_1, \ldots, \pi_{n-1}$, a set of k sequence elements $s = \{s_0, s_1, \ldots, s_{k-1}\}$, the sequence of observations as $x = x_0, x_1, \ldots, x_{n-1}$ (each one drawn from s), transition probabilities between states i and j as t_{ij}, and emission probability of element s_j by state π_i as e_{ij}, the probability of the sequence with initial state π_0 will be (Eq. 12.10):

$$p(x, \pi | t, e) = p(\pi_0) e_{\pi_0, x_0} \cdot \prod_{i=1}^{n-1} t_{\pi_{i-1}, \pi_i} e_{\pi_i, x_i} \qquad (12.10)$$

12.6.2   An Example

Consider the case of a sequence of coin tosses. Each toss could be made with either a fair coin $(p(H) = p(T) = 0.5)$ or one biased strongly towards heads $(p(H) = 0.9, p(T) = 0.1)$. Furthermore, the biased coin is substituted for the fair coin with probability $= 0.2$ after a toss, and fair for biased with probability 0.6 (Fig. 12.13). Suppose the tosser makes 20 tosses and the sequence THTHHTTTHHHHHHHHHHTT results.

If only the fair coin were tossed, the probability of the sequence would be 0.5^{20} ($\ln pr = -13.86$); if only the biased coin were used, $\ln pr = -17.49$ ($0.9^{13} \cdot 0.1^7$). The maximum likelihood estimation of coin type would then be that the (single) coin was fair with a likelihood ratio of 37.7.

HMM analysis allows more nuanced scenarios to be evaluated. The single fair and single biased coin probabilities require additional factors. The first is the probability of the starting states (here 0.5) and the 19 transitions after the first toss, in each case fair following fair, or biased following biased. The probability of the all fair case becomes $0.5 \cdot 0.5^{20} \cdot 0.8^{19}$ for a $\ln pr = -18.79$. For the all biased scenario, we have $0.5 \cdot 0.9^{13} \cdot 0.1^7 \cdot 0.4^{19}$ for a $\ln pr = -35.59$. We are not, however, limited to these two scenarios. We can also evaluate sequences where

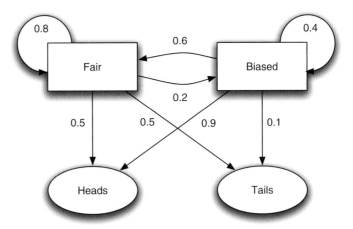

Figure 12.13: A trellis diagram of the parameters of a coin toss experiment with two coins, one fair and one foul.

coins were switched from fair to biased and vice-versa. Consider (fair coin = F, biased = B) sequences FFFFFFFFBBBBBBBBBBBBB (one transition from fair to biased) and FFFFFFFFBBBBBBBBBBBBFF (two transitions, one in each direction):

$$\text{one transition} \qquad 0.5 \cdot 0.5^8 \cdot 0.9^{10} \cdot 0.1^2 \cdot 0.8^7 \cdot 0.2 \cdot 0.4^{11} = e^{-25.15}$$

$$\text{two transitions} \quad 0.5 \cdot 0.5^8 \cdot 0.9^{10} \cdot 0.5^2 \cdot 0.8^8 \cdot 0.2 \cdot 0.4^9 \cdot 0.6 = e^{-20.83}$$

This example can easily be extended to a more complex case by adding a third coin ($\frac{1}{6}$ state transitions among them) biased strongly towards tails (Fig. 12.14). This would result in an additional scenario being evaluated where the coin state (biased head $\rightarrow$ H; biased tail $\rightarrow$ T) exactly matched the observations with six state transitions (Table 12.1). Based on the maximum probability scenario (three coins), we can re-evaluate the state transition probabilities noting that there were zero transitions between fair and the two biased coins (Fig. 12.15). If we do this, the (ln) probability of the three coin scenario jumps to -14.39 and would account for over 85% of the posterior decoding (relative probability given the restricted set of scenarios).

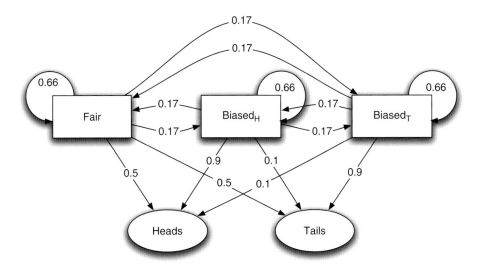

Figure 12.14: A three coin trellis diagram.

Scenario	$\ln pr$	Posterior Decoding
Three coins (six transitions)	-16.27	91.65 %
No transitions (fair coin)	-18.79	7.374 %
Two transitions (two coins)	-20.83	0.9589 %
One transition (two coins)	-25.15	0.0001275 %
No transitions (biased coin)	-35.59	0.0000000037289 %

Table 12.1: HMM coin toss scenarios.

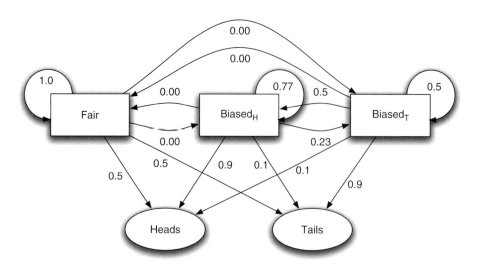

Figure 12.15: A three coin trellis diagram revised to reflect observed transitions.

12.6.3   Three Questions—Three Algorithms

Generally, we are presented with a Markov process with unknown parameters. There are three questions we might ask of the system: 1) given model parameters, what is the sequence of hidden states? 2) given model parameters, what is the probability of the observed sequence? and 3) given the output sequence, what are the parameters? These questions are addressed by the Viterbi, Forward–Backward, and Baum–Welch algorithms.

Viterbi Algorithm

Andrew Viterbi

The Viterbi (1967) algorithm is a dynamic programing procedure to determine the highest probability path of states π^* from an observed sequence x given transition probabilities t, and emission probabilities e (Eq. 12.11).

$$\pi^* = \underset{\pi}{\operatorname{argmax}}\, p(x, \pi | t, e) \tag{12.11}$$

Recalling Equation 12.10, the central recursion is derived from the observation that the highest probability path ending in state v_j at position i with observation x_i will be the probability of the total path to position $i - 1$ multiplied by the transition probability between the states in positions $i - 1$ and i, multiplied by the emission probability of x_i given state v_j maximized over the states in position $i - 1$ (Eq. 12.12).

$$p(v_{j,i}) = e_{v_j, x_i} \cdot \underset{k\ \text{states}}{\operatorname{argmax}} \left[p(v_k, i - 1) \cdot t_{v_{k,i-1}, v_j} \right] \tag{12.12}$$

Algorithm 12.1: Viterbi Algorithm

Data: Input sequence observations x of length n and states v (size k)

Data: Initial probabilities of states $p(v)$, emission probabilities e_{ab} of states (a) and elements (b) in x, transition probabilities between states t_{ab} $(a \rightarrow b)$.

Result: The probability of the optimal sequence of states π^*. The sequence itself can be determined by a traceback on *direction*.

Initialize first row of *prob* matrix;

for $i = 0$ **to** $k - 1$ **do**
$\quad \mid \quad prob\,[0, i] \leftarrow p(v_i) \cdot e_{i,x_0}$;
end

Update remainder of matrices *prob* and *direction*;

for $i = 1$ **to** $n - 1$ **do**
$\quad$ **for** $j = 0$ **to** $k - 1$ **do**
$\quad\quad$ $prevprob \leftarrow 0$;
$\quad\quad$ **for** $l = 0$ **to** $k - 1$ **do**
$\quad\quad\quad$ **if** $t_{v_l,v_j} \cdot prob\,[i - 1, l] > prevprob$ **then**
$\quad\quad\quad\quad$ $prevprob \leftarrow t_{v_l,v_j} \cdot prob\,[i - 1, l]$;
$\quad\quad\quad\quad$ $direction\,[i, j] \leftarrow l$;

$\quad\quad$ **end**
$\quad\quad$ $prob\,[i, j] \leftarrow e_{v_j,x_i} \cdot prevprob$;
$\quad$ **end**
end

return $\mathrm{argmax}_k\, prob\,[n - 1]\,[k]$

The algorithm (Alg. 12.1) proceeds by initialization, recursion, and traceback, as with other dynamic programming procedures such as the Needleman–Wunsch algorithm for string matching (Alg. 8.1).

1. Establish a matrix of n path position columns and k state rows.

2. Initialize the first column with the initial probabilities of the states multiplied by its emission probability of observation x_0.

3. Apply Equation 12.12 to each successive column keeping for each state in each column a pointer to the state in the previous column that yielded the highest probability (k in Eq. 12.12).

4. The maximum value in column $n - 1$ is the probability of the maximum probability path π^*.

5. Traceback beginning with the maximum value in column $n - 1$ to produce π^*.

As with other probabilistic algorithms, it is often convenient to work with logarithms of probabilities to avoid awkwardly small numbers.

The time complexity of this algorithm for k states and n observations is $O(k^2 n)$, but since in general $k \ll n$, the algorithm is in essence linear with n.

Forward–Backward Algorithm

The Viterbi algorithm found the single sequence of states with maximum probability, but there are an exponential number of paths (in the three coin 20 toss example there are 3^{20}), each with some non-zero probability. As with probabilistic sequence alignment, we may want to know what the total probability of a sequence of observations is given all possible state paths. This is done via a simple modification of the Viterbi algorithm where the maximization over states in the previous position (Eq. 12.12) is summed (Eq. 12.13)

$$p^F(v_{j,i}) = e_{v_j,x_i} \cdot \sum_k \left[p^F(v_k, i-1) \cdot t_{v_{k,i-1},v_j} \right] \qquad (12.13)$$

resulting in the *Forward* algorithm.

In order to determine the posterior probability that a particular observation x_i was emitted from state $\pi_i = k$, we need to determine the probability of $\pi_i = k$ given the sequence from i to the beginning and end of the entire sequence. The Forward algorithm yields the former and, surprisingly enough, the *Backward* algorithm provides the latter.

The Backward algorithm is identical to the Forward except that it begins at the end of the sequence $(i = n)$ and moves forward (Eq. 12.14).

$$p^B(v_{j,i}) = e_{v_j,x_i} \cdot \sum_k \left[p^B(v_k, i+1) \cdot t_{v_{k,i+1},v_j} \right] \qquad (12.14)$$

The posterior probability of a state in position i given observations x is then determined from the product of the Forward and Backward probabilities conditioned on the total probability from the Forward algorithm (Eq. 12.15, Alg. 12.2).

$$p^{PP}(\pi_i = v_k | x) = \frac{p^F(v_{j,i}) \cdot p^B(v_{j,i})}{P(x)} \qquad (12.15)$$

Baum–Welch Algorithm

If the paths are known, or we have a training sequence of observations to fit, transition and emission probabilities can be estimated as simple proportions of particular events to the total number. These will be the maximum likelihood estimators for these parameters given the training sequence.

If the state paths for the training sequences are unavailable or unknown, such a direct method cannot be used alone, but can be used in concert with an iterative refinement of parameter estimation (Baum et al., 1970). Initial values are set, the Forward and Backward algorithms are performed on the training sequences, and then the transition and emission frequencies estimates

Lloyd R. Welch

Algorithm 12.2: Forward–Backward Algorithm

Data: Input sequence observations x of length n and states v (size k)

Data: Initial probabilities of states $p(v)$, emission probabilities e_{ab} of states (a) and elements (b) in x, transition probabilities between states t_{ab} $(a \rightarrow b)$.

Result: Matrix of the posterior probability of each state (p^{PP}) in v at each position π_i in the sequence determined by Eq. 12.15.

Determine Forward probabilities;

Initialize first row of forward, p^F, matrix;

for $i = 0$ **to** $k - 1$ **do**

$\quad \mid \quad p^F[0, i] \leftarrow p(v_i) \cdot e_{i, x_0}$;

end

Update remainder of matrix p^F;

for $i = 1$ **to** $n - 1$ **do**

$\quad$ **for** $j = 0$ **to** $k - 1$ **do**

$\quad \quad$ $sumprob \leftarrow 0$;

$\quad \quad$ **for** $l = 0$ **to** $k - 1$ **do**

$\quad \quad \quad \mid$ $sumprob \leftarrow sumprob + \left(t_{v_l, v_j} \cdot p^F[i - 1, l] \cdot e_{v_j, x_i}\right)$;

$\quad \quad$ **end**

$\quad \quad$ $p^F[i, j] \leftarrow sumprob$;

$\quad$ **end**

end

$TotalProb \leftarrow \text{argmax}_k \, p^F[n - 1][k]$;

Determine Backward probabilities;

Initialize last row of forward, p^B, matrix;

for $i = 0$ **to** $k - 1$ **do**

$\quad \mid \quad p^B[n - 1, i] \leftarrow p(v_i) \cdot e_{i, x_{n-1}}$;

end

Update remainder of matrix p^B;

for $i = n - 2$ **to** 0 **do**

$\quad$ **for** $j = 0$ **to** $k - 1$ **do**

$\quad \quad$ $sumprob \leftarrow 0$;

$\quad \quad$ **for** $l = 0$ **to** $k - 1$ **do**

$\quad \quad \quad \mid$ $sumprob \leftarrow sumprob + \left(t_{v_l, v_j} \cdot p^B[i + 1, l] \cdot e_{v_j, x_i}\right)$;

$\quad \quad$ **end**

$\quad \quad$ $p^B[i, j] \leftarrow sumprob$;

$\quad$ **end**

end

Determine Posterior probabilities;

for $i = 0$ **to** $n - 1$ **do**

$\quad$ **for** $j = 0$ **to** $k - 1$ **do**

$\quad \quad \mid$ $p^{PP}[i, j] \leftarrow \frac{p^F[i,j] \cdot p^B[i,j]}{TotalProb}$;

$\quad$ **end**

end

return p^{PP}

---

from each of these sequences are tallied and used to calculate new parameter values. These values are used on the training sequences repeatedly, each time tallying transitions and emissions and re-estimating parameters until further improvements in the probabilities of training sequences cease.

1. Set initial parameters.

2. For each training sequence:

   Calculate $p^F(x)$ and $p^B(x)$.

   Tally state transitions and emissions implied by $p^F(x)$ and $p^B(x)$.

3. Re-estimate model parameters based on the even tallies of the training sequences.

4. Determine $p(x)$ based on the new parameters.

5. Repeat 2–4 until improvements in $p(x)$ are below a predetermined threshold.

12.6.4   HMM Alignment

Pairwise HMM alignment—As with sequence of coin tosses, HMM can be used to align sequences through maximizing the probability of the alignment given element matches (substitution or identity), insertions, and deletions. A trellis diagram is constructed (Fig. 12.16) with all of these events and their probabilities (match $= 1 - 2\delta$, insertion $=$ deletion $= \delta$, and affine insertion $=$ affine deletion $= \epsilon$). The Viterbi algorithm is also used here to determine the maximum

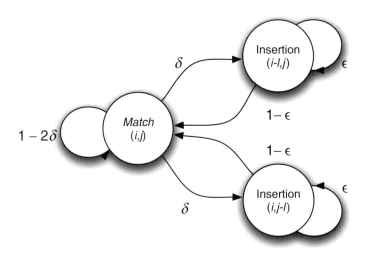

Figure 12.16: A trellis diagram for pairwise alignment with match probability $= 1 - 2\delta$, insertion $=$ deletion $= \delta$, and affine insertion $=$ affine deletion $= \epsilon$.

probability alignment with the recurrence v^M for match between the two sequences (X and Y), v^I for insertion and v^D for deletion:

$$v_{i,j}^M = p_{x_i,y_j} \cdot \max \begin{cases} (1-2\delta)v_{i-1,j-1}^M \\ (1-\epsilon)v_{i-1,j-1}^I \\ (1-\epsilon)v_{i-1,j-1}^D \end{cases} \quad (12.16)$$

$$v_{i,j}^I = q_{x_i} \cdot \max \begin{cases} \delta v_{i-1,j}^M \\ \epsilon v_{i-1,j}^I \end{cases} \quad (12.17)$$

$$v_{i,j}^D = q_{y_i} \cdot \max \begin{cases} \delta v_{i,j-1}^M \\ \epsilon v_{i,j-1}^D \end{cases} \quad (12.18)$$

When this recursion is examined in its additive log-transformed form, δ is the opening and ϵ the extension or affine gap cost of the Gotoh (1982) algorithm discussed in Chapter 8. The same dynamic programming procedure can be used for HMM alignment as for standard string alignment given that the probabilities are log transformed.

Also, as with the coin toss example above, the posterior probability of a given match or indel position can be determined using the Forward–Backward algorithms. The Baum–Welch procedure can be used to estimate the alignment parameters δ and ϵ.

Multiple HMM Alignment—Pairwise alignment can be extended to multiple sequence alignment (MSA) in a straightforward manner via the use of *Profile* HMM. A profile HMM (Krogh et al., 1994) is the model of a group of aligned sequences that can be used (via the Viterbi or dynamic programming approaches) to align groups of sequences.

Anders Krogh

The profile HMM is built up in segments. A trusted alignment(s) (usually human-derived or edited) is used as the basis for determining the emission and transition probabilities. First, ungapped regions are examined. Each aligned position, or "block" of multiple positions, is used to create the match emission probabilities. These may vary over the length of the sequence, yielding more specific probability statements (position specific score matrix—PSSM) than those that are constant over the entire length of the sequences. Insertions are modeled as the state I, allowing transitions from the match state M_i. I can transition back to the succeeding match state, M_{i+1}, or to itself (insertion length <1). This sort of model corresponds to an affine gap model. Deletions are somewhat different beasts, since a single deletion could, in principle, lead to any other downstream sequence element, necessitating large numbers (order n^2) of transitions. To handle this more cleanly, the "dummy" state D is added analogously to the insertion, with the limitation that these states can only transition back to match states or to the succeeding dummy state. With these three elements, the profile HMM is complete (Fig. 12.17). As with the PSSM, insertion and dummy (deletion) transitions and emission probabilities can vary over the length of the sequence, giving classes of sequences specific profiles. Of course, the alignments

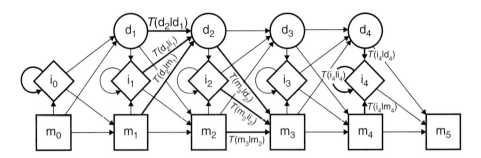

Figure 12.17: A trellis diagram for profile HMM with match, insertion, and "dummy" states for repeated deletions (Krogh et al., 1994).

used to construct and train the HMMs (*e.g.* BAli-Base, Thompson et al., 2005) will have pervasive influence on their use in subsequent analysis.

Profile HMMs are used a great deal in the identification of gene families and functional motifs. They can also be used as any HMM in the pairwise procedure above to create HMM-based MSAs (Hughey and Krogh, 1996). The operation follows the same general path as the progressive alignment of profile alignments discussed in Chapter 8.

12.6.5 Bayesian Tree Alignment

The basic operation of the TAP (Chapter 8, Sect. 10.9) in a Bayesian context comes down to the posterior probabilities of edges. A profile HMM can be used to align any pair of sequences, such as the ancestor and descendent vertices of an edge. With a suitable sequence change model (such as TKF91), the profiles can be constructed and emission and transition probabilities determined as a function of the usual parameters (edge weight and substitution model). An additional complexity in Bayesian evaluation of trees comes in the treatment of vertex medians. The posterior probability of the entire tree cannot be dependent on specific, optimal, median sequence assignments, but must integrate over all such medians (or approximate via MCMC or some other sampler) to determine the overall posterior of the tree. This (among other factors) adds a great deal of time complexity to Bayesian TAP analysis. Fleissner et al. (2005) cite this as the motivation for their approach of using only the MAP estimates of parameters. In avoiding the integration component of Bayesian TAP, they are able to decrease execution time vastly. This approach is out of step with the majority of Bayesian TAP procedures (see below), but does have enhanced scalability to recommend it. The relationship between MAP values based on maximizing parameters, and those by integration is, however, unclear.

12.6.6 Implementations

The complete General Tree Alignment Problem (GTAP) in the Bayesian framework is extremely onerous. Implementations rely on at times severe restrictions on the analysis with respect to the dimensionality of the problem (progressive HMM) to fixing edge parameters to employing single distance-based tree topologies. A few of these implementations are discussed below.

Holmes and Bruno (2001) directly extended the profile HMM of Krogh et al. (1994) onto a tree of n leaves with $2^n - 1$ emission states (Figs. 12.18 and 12.19) in their software HANDEL.

SATCHMO (Edgar and Sjölander, 2003) adds a tree component to profile HMMs by recreating the (Feng and Doolittle, 1987) progressive algorithm but based on HMMs as opposed to string alignment cost. SATCHMO determines the pairwise probability between pairs of sequences (and profile HMMs), joining the closest pair at each turn and creating new profiles. This is in essence UP-GMA (Sect. 9.5.1) using HMM probabilities as a similarity measure. SATCHMO

Kimmen Sjölander

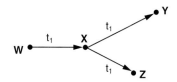

Figure 12.18: The basic median diagram of Holmes and Bruno (2001) with time parameters for the edges to be used in the TKF (Thorne et al., 1991) model.

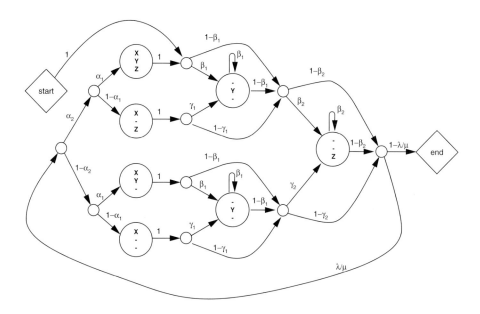

Figure 12.19: The trellis diagram of Holmes and Bruno (2001) superimposed on the tree of Figure 12.18. The α_i parameters correspond to the edge parameters (t_i) of Figure 12.18.

Gerton Lunter

also provides information on the "alignability" of subtree alignments by affinity scores.

Löytynoja and Milinkovitch (2003), in ProAlign, made a number of simplifying assumptions to create a rapid HMM approach directly as MSA. ProAlign is also a progressive alignment approach using HMMs. ProAlign, however, uses a relatively simple 5-state (A,C, G, T, and gap) Neyman (1971) model with Neighbor-Joining (Saitou and Nei, 1987) edge time parameters.

In an alternate approach, Lunter et al. (2005) developed a MCMC method for a fully Bayesian (topology as well as edges and substitution parameters) TAP procedure. The BEAST software produces a MAP tree based on the TKF91 sequence substitution model. As with other Bayesian methods, BEAST calculates the posterior probabilities of aligned positions in their MSA output (Fig. 12.20).

BaliPhy (Redelings and Suchard, 2005) takes an approach similar to BEAST in using MCMC for Bayesian joint estimation of tree and MSA. BaliPhy (like all other procedures) makes simplifying assumptions, in this case the indel model is constant over all edges (TKY91). BaliPhy also employs Metropolis–Hastings chain mixing to improve results. The overall approach is very time consuming however, requiring hundreds of hours (pc-level computer) to analyze 12 protein sequences of length 400.

12.7   Rearrangement

The Bayesian analysis of genomic rearrangements is in the earliest stages. Genomic rearrangement models (Sect. 10.13.2) are varied, complex, and at present not very biological. The identification of realistic scenarios of rearrangement

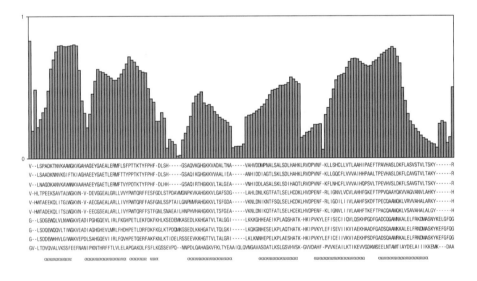

Figure 12.20: Posterior decodings of aligned positions from Lunter et al. (2005).

are elusive, and statistical models of their properties even more distant. Larget et al. (2004, 2005) have produced BADGER to take the mainly parsimony-based rearrangement distance methods and describe them in a Bayesian framework.

BADGER uses an MC^3 approach based on Poisson priors and random permutations of locus synteny to establish likelihoods and posterior probabilities of rearrangement scenarios. This approach is a descriptive one, not based in any particularly supported view of genomic rearrangement, but has the virtue of practical execution times of days for nearly 100 metazoan mitochondrial genomes of 40 or so loci.

Mark Suchard

12.8 Criticisms of Bayesian Methods

Criticisms of Bayesian methods (above and beyond those of likelihood) center on prior probabilities. Not only are the specific shape and distributional details at issue, but also their fundamental existence. If we take the case of time parameters on edges of trees, what is meant by a prior distribution on edges? Is it a prior on that edge? Is it specific to that tree or clade? Or is the prior more general, describing the distribution of edges over the entire tree, or all trees? As mentioned above, simply by increasing the number of taxa, the expected edge time parameter is reduced. Is there a single model—or multiple, perhaps one for each tree (to maximize MAP)? These questions apply equally well to all manner of other distributional assumptions.

The specific distributions themselves can carry awkward baggage as well. The uniform distribution is often used to describe ignorance, or lack of preference among outcomes. Yet this distribution must be arbitrarily bounded, or probabilities of all scenarios go to zero. Other distributions may be used, but what is their justification in the absence of empirical knowledge? In historical science, there is no frequentist experience to draw on. All events are unique. The very notion of a distribution in such a situation is highly problematic. Bayesian methods draw their strength from the integration of information over alternate scenarios. If, however, there is only one scenario—one history, is such an integration valid?

Brent Larget

Perhaps systematics is uniquely difficult in the Bayesian framework. The conflation of natural law and analytical procedure makes any statistical statement at least as much about inferential method as nature. Do our models describe events—or our ability to perceive them?

12.9 Exercises

1. If we have a tree T with n leaves and follow a Yule process, how many trees are derivable from T with $n + 1$ leaves? If we were to follow a uniform model, how many trees are derivable?

2. We observe the sequence of coin tosses: HHTTTHHHHHHHHTT, and are told that one of three possible coins was used: a fair coin ($pH = pT = 0.5$),

a coin biased towards heads ($pH = 0.8$), or one biased towards tails ($pT = 0.8$), and that these three coins had the prior probabilities of 0.5 for the fair coin and 0.25 for each of the biased coins. What are the expected transition probabilities (temperature aside) in a Markov Chain among these three states? What are the expected frequencies of the three states after a long chain?

3. If the prior distribution of tree topologies with 10 leaves is uniform, what is the prior probability of a clade with 2 leaves? 5 leaves? 8 leaves?

4. For the three coins in exercise 2, if the probability of the same coin being used in successive tosses is 0.5 and the probability of switching to each of the other two is 0.25 (the initial coin choice follows the prior), draw the trellis diagram for this Hidden Markov Model.

   If only a single coin were used, determine the probabilities of the coin toss sequence for the fair and two biased coins.

   If the sequence of coins used were FFFFFHHHHHHHHHTT (F=fair, H = biased towards heads, and T = biased towards tails), determine the probability of the coin toss sequence.

5. Given the HMM trellis of 12.14, and the coin toss result HHTTTHHHH-HHHHTT, assuming the three coin states have equal initial probability, determine the maximum probability state path (Viterbi) and the probabilities of each state at each position (Forward–Backward).

6. Draw a general (no numbers) profile HMM trellis diagram for a sequence model with sites that follow GTR (no indels), sites that follow 5-State Neyman (A, C, G, T + gaps), and sites that are invariant (no indels either).

Chapter 13

Comparison of Optimality Criteria

As discussed in the four previous chapters, there is a diversity of methods to reconstruct phylogenetic trees from comparative data. In general, most forms of such comparative data (*e.g.* qualitative anatomical features, DNA sequences) are amenable to analysis by any of the currently applied methods. In this chapter, the relative merits of the methods are discussed, as are three bases used to distinguish among the approaches: epistemology, statistical behavior, and performance.

The approaches discussed here are of four general types: distance analysis, parsimony, or minimization, maximum likelihood, and posterior probability. The term "posterior probability" is used in this chapter to refer to the Bayesian approaches that maximize the posterior probability of trees (MAP or Topology–Bayes), as opposed to their component parts (clade posteriors or Clade–Bayes). As discussed in Chapter 12, clade-based posteriors do not offer an optimality value *per se*, hence are not directly comparable to the methods discussed in this chapter.

There is no single answer to "what is the best method?" in same way there is no single answer to "what is the best tree?" A criterion must be specified in both cases. Given an optimality criterion, we can rank competing hypotheses and identify the best tree or trees (at least of those we have found). The same procedure applies to methods, a criterion must be proposed in order to choose a method; and the criterion justified.

13.1 Distance and Character Methods

Before discussing the main points of comparison among methods, a general distinction needs to be made between methods that rely on overall dissimilarity (or similarity) and those that directly operate on individual character observations (*e.g.* anatomical features, molecular sequence data). Distance methods (Chapter 9) operate on the degree of pairwise difference between taxa. The distances may be raw, scaled, or normalized transforms of dissimilarity. When trees are constructed from these data, by whatever criterion, no attempt is (or can be) made to identify medians at tree vertices. No hypothetical ancestral state reconstructions are proposed, no global homology statements erected, and no distinction between primitive and derived similarity made. In contrast, character-based methods are alike in that they make explicit statements about median ancestral reconstructions. The edges of trees then represent the transformations between ancestral and descendent states (assuming the tree is directed or rooted).

With parsimony, the differences between these node states are summed, yielding the parsimony score. Likelihood multiplies the probabilities of the same edge transformations given a model and time parameter. Bayesian methods also do this, integrating over the distribution of model and time parameters.

Whatever other admirable mathematical qualities distance analysis may have, it does not yield the specific events that are reconstructed to have occurred on tree edges. The reconstruction of edge events forms much of the basis for discussion in the comparative biology literature. Biologists are interested in, and largely motivated by, the desire to discover patterns and understand the general and specific nature of these events. The entirety of historical evolutionary change is contained in these edge transformations. Of course, character events can be reconstructed *ex post facto* on a distance tree, but by what means? This brings us back to character-based methods. Furthermore, since the distance criterion differs from those of the character methods, any distance tree result is unlikely to be optimal for the character-mapping criterion, again returning us to character-based analysis *ab initio*. For this reason, most biologists concerned with tree-based analysis of variation employ character-based methods[1].

13.2 Epistemology

One of the earliest and most fundamental distinctions among systematic methods was made on the basis of epistemology. That is, how do we "know" things, how are scientific hypotheses proposed and tested, and what is the relationship between observation and inference? A fundamental distinction has been made between a deductive (or more precisely hypothetico-deductive) process and the inductive approach embodied in statistical estimation procedures as applied to systematic problems (see Chapter 4 for general discussion of modes of inference).

---

[1]There are several biological areas that contain counter examples including the medical analysis of human pathogens and evolutionary studies of non-eukaryotic organisms. These cases are presumably due more to tradition than any view of the superiority of the method.

13.2.1 Ockham's Razor and Popperian Argumentation

Platnick and Gaffney (1977), Gaffney (1979), Farris (1983), and Kluge (1997) among others, have proposed and defended the proposition that tree reconstruction based on the minimization principle of parsimony (*i.e.* Ockham's Razor) is not only desirable on its own merits, but is also the method that conforms most closely with the ideas elucidated by Popper (1959, 1963, 1972, 1983). The principle of minimal causative explanation has great power and elegance on its own. For a method that makes no specific process claims, parsimony has been remarkably successful in providing a means to understand the enormous diversity of biological pattern that confronts us. The minimization approach of parsimony offers a specific method to achieve the base-line goal of choosing genealogical explanations of variation, and of identifying non-conforming observations that require additional modes of causality. More than the mere desire to avoid epicycles, parsimony became a basis for hypothesis testing through the union of the ideas of Ockham and Popper.

Popper (following Hume, 1748) emphasized that inductive methods could never be satisfactory, since proof would require an exhaustive examination of cases. In place of this verification procedure, Popper proposed that hypotheses should be tested by searching for contradictory or falsifying observations since, in principle, a single falsification would reject the hypothesis. This was not only absolute but efficient.

As discussed in Chapter 4, this hypothetico-deductive approach is not applied naïvely (since all non-trivial systematic hypotheses are likely to be falsified by at least one observation), but such that the hypothesis that is "least" falsified is most favored (Lakatos, 1970). This application of falsification creates the link with Ockham's razor, justifying and underpinning the use of parsimony in systematic analysis (Farris, 1983). Any parsimony suboptimal result, as generated by a method based on an alternate criterion, breaks this connection with falsification and the hypothetico-deductive method. Furthermore, the greater the number of observations brought to bear on the problem, the greater the number of opportunities to falsify the hypothesis, and the more severely it has been tested (Eq. 4.3).

Imray Lakatos
(1922–1974)

The complement of Popper's falsification is explanatory power. Cladograms and trees "explain" variation through genealogical inheritance from ancestor to descendant. Observed variation that is consistent with a given tree explanation will have an overall minimal number of changes (cost), with non-minimal changes (homoplasies) not as well explained by that tree as they would be by another. The tree that minimizes the number of *ad hoc* homoplastic changes over the data set best explains the variation in terms of genealogy. As Farris (1983) pointed out, not only does the most parsimonious tree explain the variation, but does so without the requirement of other complex aspects required by statistical explanation, such as overarching process models and the parameters associated with edge lengths. Parsimonious trees have maximal explanatory power with regard to observed variation, and they do this without the accessory explanatory elements of other methods (*e.g.* model and tree parameters).

13.2.2 Parsimony and the Evolutionary Process

When discussing the merits of parsimony as a phylogenetic procedure, it is important to keep clear the distinction between parsimony as an inferential procedure and parsimony as a model of evolution. As a philosophical basis for testing hypotheses, parsimony does not assume that evolution occurs parsimoniously. In fact, all non-trivial data sets exhibit non-parsimonious (*i.e.* homoplastic) events. Neither, in this context, does parsimony assume that homoplasy is rare, or that rates of evolution are fast or slow (Farris, 1983; Sober, 1988, 2004) (but see below). The epistemological basis of parsimony as a process of discovery and hypothesis testing in science is general and useful. As stated by Wiley (1975), systematic analysis "must be done under the rules of parsimony, not because nature is parsimonious, but because only parsimonious hypotheses can be defended by the investigator without resorting to authoritarianism or apriorism."

Kluge (*e.g.* Kluge, 2005) has taken a different tack in that he has proposed that the simple model of evolution as descent with modification (in essence, a minimal evolutionary assumption that offspring resemble their parents more than non-parents, but not exactly) leads to parsimony as an evaluation principle. Again, such a statement makes no assertion of rates of change; rather it is perhaps related to the "evolutionary path" model of Farris (1973a) (Section 11.3.1).

13.2.3 Induction and Statistical Estimation

One of the strongest criticisms of statistical approaches in systematics is derived from the fact that systematics is a historical science. The events whose explanation are sought are unique to time and place. They are historical singulars (Kluge, 1998). As such, notions of "average" behavior inherent in probabilistic models is problematic. If each character observation is a unique object, it cannot be a random sample drawn from a parameterized distribution expressed as a model of change (*e.g.* GTR+I+Γ). Interestingly, when characters (and edges) are treated as having uncorrelated evolutionary rates (No-Common-Mechanism; Tuffley and Steel, 1997), parsimony and likelihood become exactly concordant in tree choice. This appears to express neatly an important difference not necessarily between statistical and minimization-based methods, but between treating character changes as a class as opposed to historical individuals.

By this rationale, parsimonious trees have explanatory power in minimizing *ad hoc* hypotheses, and increased observation will increase the severity of test, but would not reduce any measure of "sampling error," since there is no sample distribution that is measured.

13.2.4 Hypothesis Testing and Optimality Criteria

Much of the argumentation and philosophical constructs surrounding hypothesis testing in systematics were developed in the context of parsimony. However, Popper (1959) himself defined support, severity of test, and corroboration in terms of likelihood functions involving the probabilities of evidence (data) given a hypothesis and background knowledge. He was explicit in saying his probability was defined as "relative frequency" and the probability of the evidence given a hypothesis, "Fisher's likelihood function." Popper's optimal hypothesis is that which maximizes $p(e|h, b)$, the probability of the data given the hypothesis and background knowledge—in short the likelihood.

As discussed in Section 4.1.7, at least in my opinion, hypothetico-deductive inference can proceed via a falsification step based on any objective (and transitive) optimality function. As long as hypotheses can be competed in a mathematically objective manner, hypotheses can be tested, and non-optimal solutions falsified. The key question is the appropriateness of the chosen function.

13.3   Statistical Behavior

There is a large literature concerning the statistical properties of systematic methods, centered on the estimation behavior of various optimality criteria. When contrasting approaches to phylogenetic reconstruction are treated as alternate estimators, their behavior is most often discussed in light of three aspects: consistency, efficiency, and robustness. There are many cases of provable results, as well as demonstrations on contrived or simulated data. One of the most pressing issues that faces systematists is whether the conditions under which these results are generated or proven apply to the empirical cases encountered in the course of research.

13.3.1   Probability

Before diving into the hurly-burly, it is worth revisiting the ideas embodied by the concept of probability and their application to historical science. As discussed more fully in Chapter 6, there are alternate interpretations of probability. In the context of the arguments discussed here, probability can have a logical or degree of belief interpretation or, alternately, it may signify truth statements about the natural world akin to the frequency of repeated events. In the first case, probabilities reflect relative confidence in an outcome. A statement such as "this coin has an 80% chance of being fair" would be an example of the first interpretation. The coin is either fair or not; the probability attached is similar to a betting or odds ratio often used in Bayesian analysis (DeGroot and Schervish, 2006). By contrast, in the second case, probabilities imply a statement about what is ontologically "real." "This coin has an 80% chance of yielding heads on a given toss," offers a property of the coin and a statement regarding physical reality.

Historical science does not easily permit a frequentist interpretation. What is to be our course of action if presented with the question, "What is the probability that Octavian and Caesar Augustus were the same person?" Such a situation cannot be re-run or repeated in any way. Either he was or he wasn't. Any notion of probability is necessarily limited to our degree of belief, based on the analysis of empirical data. Such is the case when presented with a phylogenetic statement, "are groups A and B sister taxa?" They either are or are not, and probabilistic hypothesis testing would yield a value allowing us to assess the relative degree of support of these two hypotheses—but not their reality. In short, probability statements about unique historical events do not describe nature, but rather our understanding of it[2].

This restricted, and perhaps non-consensus, interpretation of probability, however, has absolutely no effect on the formal analysis of probabilities or statistical properties of estimators (as long as the probabilities follow the basic axioms, Sect. 6.1.1). What this interpretation does do is proscribe our ability to make statements that are "true." We cannot say whether a given set of historical statements (*i.e.* a tree) is true, but we can say we have a relative degree of belief based on empirical observation and an optimality criterion.

13.3.2 Consistency

A concept central to the literature concerning the comparison of methods is the concept of statistical consistency. Informally, consistency is the behavior of an estimator $\hat{\theta}$ of a parameter θ where, as the sample size (n) grows, the difference between $\hat{\theta}$ and θ becomes arbitrarily small (Eq. 13.1). If the estimator exactly converges to the parameter value,

$$\lim_{n \to \infty} |\hat{\theta} - \theta| < \epsilon \qquad (13.1)$$

it is said to be *strongly* consistent (Eq. 13.2).

$$\lim_{n \to \infty} |\hat{\theta} - \theta| = 0 \qquad (13.2)$$

In the context here, $\hat{\theta}$ would be the reconstructed tree (or graph) and θ the "true" tree. The question posed to optimality criteria is whether optimality criteria are or are not consistent, and under what conditions.

This line of reasoning was first raised by Felsenstein (1978) (Sect. 11.1.1) when he illustrated a situation where parsimony was guaranteed to favor an incorrect tree (inconsistent), while a likelihood analysis would not. His first case was perhaps overly simple and unrealistic, but was followed by a large series of analyses of more complex models in various situations. Discussions have dealt both with the question of whether a particular method is consistent and also with whether consistency is a useful quality by which to judge methods.

Assumptions

Felsenstein first demonstrated that parsimony could be inconsistent under conditions of unequal branches and a simple evolutionary model and asserted that likelihood in the context of tree reconstruction should be consistent based on the condition of Wald (1949).

Abraham Wald
(1902–1950)

---

[2]A classic example of this form of analysis concerning the authorship of *The Federalist* papers is found in Mosteller and Wallace (1984).

Wald proved that maximum likelihood estimators would be consistent given several conditions[3]. Many of these same assumptions are found in later consistency proofs. The most important of these conditions are: 1) that the observations (data) are independent, 2) that they are identically distributed random variables, and 3) that the number of parameters to be estimated is finite, while the number of observations increases without bound. Each of these assumptions is potentially problematic in the situations posed by systematic analysis.

The independence assumption is entirely appropriate for the sorts of standard examples of statistical reasoning such as drawing balls from an urn or tossing a coin. In these cases, individual observations have no effect on the chance of another event or its interpretation. If we take DNA sequence data as an example set of observations for discussion, empirical studies in no way match this description. Empirical sequence observations are not randomly drawn from the genome, one at a time. Almost all phylogenetic analyses of sequence data draw their observations from a restricted segment of the genome (a locus or loci) defined by a common metabolic purpose (*e.g.* cytochrome oxidases) and adjacency on a chromosome. Furthermore, sequences are typically identified in a specific region or locus and all nucleotides between two biochemically defined positions (often primers) determined. These are not random samples of the genome in any way, but are observations from a highly localized, functionally related region. Multi-locus studies certainly improve on this, but even large collections of loci provided by EST analysis (*e.g.* Dunn et al., 2008, Hejnöl et al., 2009) are limited to the relatively small fraction of the genome (10^{-3} or less, if 10^{-1} of the transcriptome) that are expressed as proteins.

A second difficulty with the independence assumption is found in all comparative data. Even when entire genomes are sampled, these sequences have been in the same historical "bottle" so to speak, for their entire history. We are forced by necessity to treat comparative observations of anatomy and sequence as if they were independent, but their shared history shows this operational assumption to be false.

A third problem with the independence assumption is derived from the length-variable nature of many loci used in molecular systematics studies. The "observations" used in systematic analyses are often aligned positions. These are constructed from raw observation and are, in fact, highly inferential objects. When sequences of an organism are determined, it is without reference to sequences in other organisms. The raw sequences undergo an analytical process to determine which nucleotide sequence elements in one organism correspond to those in another. Alignment and optimization of length-variable sequences infer sequence correspondences via the optimization of a function based on parameters for the transformations between sequences (*e.g.* transition, transversions, insertion, and deletions) and an optimality criterion (*e.g.* Sum-of-Pairs distance, Sect. 8.5). The correspondences between any given pair of sequence elements is

---

[3]Wald also applied his statistical acumen in the British Air Ministry during the Second World War urging the counter-intuitive path of armoring locations on bombers that did *not* have bullet holes. This was because the damage was only found on bombers that came back, not those that suffered fatal damage and left no record—an extreme case of survivor bias.

highly dependent on those elements adjacent to them and in the remainder of the sequences. Hence, comparative sequence data are highly interdependent and obviously do not satisfy the definition of independence.

The identical distribution assumption likewise presents difficulties. Most analyses of consistency assume a single evolutionary stochastic model for sequence change. This may include variable rates among the aligned positions (if such exist) and tree edges, but a unitary model is assumed. Even within a single locus, there are well-described varying dynamics due to the metabolic role various components of products of the locus play. Obvious examples include the "stem" and "loop" regions of structural RNAs and transmembrane regions of cytochrome oxidases. More complex models can, of course, be constructed, but this entails the problems associated with estimating additional parameters with limited data.

The final assumption concerns limit behavior as the input data increase without bound. The data presented by biological problems are finite, and in many cases, relatively small. If a stochastic model is to be applied to a single locus (much less to a subregion), sequence data sizes will be limited to hundreds to low thousands of base-pairs. It is unclear if these sample sizes are sufficient to ensure asymptotic behavior. An entire transcriptome might constitute 10^4 loci, but it would be difficult to justify a single stochastic model for such a set of metabolically diverse genomic components. Entire genomes vary from thousands of base-pairs in viruses to billions in vertebrates, setting a hard limit on sample size, and as with transcriptome analysis, a single model is unlikely to be satisfactory.

The assumption violations discussed above do not impede the use of statistical methods as optimality criteria for the relative ranking of hypotheses, but they do bring into question the application of consistency of estimators in real-world systematic analysis.

Distances

As mentioned earlier (Chapter 9), one of the more vexing aspects of distance analysis is negative branches. These appear in analysis using a variety of branch-length estimation procedures including least-squares (Fitch and Margoliash, 1967), greatest-lower-bound (Farris, 1972), and Neighbor-Joining (Saitou and Nei, 1987). The analytical meaning of such edges with respect to the methods is clear—the reconstruction assumptions are violated. Biologically, of course, they are meaningless. They also have clear implications for the behavior of these edge lengths as estimators of amount of evolution. Since there can be no such thing as "negative" evolution, the minimum value for any branch length must be zero. These estimators, then, must be inconsistent.

The tree topology itself, nonetheless, could still be consistently estimated through minimization of the sum of these branch lengths. Gascuel et al. (2001) examined this issue and showed that for the most "reliable" method of branch length estimation, generalized-least-squares, as well as for the widely used weighted-least-squares, Minimum Evolution (ME) tree reconstruction was not

1	AABAABB		
2	ABABBAA	$d(1,2) = 6$	$d(2,4) = 6$
3	BACCCAA	$d(3,4) = 6$	$d(1,4) = 5$
4	BCAAACC	$d(1,3) = 6$	$d(2,3) = 5$

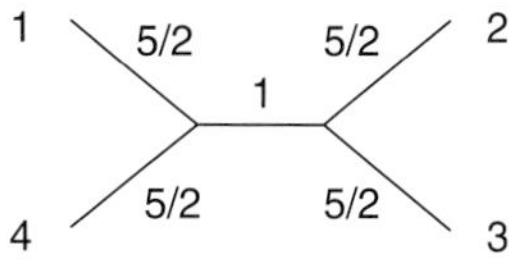

Figure 13.1: The example of Huson and Steel (2004) showing the underlying data (left) perfectly compatible with the tree 12|34, the distance matrix derived from the data (center), and the perfectly additive tree 14|23 (right).

generally consistent (even under a variety of methods employed to avoid negative branches).

Huson and Steel (2004) examined more general conditions and found that there were cases where the underlying character data were homoplasy-free on one tree, but when transformed into (uncorrected) distances were still perfectly treelike and clocklike, yet on a different tree. The distances are compatible, additive, and entirely misleading (Fig. 13.1). This is a mathematical instantiation of the problems presented by plesiomorphic and autapomorphic characters that Hennig described.

In addition to this result, Huson and Steel (2004) show a similar result using stochastic models, where the inconsistent result would be unsuspected because the distances appear perfectly tree like. This effect can also occur when distances are transformed by an incorrect model. In a similar vein, Steel (2009) showed that distance-based reconstruction can be inconsistent when rate variation is modeled by the gamma distribution, but the shape parameter unknown.

Parsimony

Felsenstein (1978) showed that parsimony (under a specific 2-state model) was inconsistent in cases where there was a short central edge connected to two long and two short pendant edges (Fig. 13.2, $p_1 = p_3$ and $p_2 = p_4 = p_5$). Under these conditions, the random match similarity of the terminals with long branches (p_1 and p_3) overwhelmed the parsimony signal of the short central branch (p_5). Most of the conditions under which parsimony has been shown to be inconsistent center on this phenomenon.

Penny et al. (1991) generalized Felsenstein's four-taxon case to arbitrary edge probabilities. With $\omega_i = (1 - 2p_i)$ of Fig. 13.2, parsimony will be consistent if and only if

$$\omega_5 < \min \left(\frac{\omega_1\omega_2 + \omega_3\omega_4}{\omega_1\omega_3 + \omega_2\omega_4}, \frac{\omega_1\omega_2 + \omega_3\omega_4}{\omega_1\omega_4 + \omega_2\omega_3} \right) \tag{13.3}$$

These conditions will always be met if change is clocklike and also apply to distance methods.

A more general case, unlimited in number of states or taxa, was analyzed by Steel (2000). Under the Neyman (1971) model, Steel proved that when overall transformation rates are low and fairly similar among branches, parsimony

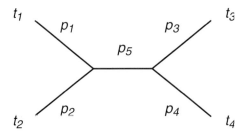

Figure 13.2: The example of Penny et al. (1991) with terminal taxa $t_{1...4}$ and probabilities of change along edges $p_{1...5}$.

would be consistent. More precisely, for edge probabilities p_i (over the entire edge as in Felsenstein, 1978 not instantaneous) on a tree $T = (V, E)$, two relations must hold (Eq. 13.5) as the sufficient condition for parsimony to be consistent.

for

$$p_{sum} = \sum\nolimits_{\forall e \in E} p_e$$

$$p_{min} = \min\nolimits_{\forall e \in E} p_e \qquad (13.4)$$

parsimony is statistically consistent if:

$$p_{sum} < 1$$

$$p_{min} \geq \frac{p_{sum}^2}{1 - p_{sum}} \qquad (13.5)$$

Equation 13.5 suggests that as a tree becomes larger, parsimony will have an increasingly difficult time maintaining consistency (with mean $\mu t = 0.1$, the limit would be about 12 taxa for binary data and 22 for 4-state). However, Steel and Penny (2004) have shown that if the (non-additive) state set (r) is sufficiently large ($r \geq 4^{nk}$, for n taxa and k characters), the parsimony and maximum average likelihood tree are identical, hence parsimony would be consistent.

Lastly, several model-based "corrected" forms of parsimony (Steel et al., 1993; Penny et al., 1996) have been shown to be consistent. Of course, the generating model of the sequences and tree must be the same as that used in the correction.

Likelihood

In discussing the consistency of maximum likelihood tree reconstruction, it is important to keep clear the distinctions among the various flavors of likelihood that have been applied to systematic problems (Chapter 11). The strengths and weaknesses of likelihood methods stem from the model upon which the analysis is based. A rigid requirement for likelihood methods to be consistent is that the model used for reconstruction is the same as that which generated the tree in the first place. In general, likelihood will be inconsistent otherwise. This could be as simple a case as where a single incorrect model is chosen, or where the reconstruction assumes a single model, and the "true" scenario was based on two—even if they were generated by the same tree (Chang, 1996). To ensure consistency, the likelihood model must accurately depict the reality of historical change. This model must be correct for all the lineages/edges (ancestral as well as extant) over the entire time and environmental conditions suffered by the taxa since their root origin. Of course, as with all historical phenomena, this information is unavailable, hence consistency can never be assured for any empirical analysis. Simulation offers some means of examining performance in light of "true" trees, but simulations are human constructs, limited, as nature is not, by our imagination.

Maximum Average Likelihood (MAL)—As mentioned above, Felsenstein (1978) cited Wald (1949) in his assertion that MAL was consistent. One of the key points in Wald's proof is that the parameters to be estimated must be finite, not growing with the data size. As long as unique parameters are not estimated for each observation (*e.g.* an aligned nucleotide sequence position), this condition should hold. Several likelihood models violate this condition (*e.g.* No-Common-Mechanism or NCM, Tuffley and Steel, 1997), hence, do not meet the condition Wald described (although some NCM methods are still consistent [Steel, 2011b]).

Using a common mechanism model as described by Felsenstein (1981), Yang (1994a) proved that MAL was consistent in the estimation of tree topology. The estimation of edge parameters and other aspects of the full model was shown also to be consistently estimated under MAL by Chang (1996).

Steel (2011b) investigated the properties of several NCM models showing that subtle differences between models and reconstruction methods resulted in variants which were (NCM-N_4[4] using linear invariants, NCM with a molecular clock, NCM-N_∞) and were not (NCM-$N_{r<1}$, other than NCM-N_4 earlier) consistent.

Most-Parsimonious Likelihood (MPL—also referred to as ancestral maximum likelihood)—In this form of likelihood, the node assignments are made based on the single most likely state as opposed to the averaging of node assignments over all possible states (Sect. 11.3.1; Barry and Hartigan, 1987). Mossel et al. (2009) show that MPL has a "shrinkage" problem with respect to short edges on a tree, implying that MPL is not statistically consistent. This result is supported by the proof (Steel and Penny, 2004) that when taxon sampling densities are high enough, the MPL result is the parsimony result.

Evolutionary Path Likelihood (EPL)—The Farris (1973a) form of likelihood specifies not only the tree topology and node state reconstructions, but each intermediate conditions along each edge (Sect. 11.3.1). Given that Farris proved that the EPL tree is precisely the parsimony tree under very general evolutionary condition (*e.g.* no rate restrictions), EPL would have all the strengths and weaknesses of parsimony, including inconsistency in the circumstances described above.

---

[4]NCM-N_i signifies the No-Common-Mechanism model with i character states.

Maximum Integrated Likelihood (MIL)—This form of likelihood can be employed when prior distributions on model and edge parameters are known (Sect. 11.2.1). In essence, MIL is a Bayesian analysis where the prior distribution on trees is uniform. As a special case of the general Bayesian framework, MIL has been shown to be consistent, given, of course, appropriate priors and model (Steel, 2011a).

Bayesian Posterior Probability

Given the derivation of Bayesian methods in likelihood approaches, it was initially thought that the approach of maximizing posterior probability of trees[5] would be statistically consistent. This was thrown into doubt in the situation where data were evenly supportive of multiple topologies yet a specific tree was favored among those that should have been equal (Yang and Rannala, 2005; Lewis et al., 2005; Kolaczkowski and Thornton, 2009). This "Star Paradox" was thought to suffer from long-branch-attraction type problems, rendering the method inconsistent. The method was later proven to be asymptotically consistent by Steel (2011a), and the previous instances likely due to finite data.

Does Consistency Matter?

As mentioned above, Felsenstein (1978) first urged the importance of statistical consistency in systematic analysis. Certainly, the idea of a procedure that is guaranteed to converge to the "true" result would seem to be desirable, and one that did not would be undesirable. This sentiment has been expressed by many statisticians (Fisher, 1950; Neyman, 1952; Kendall and Stuart, 1973) as the only reasonable course for estimators. Others, however, did not feel this way. Edwards (1972) stated that asymptotic behavior was irrelevant, and that the only operation that mattered was the relative ranking of hypotheses.

Traditionally, researchers who favor ML have emphasized the centrality of consistency, while those more interested in parsimony have downplayed its importance. Either way, the relevance of consistency in real world analysis is unclear for two reasons. The first concerns the requirement of a match between generating ("true") model and that used during ML reconstruction. The second reason is finite data size.

Models used currently in systematic ML analysis are clearly simplifications of the myriad forces molding the evolution of creatures in time and space. This simplification makes analysis tractable, and is even desirable to a certain extent, since any tree can be made to match the data perfectly by sufficient manipulation of a large enough collection of parameters (the solution becomes non-identifiable, violating one of Wald's assumptions; Steel et al., 1994; Yang et al., 1995). Nonetheless, this invalidates the applicability of proven consistency results to any empirical analysis. A stark example of this invalidation is the use of stochastic models that ignore insertions and deletions in length-variable

---

[5] As opposed to that of clades.

sequences. An analysis that treats gaps as missing data of necessity does not model the "real" process completely. In many cases, very slight differences in models and analytical assumptions can have large effects. As stated by Steel (2011b) when discussing varieties of NCM models,

> This brings into question the robustness of any consistency results to even slight model misspecification and suggests that other statistical considerations (*e.g.* bias, efficiency) may override consistency issues.

A second issue, as discussed above, is the finite data available to research (*e.g.* even whole genomes are finite, if large). Asymptotic behavior is irrelevant if available data are insufficient to show it, and finite-sample analysis need not reflect large-sample behavior (Kim, 1998). An example of this may be the "Star Paradox" (above) of Bayesian methods identified in finite cases (Kolaczkowski and Thornton, 2009) but shown not to exist asymptotically (Steel, 2011a).

The real-world inapplicability of consistency proofs does not impugn likelihood methods in any way; it simply signifies that consistency is not a basis to favor likelihood over other methods (*e.g.* parsimony). Likelihood remains a competitive and defensible optimality criterion for use in hypothesis testing. As Sober (1988) wrote,

> Likelihood describes which hypotheses are best supported by the evidence. When the evidence is misleading, the best-supported hypothesis will be a false one. A rule of inference that correctly conveys the evidential meaning of observations *ought* to point to a falsehood when the evidence is misleading. When it does so, it correctly captures what the evidence is saying. (italics original)

13.3.3 Efficiency

Informally, efficiency is the amount of data required by an estimator to produce a value very close to the actual parameter. A more efficient estimator will require fewer data to generate a specific quality of result than one that is less efficient. Likelihood methods are often asymptotically efficient (Yang, 1997), meaning that they will produce good estimates, given enough data.

In an environment of finite data sets, efficiency can be an important issue. An inefficient method may return inferior results on a finite data set when asymptotically it would be superior. Gaut and Lewis (1995) and Yang (1996, 1997) present cases where likelihood reconstruction with the correct model (under simulation) is out-performed by both likelihood with "incorrect" models and parsimony.

An extreme case of this has been shown to occur by Steel and Penny (2000), where parsimony requires only 16 residues to reconstruct a tree correctly with 0.99 probability, whereas maximum average likelihood could require 10^{10}. This is not to say that parsimony methods are uniformly or generally more efficient than likelihood or Bayesian methods. However, efficiency is clearly an important factor to consider in real-world analysis of finite data.

13.3.4   Robustness

Robustness refers to the performance of an estimator when faced with violations of its assumptions (distribution, independence, *etc.*). In the context of systematic analysis, the main issue is model violation. Given that phylogeny estimation exists in an environment of finite data and unknown (and likely complex) generating models, methods overly sensitive to violations of their assumptions will be unsatisfying tools when faced with real data.

Steel (2011b) examined the conditions for consistency among a set of NCM models and found that relatively slight alterations of conditions had marked effects on results (whether the model was consistent or not). Yang (1997) demonstrated similar effects for more typical common-mechanism models, disconcertingly finding "wrong" models outperforming the right. As classified by Sanderson and Kim (2000), likelihood (and Bayesian) methods are parametric, in that they rely on stochastic models based on a specific set of parameters. They contrast this with non-parametric phylogeny reconstruction methods such as parsimony. Parametric methods will always have more power, but be intimately connected to the assumptions of their models and will be perturbed more easily than non-parametric methods that operate well over a broad area of distribution and parameter space. An example of this fragility is shown in simulations by Kolaczkowski and Thornton (2004, 2009) (and potentially real data by Pickett et al., 2005), where parsimony performed significantly better than likelihood under conditions of heterotachy over edges (Fig. 13.3).

Sanderson and Kim (2000) state that parsimony is a robust estimation technique well suited to situations where "true" models are unknown. They also discuss the computational burden of likelihood estimation being so much greater than that for parsimony (in general, NCM, of course, is an exception with low complexity of calculation), and that parsimony tree space will always be better explored than that of likelihood. Hence, although given enough time, likelihood may perform better, time, like data availability, is finite.

As mentioned earlier, transcriptome data sets present on the order of 10^4 expressed loci, each with its own metabolic role. The task of constructing models able to deal with the varying evolutionary dynamics and history of such a collection of sources of information, while avoiding non-identifiability, is daunting. Most likely, the simple models we use today will be inadequately robust to deal with such genome scale diversity. It may be that only non-parametric systematic analysis will be sufficiently robust to handle such problems.

13.4   Performance

An additional, and perhaps utilitarian, means of comparing methods is their performance on real and simulated data sets. Unfortunately, when dealing with actual problems and data, we cannot know the true (in the historical sense) patterns and processes that resulted in the diversity we now see. On the other hand, analysis based on real-world data is subject to all the effects and defects our methods seek to understand. This would include (but is not limited to) evolutionary processes varying over time, space, and taxon, and finite,

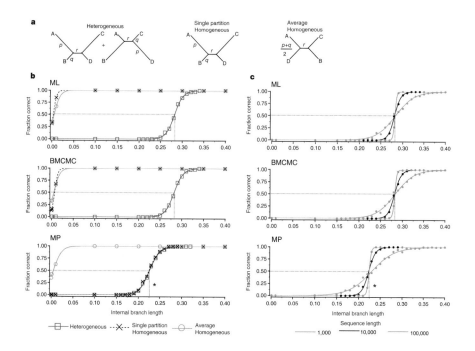

Figure 13.3: Results of Kolaczkowski and Thornton (2004) showing the superior performance of parsimony over likelihood and Bayesian methods under a condition of heterotachy. ML = maximum likelihood, BMCMC = Bayesian MCMC, and MP = parsimony. See Plate 13.3 for the color Figure.

incomplete data due to inherent limitations on data quantity, extinction, and technical difficulties. Simulated data, on the other hand, do not share these problems. Conclusions based on simulations, however, are bound ineluctably to the conscious and unconscious assumptions, opinions, views, and experiences of their creators—but they are known[6]. Replicate analyses can be performed, ancestors observed, extinction and sampling precisely determined. Empirical data are the real thing, but they are uncontrolled. Simulated data are controllable, but synthetic.

Due to this distinction, performance comparisons of phylogenetic methods can argue at cross purposes, and several important factors are only amenable to one or the other form of examination.

13.4.1 Long-Branch Attraction

Since Felsenstein's initial case, Long-Branch Attraction (LBA) has been prominent in comparison of methods, most usually in criticisms of parsimony. LBA is an expression of a lack of statistical consistency in the asymptotic case. As

---

[6]A biological "simulation" was undertaken by Hillis et al. (1992) with mutagenized T7 bacteriophage. Unfortunately (or not), all methods performed perfectly, including the consensually pathological UPGMA.

has been shown, distance and common-mechanism average likelihood methods can also be inconsistent in fairly simple cases of either incorrect or insufficient stochastic models (Chang, 1996).

Performance of methods in real-world situations will always be based on finite data, and many studies have addressed this reality via simulation. Typically, a small set of sequences (4–10 or so, compact enough to be analyzed easily) are evolved via some model under a variety of branch length scenarios (*e.g.* Swofford et al., 2001) and then subjected to analysis via a set of methods. The result is compared to the "true" history of the simulation.

Simulations have repeatedly shown LBA for parsimony (*e.g.* Swofford et al., 2001; Pol and Siddall, 2001), but this phenomenon is not limited to parsimony. Pol and Siddall (2001) showed in a 10-taxon case with 1kb of simulated, aligned sequence data that likelihood was prone to this problem as well (Fig. 13.4). This result holds regardless of whether likelihood analysis is performed using an incorrect or correct (generating) model. Additionally, likelihood also suffers from a Long-Branch Repulsion (a true long branch is not recovered—named the "Farris Zone," Siddall, 1998, or the Anti or Inverse Felsenstein Zone) not found in parsimony analyses. Parsimony outperforms likelihood in this context as the probability of change on the short branches approaches zero and the sequences on the longer edges approach randomization. With k sites, the probability of parsimony reconstructing the tree correctly approaches $1 - (\frac{3}{4})^k$ while that for likelihood will be no more than $\frac{2}{3}$ (Steel and Penny, 2000).

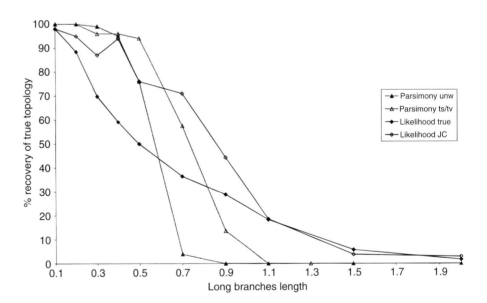

Figure 13.4: Figure of Pol and Siddall (2001) showing the failure of both parsimony and likelihood to identify the correct tree as branches show increasing differences in length.

As mentioned above, a similar case of parsimony outperforming both likeli-hood and Bayesian methods with simulated varying rates of change is shown by Kolaczkowski and Thornton (2004). This was observed with even a small frac-tion of characters suffering heterotachy. Kolaczkowski and Thornton (2009) later showed that LBA could also occur in Bayesian analyses under similar conditions. Simulations have shown that parsimony can also be hindered by heterotachy, but not as severely as statistical methods (Philippe et al., 2005).

So what are we to do about LBA? All methods can, under some conditions, exhibit this behavior. However, the only way to recognize the phenomenon is by knowing the actual history of a group. Yet, we cannot know this history for real data. What if, as the data would suggest, a "long" branch merely denoted well-supported sister taxa? If, for example, a pair of *Panorpa* species were included in an analysis of insect orders (Wheeler et al., 2001), the edge joining these taxa would show a great deal of change—a long branch. Should we doubt such a grouping?

13.4.2 Congruence

In my mind, one of the most powerful tests of any phylogenetic hypothesis is congruence. Whether that congruence is among characters of a single analysis or between multiple data partitions, we expect that alternate sources of infor-mation should agree about phylogenetic conclusions (assuming a single history, horizontal and reticulate evolution aside). In this way, data and methods are predictive (or more precisely retrodictive), in that they anticipate new obser-vations. Comparison with "known" phylogenies is really an extreme form of congruence where the comparison is between a result and the previous body of evidence supporting a phylogeny.

As far as I am aware, there are very few explicit comparisons of optimality criteria based on the congruence behavior of real data (Wheeler, 2006). This may be due to the dearth of methods available for the statistical analysis of qualita-tive data in the past, but with the implementations of Neyman (1971) and NCM (Tuffley and Steel, 1997), this problem would seem to have been ameliorated. Given the ability to measure congruence irrespective of the true relationships, this procedure should be a powerful measure of the behavior of methods[7].

13.5 Convergence

As discussed above and in previous chapters, several instances have been shown to lead to convergent results between parsimony and other methods, most com-monly likelihood. Farris (1973a) with EPL was the first to show that likelihood (that version at least) and parsimony would rank topologies in exactly the same way when intermediate ancestors were reconstructed. Goldman (1990) showed this convergence when change probabilities were invariant over edges. Later, the

---

[7]As long as the methods are label invariant, as opposed to something akin to alphabetical order, which would be perfectly—if trivially—congruent.

common edge rate parameter was relaxed and NCM provided a third general case of convergence in optimality among methods. To this has been added the case where the number of character states is large (Steel and Penny, 2004), and, for MPL, when taxon sampling densities are sufficiently great.

One of the implications of method convergence is that potentially we can identify and evaluate their underlying assumptions. Statements such as "since parsimony is statistically consistent in scenario A, parsimony assumes the conditions of A," is clearly misguided (Sober, 2004, 2005). As Sober points out, the best that can be done is to say that "since parsimony and likelihood converge in situation A, those conditions not assumed by likelihood in that scenario are also not assumed by parsimony." This reasoning serves to rule out assumptions, but is unable to rule them in, or to identify necessary and sufficient conditions for desirable behavior.

It is difficult to know if continued mathematical analysis of methods will find increasingly general cases where methods converge. It is possible that more complex and realistic models (in the vein of NCM) will identify new areas where methods choose the same optimal tree hypothesis, or they may dissolve into non-identifiability. Whether there is some shore of unification or not, areas of intersection among methods reveal a great deal about them, their strengths and weaknesses, and thus allow us to evaluate empirical results with greater clarity and precision.

13.6   Can We Argue Optimality Criteria?

How are we to choose an optimality criterion? Neither epistemology nor asymptotic consistency offer unfailing guidance. All optimality criteria can participate in hypothetico-deductive hypothesis testing, and all methods are subject to inconsistent behavior asymptotically. The empirical reality of finite character data and imperfect taxon samples reduces the impact of stochastically ideal asymptotic behaviors, leaving us to assay performance in absence of known or repeatable phenomena.

Another confounding effect is the computational complexity of systematic problems. Optimality-based tree reconstruction is, in general, an NP–hard problem. As such, we are not likely ever to have a (if there is only one) known, guaranteed optimal solution to tree reconstruction problems. Given that we cannot, for non-trivial problems, determine optimal solutions, and these would be required to satisfy at least the strong form of statistical consistency, it is unclear if consistency could ever apply to an empirical case.

One choice might be the syncretistic escape of employing multiple methods. Of course, if there is complete agreement, there is no issue. When optimality criteria disagree, however, which is the most common situation (at least for larger data sets), what to do? Alternate optimality criteria have different motivations and epistemological underpinnings (Giribet et al., 2002). An average or consensus of such varied objects is a difficult path, especially since the criteria are supporting alternate hypotheses (or the problem wouldn't exist).

Optimality criteria are analytical assumptions used to interpret historically unique events. As such, they cannot be tested empirically for accuracy in the sense of propinquity to truth. We simply cannot know this. Methods can be tested for precision in the guise of congruence, and this may help distinguish among criteria in a useful and intuitive manner. However, at the core, optimality criteria are assumptions that require support and justification when employed. Alternate assumptions lead to alternate results. Only by justifying our assumptions can we support our conclusions.

13.7 Exercises

1. Under what conditions would distance analysis be appropriate to analyze underlying character data?

2. Could Popper's ideas of support, corroboration, and severity of test be defined in terms other than Fisher's likelihood?

3. Would it be reasonable to limit the phylogenetic methods employed to those cases or those conditions, where methods converge?

4. Consider a data set comprised of anatomical data for a group of taxa composed of both living and extinct taxa, and complete genomic sequences for the extant taxa. How do the strengths and weaknesses of different methods apply to this case? Which optimality criterion is best suited to analyze these data and why?

Part IV

Trees

Chapter 14

Tree Searching

As is well-known (Foulds and Graham, 1982; Day, 1987; Roch, 2006), the search for "best" trees is an NP–hard problem. The number of possible trees is exponentially large (Eq. 2.1), hence, other than for trivial data sets, systematic analysis requires the use of heuristic tree search procedures. The discussion here is, for the most part, general to optimality criteria. Any objective criterion can be applied to a given tree to yield a cost or optimality value. This chapter is limited to the various exact and heuristic procedures used to explore tree space and identify "best" solutions.

A corollary of the exponential number of possible trees is that the set of optimal trees can be exponentially large as well. A search procedure must not only seek to find an optimal solution, but the complete (or at least representative) set of optimal solutions.

14.1 Exact Solutions

Although not generally employed, exact solution techniques can be comforting and, at times, useful. Exact solutions can be identified when leaf numbers are small, and the understanding of exact techniques can inform the design of heuristic procedures and the inference of their effectiveness. Two exact solution procedures are in common use: simple enumeration and the more efficient Branch-and-Bound.

14.1.1 Explicit Enumeration

Explicit enumeration is a brute force method of simply evaluating all possible tree topologies for a given set of taxa. This operation becomes tiresome quickly and is rarely employed for anything other than analytical purposes (Algs. 14.1 and 14.2, Fig. 14.1). This is due to the combinatorially large number of trees. Naïvely

Systematics: A Course of Lectures, First Edition. Ward C. Wheeler.
© 2012 Ward C. Wheeler. Published 2012 by Blackwell Publishing Ltd.

Algorithm 14.1: ExplicitEnumeration

Data: Input data of leaf taxa $(L_0, L_1, \ldots, L_{n-1})$ and observations (m)
      for each leaf taxon
Data: Element character cost matrix σ of pairwise distances between all
      observations (as in Chapter 8)
Result: Set of *BestTrees*, $|BestTrees| \geq 1$, trees of optimal cost
Initial tree of three leaves
$T \leftarrow (L_0(L_1, L_2))$;
$MinCost \leftarrow \infty$;
Initial tree cost set to maximum
$BestTrees \leftarrow \varnothing$;
Set of optimal trees is initially empty
$RecurseAllTrees(T, L, m, \sigma, 3)$ [Alg. 14.2];
return *BestTrees*

Algorithm 14.2: RecurseAllTrees

Data: Tree $T = (V, E)$
Data: L of Algorithm 14.1
Data: m of Algorithm 14.1
Data: σ of Algorithm 14.1
Data: i
Add each leaf taxon in turn
while $i \leq |L|$ **do**
    Each edge on tree of i leaves
    for $j = 0$ **to** $2i - 4$ **do**
        Add leaf taxon i to edge j
        $T' \leftarrow maketree(T, j)$;
        All leaf taxa not yet in tree
        if $i < |L|$ **then**
            Add leaf taxon i to edge j
            $RecurseAllTrees(T', L, m, \sigma, i + 1)$
        else
            Determine cost of tree, lower is better
            $T'_{cost} \leftarrow getcost(T', L, m, \sigma)$;
            if $T'_{cost} < MinCost$ **then**
                $MinCost \leftarrow T'_{cost}$;
                $BestTrees \leftarrow \{T'\}$;
            else if $T'_{cost} = MinCost$ **then**
                $BestTrees \leftarrow BestTrees \cup \{T'\}$;
            end
    end
end

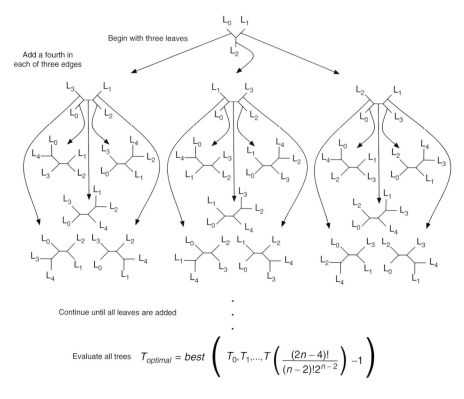

Figure 14.1: Explicit enumeration and evaluation of all trees.

plodding through space evaluating every possibility will be tractable only for a relatively small number of taxa (usually topping out at 15 or so taxa).

In the algorithmic descriptions of this chapter, the cost function of a tree, *getcost*(), is left undefined in order to be general with respect to optimality criterion.

14.1.2 Implicit Enumeration—Branch-and-Bound

The Branch-and-Bound algorithm (Algs. 14.3 and 14.4; Fig. 14.2) (Land and Doig, 1960; Hendy and Penny, 1982) seeks to reduce the number of trees examined by taking advantage of the fact that partial trees can only become more costly with the addition of more leaf taxa (assuming distances are metric and all leaves unique). If a partial tree cost is already equal to or greater than a complete tree, that entire line of trees generated by adding more taxa to that subtree, must be more costly than the current best tree known (the bound) and can be excluded from further examination. In general, the method can reduce the number of explicitly enumerated trees greatly (hence the term "implicit enumeration"), but not with a guaranteed time performance. Individual data sets may have significant or marginal reductions in numbers of trees examined. This is due to the fact

---

Algorithm 14.3: Branch-and-Bound Tree Search

Data: Input data of unique leaf taxa $(L_0, L_1, \ldots, Ln - 1)$ and
observations (m) for each leaf taxon
Data: Element character cost matrix σ of pairwise distances between all
observations
Data: Initial cost of a complete tree as $Bound$
Result: Set of $BestTrees$, $|BestTrees| \geq 1$ trees, of optimal cost
Initial tree of three leaves
$T \leftarrow (L_0(L_1, L_2))$;
$MinCost \leftarrow \infty$;
Initial tree cost set to maximum
Set of optimal trees is initially empty
$BestTrees \leftarrow \varnothing$;
$BoundRecurse(T, L, m, \sigma, 3, Bound)$ [Alg. 14.4];
return $BestTrees$

Algorithm 14.4: BoundRecurse

Data: Tree $T = (V, E)$ of Algorithm 14.3
Data: L of Algorithm 14.3
Data: m of Algorithm 14.3
Data: σ of Algorithm 14.3
Data: $Bound$ of Algorithm 14.3
Add each leaf taxon in turn
while $i \leq |L|$ **do**
    Each edge on tree of i leaves
    for $j = 0$ **to** $2i - 4$ **do**
        Add leaf taxon i to edge j
        $T' \leftarrow maketree(T, j)$;
        Determine cost of partial or complete tree
        $T'_{cost} \leftarrow getcost(T', L, m, \sigma)$;
        All leaf taxa not yet in tree
        if $i < |L|$ **then**
            Cost of partial tree less than current best tree
            if $T'_{cost} < Bound$ **then**
                Go on with search
                $BoundRecurse(T', L, m, \sigma, i + 1, Bound)$;
            end
        else
            if $T'_{cost} < Bound$ **then**
                Improve Bound
                $Bound \leftarrow T'_{cost}$;
                $BestTrees \leftarrow \{T'\}$;
            else if $T'_{cost} = Bound$ **then**
                $BestTrees \leftarrow BestTrees \cup \{T'\}$;
            end
        end
    end
end

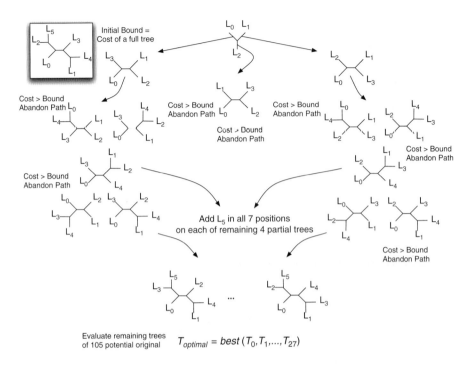

Figure 14.2: Branch-and-Bound implicit enumeration and evaluation of all trees. In this example, the initial bound from a full tree was able to exclude 77 of 105 possible trees.

that for Branch-and-Bound techniques to yield large time improvements, a large fraction of subtrees must have high cost, and that cannot be ensured. Furthermore, in the worst case, more trees (partial as well as complete) can be evaluated than for an explicit search, but that is a low probability event.

14.2   Heuristic Solutions

In absence of exact and complete solutions, we must rely on heuristic procedures. These methods explore solution space, searching for (but not guaranteeing) exact solutions. Various methods have been designed with different objectives and situations in mind. No one approach is likely to perform best (in time or quality of solution) for all data sets or all problems. For this reason, systematists use combinations of heuristic procedures when attacking real-world problems.

14.2.1   Local versus Global Optima

The central challenge faced by approximate solutions is the identification of global versus local solutions. If we use the metaphor of a landscape to represent

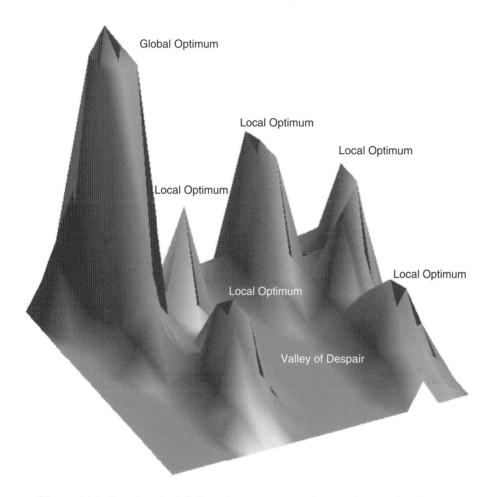

Figure 14.3: Local and global optima represented as peaks in a landscape.

the optimization challenge faced by tree searching, there are peaks and valleys of tree costs (Fig. 14.3). There can be great relief on the landscape involving lesser and greater peaks with one or more highest peaks representing areas (*i.e.* groups of trees) of optimal cost. Separating high points in the space are valleys of higher cost (lower optimality). The trick of heuristic solutions is avoiding being trapped in local, lesser peaks and navigating a path to global optima.

The tried and true techniques of Wagner addition and Branch-Swapping combined in trajectory search are effective at exploring a neighborhood of solutions and finding local optima. Other methods, involving randomization, perturbation, recombination, and suboptimal trajectories, have been designed to escape local optima while striving for global solutions.

14.3   Trajectory Search

A trajectory (local) search for an optimal solution comprises two parts. The first
step is the identification of a rough solution within a specific neighborhood of
potential solutions. The second step is the refinement of the initial solution to
identify the "best" cost solution within the neighborhood. Until the advent of
the Ratchet (Nixon, 1999), phylogenetic reconstruction in systematics consisted
solely of one or more trajectory searches involving the construction of an initial
solution (usually via the Wagner algorithm, below) and tree refinement via
branch-swapping.

14.3.1   Wagner Algorithm

The Wagner algorithm is a procedure to add leaf taxa sequentially to a growing
tree via a "greedy" (always choosing the immediately best option) procedure
(Farris, 1970). The method is named after Wagner (1961) who used a character-
based ground-plan analysis in his studies of fern relationships.

As employed today, the Wagner algorithm (Alg. 14.5; Fig. 14.4) begins with
three leaf taxa and adds each remaining leaf in turn to the existing tree by

J. Steven Farris

Algorithm 14.5: WagnerBuild

Data: Input data of leaf taxa $(L_0, L_1, \ldots, Ln - 1)$ and observations (m)
for each leaf taxon

Data: Element character cost matrix σ of pairwise distances between all
observations

Result: Single tree, *BestTree* of local optimal cost

Initial tree of three leaves

$BestTree \leftarrow (L_0(L_1, L_2))$ Add each leaf taxon in turn

for $i = 3$ **to** $|L| - 1$ **do**

$\quad$ Initial tree cost reset to maximum;

$\quad MinCost \leftarrow \infty$;

$\quad T \leftarrow BestTree$;

$\quad$ Each edge on tree of i leaves

$\quad$ **for** $j = 0$ **to** $2i - 4$ **do**

$\quad\quad$ Add leaf taxon i to edge j

$\quad\quad T' \leftarrow maketree(T, j)$;

$\quad\quad$ Determine cost of tree, lower = better

$\quad\quad T'_{cost} \leftarrow getcost(T', L, m, \sigma)$;

$\quad\quad$ **if** $T'_{cost} < MinCost$ **then**

$\quad\quad\quad MinCost \leftarrow T'_{cost}$;

$\quad\quad\quad BestTree \leftarrow T'$;

$\quad\quad$ **end**

$\quad$ **end**

end

return *BestTree*

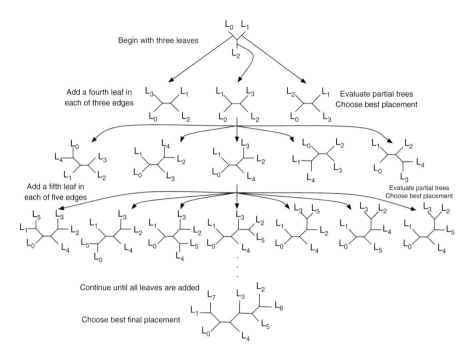

Figure 14.4: Initial tree construction procedure via the Wagner algorithm (Farris, 1970).

creating a new vertex and edge to the new leaf. Since there are $2n - 3$ edges for a tree with n leaves, the number of operations required to build a complete tree is $O(n^2)$.

Variations

The basic Wagner algorithm (Alg. 14.5) builds a single tree, adding taxa in arbitrary order (but optimal position). There are two variations that are often added to the basic algorithm. The first is in the order of taxon addition ("addition sequence"). The taxa may be added in random order (see below), or in attempts to minimize the distance between the existing partial tree and the incoming leaf taxon. Such a leaf might be chosen in an attempt to minimize the overall solution cost. An additional loop could be added, between the i and j loops of Algorithm 14.5, to choose the leaf to be added. This would add time complexity to the procedure resulting in an $O(n^3)$ operation (Farris, 1970).

A second common variation comes in expanding the *BestTree* returned to a set of "best" trees. In this case, trees that were less than or equal to the current minimum cost would be kept and built upon in later steps. The multiple trees might aid in overcoming some of the greediness of the algorithm and find better solutions. This too would incur time-complexity cost (a constant factor determined by the number of trees maintained) in keeping multiple builds going simultaneously, and ensuring only unique trees were produced.

Experience has shown that random addition sequences (RAS; used multiply with refinement—below) is the most useful Wagner Build procedure. Rather than trying to generate particularly optimal trees at this stage (with higher time complexity), the random-order builds offer a useful spread of initial trees for later refinement procedures in quadratic time.

14.3.2 Branch-Swapping Refinement

David Swofford

Michael Steel

The most common form of tree search refinement is known as *Branch-Swapping*. This is because the basic procedure removes branches (vertices and their subtrees or splits) from trees and adds them back to the remaining subtree edges (save where the original break took place). There are three flavors of branch-swapping in common use that are referred to (in order of increasingly large explored neighborhoods) as Nearest-Neighbor-Interchange (NNI), Subtree-Pruning and Regrafting (SPR), and Tree-Bisection and Regrafting (TBR). These tree manipulations have been present in implementations as far back as PHYSYS (Mickevich and Farris, 1980) and early versions of PAUP (Swofford, 1990; and earlier) but were not explicitly documented. The discussion here is based on the formalism of Semple and Steel (2003).

Nearest-Neighbor-Interchange

Nearest-Neighbor-Interchange (NNI) (Camin and Sokal, 1965; Robinson, 1971) performs the most limited set of tree rearrangements. If we have a tree ($T = (V, E)$) with internal (*i.e.* non-pendant) edge $e = \{u, v\}$ and an adjacent edge $e' = v, v'$, NNI is accomplished by first removing e and creating two partitions (subtrees) of T, then adding a new edge $e'' = \{u, v''\}$ where v'' is a new vertex created on an edge incident on v', $\{v'', v'\}$ and contracting all vertices of degree two (Fig. 14.5; Alg. 14.6). NNI produces $2(n-3)$ rearrangements on a given tree (two per tree partition). Sankoff et al. (1994) expanded the number of rearrangements in NNI to a fixed number > 2 for each edge deletion in their "window" approach (the "window" is defined around the original position of the pruned edge). Since the window is a constant value (they tested out to 8) the cost of the "window" approach is still linear in size of leaf set, but with a larger constant factor. Unsurprisingly, with a larger neighborhood, this approach was more effective than simple NNI.

Subtree Pruning and Regrafting

Subtree Pruning and Regrafting (SPR) expands the number of potential rearrangements of a given tree by enlarging the number of edges for re-addition (regrafting) of T_0 by T_1 (of Alg. 14.6). Instead of restricting the re-addition edges

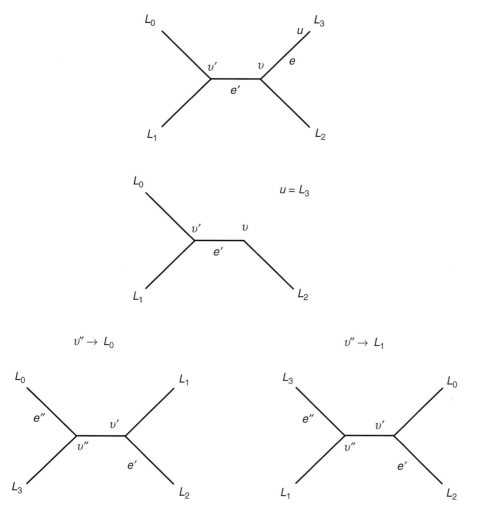

Figure 14.5: Nearest-Neighbor-Interchange (NNI) rearrangement. Original tree above, partitioned tree center, and rearranged trees below.

to those incident on v', all edges in T_0 are available (Fig. 14.6, Alg. 14.7). With the added target edges, the neighborhood of SPR rearrangement trees is expanded over that of NNI by a factor of $2n - 7$ for a total of $2(n - 3)(2n - 7)$ (Allen and Steel, 2001).

Tree-Bisection and Regrafting

Tree-Bisection and Regrafting (TBR; Swofford, 1990) or "Branch-Breaking" Farris, 1988) further enlarges the rearrangement neighborhood by allowing new

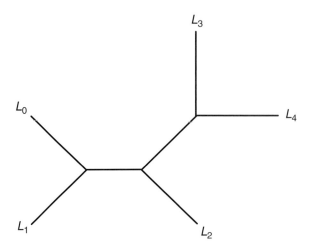

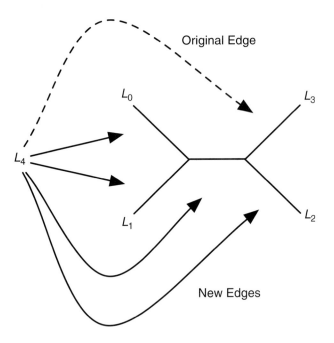

Figure 14.6: Subtree Pruning and Regrafting (SPR) rearrangement. Original tree above, and rearrangement positions below.

Algorithm 14.6: NearestNeighborInterchange

Data: Input tree $T = (V, E)$ of $|L| > 3$ leaf $(L \subset V)$ taxa
Result: Set of Nearest-Neighbor-Interchange trees ($NNITrees$)
$NNITrees \leftarrow \varnothing$;
Each edge in turn
for $i = 0$ **to** $|L| - 3$ **do**
    Internal edge $e_i = \{u, v\}$;
    Adjacent edge $e_i' = \{v', v\}$;
    Delete edge and split tree
    $T_0 \leftarrow T \setminus e_i$; subtree without u
    $T_1 \leftarrow T \setminus e_i \setminus T_0$; remaining subtree from u
    Each edge incident on v' in T_0 other than e'
    for $e_j = \{u', v'\} \in E_0$ *incident on v' and $e_j \neq e_i'$* **do**
        $v'' \leftarrow$ new vertex on e_j between u' and v';
        Reconnect T_1 to T_0 via edge $\{u, v''\}$
        Create new edges
        $e'' \leftarrow \{u, v''\}$;
        $e_j' \leftarrow \{u', v''\}$;
        $e_j'' \leftarrow \{v'', v'\}$;
        $T' \leftarrow T_0$;
        $NNITrees \leftarrow NNITrees \cup \{T'\}$;
    end
end
return $NNITrees$

---

edges to be created between all edges in the two tree partitions (Fig. 14.7, Alg. 14.8). TBR tests the "rerooting" of T_1 as it is added to T_0 by creating new nodes on each of the edges of T_1 and adding them to each edge in T_0. Although the exact number of tree rearrangements depends on the tree shape, the additional edge regraft options $(2 \cdot |T_1| - 3)$ make the overall operation $O(n^3)$.

Enhanced and expanded TBR-type rearrangements are possible with k-partitions of the original tree instead of the two specified in TBR. These would expand the neighborhood of rearranged trees but at considerable cost. For k partitions of the original tree, there would be $O(n^{3(k-1)})$ rearrangements. This was implemented in Goloboff (1999b).

The various edge locations available for regrafting between tree partitions (as referred to in Algs. 14.6 to 14.8) are summarized in Figure 14.8.

14.3.3 Swapping as Distance

The size of the rearrangement neighborhoods defines a metric distance between trees. These metrics are often used to measure the dissimilarity of trees in

---

Algorithm 14.7: SubTreePruningandRegrafting

Data: Input tree $T = (V, E)$ of $|L| > 3$ leaf ($L \subset V$) taxa
Result: Return set of Sub-Tree Pruning and Regrafting Trees, SPR
            neighborhood ($SPRTrees$)

$SPRTrees \leftarrow \varnothing$;
Each edge in turn
for $i = 0$ **to** $2 \cdot |L| - 1$ **do**
    | Edge $e_i = \{u, v\}$;
    | Delete edge and split tree
    | $T_0 \leftarrow T \setminus e_i$; subtree without u
    | $T_1 \leftarrow T \setminus e_i \setminus T_0$; remaining subtree from u
    | Each edge in T_0;
    | **for** $e_j = \{u', v'\} \in E_0$ *and* $e_j \neq e_i$ **do**
        | $v'' \leftarrow$ new vertex on e_j between u' and v';
        | Reconnect T_1 to T_0 via edge $\{u, v''\}$
        | Create new edges
        | $e'' \leftarrow \{u, v''\}$;
        | $e'_j \leftarrow \{u', v''\}$;
        | $e''_j \leftarrow \{v'', v'\}$;
        | $T' \leftarrow T_0$;
        | $SPRTrees \leftarrow SPRTrees \cup \{T'\}$;
    | **end**
end
return $SPRTrees$

---

terms of the number and type of rearrangements required to "edit" one tree into another. Unfortunately, each of NNI (Kriwánek, 1986), SPR (Bordewich and Semple, 2005), and TBR (Allen and Steel, 2001) distances are NP–hard calculations. There are useful heuristic algorithms (*e.g.* Bonet et al., 2006; Goloboff, 2008; Bordewich et al., 2007), but exact minimum distances are likely to be unknown.

14.3.4   Depth-First versus Breadth-First Searching

As defined above, both SPR and TBR branch-swapping are *Depth-First* search procedures. That is, each subtree (defined by a deletion of a particular edge) is added back to each possible alternate edge before the next edge-defined subtree is examined. Since the addition points cover adding back the subtree at all distances (in terms of number edges) from its original position, this is termed Depth-First. An alternative would be to add back subtrees only to edges a fixed distance (k) from their original location, cycling through all possible subtrees before adding each one again to the next further distance. In this way,

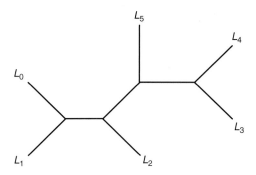

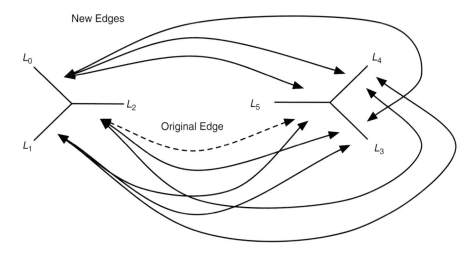

Figure 14.7: Tree-Bisection and Regrafting (TBR) rearrangement. Original tree above, and rearrangement positions below.

all tree modifications at a fixed distance are examined before moving out to the next distance (Fig. 14.9). NNI is, in essence, a Breadth-First search limited to $k = 1$.

Both approaches, upon completion, will examine the same set of modified trees, but in a different order. If intermediate improvements are immediately used as new starting points for continued swapping (abandoning the initial tree, as opposed to completing all rearrangements of the initial tree), the trajectory of visited trees can be altered and different heuristic results obtained. For a

---

Algorithm 14.8: TreeBisectionandRegrafting

Data: Input tree $T = (V, E)$ of $|L| > 3$ leaf ($L \subset V$) taxa
Result: Set of Tree Bisection and Regrafting Trees (*TBRTrees*)
$TBRTrees \leftarrow \varnothing$;
Each edge in turn
for $i = 0$ **to** $2 \cdot |L| - 4$ **do**
    Edge $e_i - \{u, v\}$,
    Delete edge and split tree $T_0 \leftarrow T \setminus e_i$; subtree without u
    $T_1 \leftarrow T \setminus e_i \setminus T_0$; remaining subtree from u
    Each edge in T_0;
    for $e_j = \{u', v'\} \in E_0$ *and* $e_j \neq e_i$ **do**
        $v' \leftarrow$ new vertex between u_i and v_i of e_j;
        Each edge in T_1;
        for $e'' = \{u'', v''\} \in T_1$ **do**
            Define new 'root,' r for T_1 (edge to connect to T_0)
            $r \leftarrow$ new vertex between u'' and v'';
            Add back T_1 via edge $e' = \{r, v'\}$
            $e' \leftarrow \{r, v'\}$ in T_0;
            $T' \leftarrow T_0$;
            $TBRTrees \leftarrow TBRTrees \cup \{T'\}$;
        end
    end
end
return *TBRTrees*

Enrico Fermi
(1901–1954)

The term "Monte Carlo" optimization based on randomization techniques originated with Enrico Fermi in the Manhattan Project of WWII.

given depth, breadth-first searches are linear with number of leaves (with constant factor—$O(kn)$), hence can be very useful for refining large trees, or as a component in combined search strategies. The "window" approach of Sankoff et al. (1994) is such an application.

14.4 Randomization

As mentioned above (Sect. 14.3.1), a useful variation in the basic Wagner algorithm is to add leaf taxa to the growing tree in random order (a "Monte Carlo" type randomization technique). The idea is that the various randomizations place the initial solutions in different areas of the optimality landscape (Fig. 14.3), allowing the refinement tree rearrangements to have a better chance of successfully hill climbing to an optimal solution. The coupling of such a "random addition sequence" with TBR refinement has been termed "RAS+TBR" and, with suitable numbers of repetitions, was the dominant technique of phylogenetic tree searching until 1999. Increasing the number of RAS+TBR replicates would be performed until stability of result or exhaustion intervened. This approach was effective for the size of the data sets then current, but was insufficient for the large molecular data sets even then beginning to appear (Chase et al., 1993).

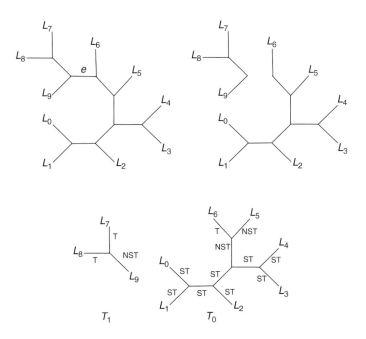

Figure 14.8: Edges available for rearrangement after deletion of edge e are labeled as "N" for NNI, "S" for SPR, and "T" for TBR. T_0 and T_1 are as in Algs. 14.6 to 14.8. Top left is the original tree, top right, the tree partitions, bottom the rearrangement edges.

14.5 Perturbation

In general, the RAS+TBR technique is effective for data sets with fewer than 100 taxa. Assuming sufficient randomizations are performed (typically in the hundreds), investigators achieved stable analytical results (in terms of best cost found), if not guaranteed optimality. In the 1990s, larger data sets, especially in botany, were produced with hundreds (Chase et al., 1993) and even thousands (Soltis et al., 2000) of taxa. These data sets required months of analysis on then standard hardware and yielded clearly unsatisfactory results (Rice et al., 1997).

Kevin Nixon

The basic problem was that searches were stuck in isolated local optima, and the Wagner randomizations were insufficient to identify good enough starting points for hill climbing (in the form of branch-swapping) to a stable solution (there were also implementation shortcomings detailed by Nixon, 1999). Nixon (1999) proposed a perturbation-based method, the "Parsimony Ratchet" or more simply the "Ratchet," that dramatically improved the speed and effectiveness of tree searching. In short, the ratchet is a refinement technique that alternates hill climbing on a perturbed (reweighted) data set with refinement of the original data. Multiple cycles of reweighting + branch-swapping and original weighting + branch-swapping allowed the escape of local optima ("islands" in

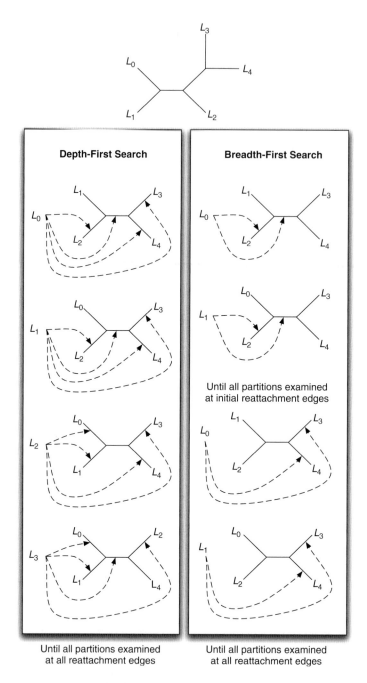

Figure 14.9: Depth-First (left) and Breadth-First (right) search orders on a tree (top). The Depth-First order evaluates all reattachment points of a subtree before moving on to the next tree partition, whereas the Breadth-First search examines reattachments of subtrees at fixed distances from their original point, each round examining each tree partition in turn before moving on to the next, more distant reattachment edges.

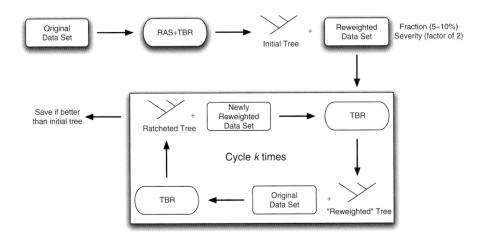

Figure 14.10: Ratchet refinement procedure of Nixon (1999).

the parlance of Maddison, 1991) and the identification of more globally optimal solutions (Fig. 14.10).

The idea behind the reweighting was that the reweighted characters have a distorted optimality landscape compared to that of the original. Hill climbing on this new surface may form a link between local and more global solutions. This distortion would be in proportion to both the fraction of characters reweighted and the severity of weight change. Nixon found the best performance with a relatively low percentage (5–10%) of the characters affected and with a mere doubling of their weights (other problems may perform better with other parameters).

The steps of the Ratchet are as follows:

1. Define an initial solution (*e.g.* via RAS+TBR).

2. Randomly reweight some fraction of characters by some factor (usually 5–10% of characters by 2).

3. Refine using Branch-Swapping (usually TBR).

4. Return weights to their original values.

5. Refine using Branch-Swapping (usually TBR).

6. Repeat 1–5 k times.

Given that the Ratchet still relies on branch-swapping refinement (Alg. 14.9), it will have the time complexity of the swapping regime (*e.g.* $O(n^3)$ if TBR is used), with a constant factor for the number of cycles performed ($O(kn^3)$).

Algorithm 14.9: RatchetRefinement

Created via RAS+TBR or other method

Data: Input tree $T = (V, E)$

Data: Input data matrix M

Typically all weighted to unity

Data: Input weight regime W

Data: Input number *cycles* of ratchet cycles to be performed

Data: Input fraction *fraction* of characters to be reweighted

Data: Input reweight *severity* factor

Result: Single ratchet refined tree (*RatchetTree*) if lower cost tree than input is found, $\varnothing$ if not.

$RatchetTree \leftarrow \varnothing$;

$mincost \leftarrow T_{cost}$;

Each replicate;

for $i = 0$ **to** $cycles - 1$ **do**

    Randomly perturb data weight set

    $W' \leftarrow perturb(W, fraction, severity)$;

    Perform TBR branch-swapping on input tree with new weight set

    $T' \leftarrow TBR(T, M, W')$;

    Perform TBR branch-swapping on perturbed tree with original weight set

    $T'' \leftarrow TBR(T', M, W)$;

    Check for better tree

    if $T''_{cost} < mincost$ **then**

        $RatchetTree \leftarrow T''$;

        $mincost \leftarrow T''_{cost}$;

    end

end

return *RatchetTree*

The Ratchet had an immediate and significant effect on empirical studies. The Chase et al. (1993) data set was analyzable in hours, rather than months, and better (16218 versus 16225 parsimony steps) trees were found. Giribet and Wheeler (1999) reanalyzed their metazoan data and also found shorter trees (7032 versus 7028) in greatly reduced time. The method was extended to likelihood techniques (Voss, 2003; Wheeler, 2006). Ratcheting is now a standard component of phylogenetic tree searching.

Several variations of ratchet procedures have been explored centering around the types of perturbation (Varón et al., 2010, allows perturbing indel and other parameters as well as character weight) and character selection (complementary reweighted character sets over pairs of cycles) operations.

14.6 Sectorial Searches and Disc-Covering Methods

Sectorial Searches (Goloboff, 1999a; Goloboff et al., 2003) and Disc-Covering Methods (Huson et al., 1999; Roshan et al., 2004) are both methods that subdivide data sets into sectors, analyze these sectors separately, combine the sectors into a complete tree, and refine that complete tree with branch-swapping. Beneath this surface similarity lie important differences, however, leading to significant differences in effectiveness and performance.

14.6.1 Sectorial Searches

Goloboff (1999a) proposed Sectorial Searches (SS) along with a variety of other methods to address the problem of composite optima in large data sets. These data sets are large enough (hundreds to thousands of taxa) that subtrees of the problem had their own optimization challenges and it was very unlikely that RAS+TBR would be able to achieve optimal arrangement of all the subtrees simultaneously. Goloboff defined *sectors* of trees (Fig. 14.11) based on an initial

Pablo Goloboff

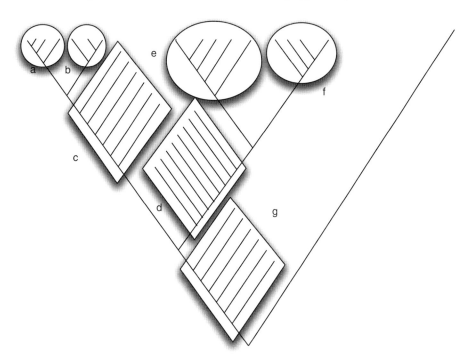

Figure 14.11: A tree showing several potential sectors (a–g).

tree or trees. These sectors could be determined randomly (RSS) or by consensus among several trees (CSS), focusing in on areas of disagreement among candidate solutions. Later (Goloboff et al., 2003), balanced sectors (XSS) were defined to create similarly sized subunits.

In Goloboff's procedure, the sectors are used to define a new, reduced data set based on the root state condition of each sector. This reduced data set is then searched using some number of RAS+TBR replicates—akin to a Breadth-First search around edges incident upon sectors. The resulting reduced tree is then reincorporated into the original tree and adjusted to reflect the relationships among the sectors after the reduced data search. If this overall tree is different from the original (or sufficiently so; Goloboff advocated requiring multiple differences), a round of TBR branch-swapping is performed on the entire tree. If this new tree is superior, it is kept. Multiple rounds of SS may be performed using improved trees as starting points and randomizing sector selection (Fig. 14.12, Alg. 14.10). Sectorial searches, in concert with Ratchet and other methods (*e.g.* simulated annealing, genetic algorithm; see below), have been extremely effective in searching large data sets (Giribet, 2007).

14.6.2   Disc-Covering Methods

Huson et al. (1999) first proposed DCM for distance matrix analysis. The method was improved through DCM2 and finds its most recent incarnation in Rec-I-DCM3 (Roshan et al., 2004). The method is a meta-heuristic, described as a "boosting" procedure relying on other methods (such as TNT,

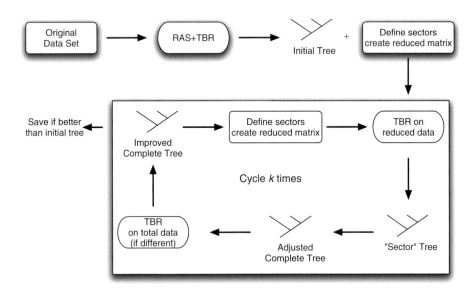

Figure 14.12: Sectorial Search refinement procedure of Goloboff (1999a).

Algorithm 14.10: SectorialSearch

Created via RAS+TBR or other method;
Data: Input tree $T = (V, E)$
Data: Input data matrix M
Data: Input number $SScycles$ of sectorial searches to be performed
Result: Single sectorial search tree ($SSTree$) if lower cost tree than input
        is found, $\varnothing$ if not
$SSTree \leftarrow \varnothing$;
Each replicate;
for $i = 0$ **to** $SScycles - 1$ **do**
    Define set of sectors in T (either randomly—RSS—or by
    consensus—CSS—of several input trees) and create a reduced data
    set from M
    $S \leftarrow defineSector(T, M)$;
    Perform $RAScycles$ of RAS+TBR on reduced data set S
    $T^S \leftarrow RASTBR(RAScycles, S)$;
    Adjust T to conform to relationships of sectors in T_S
    $T' \leftarrow makeTree(T, T_S)$;
    If there were changes in the overall tree
    if $T \neq T'$ **then**
        Perform TBR branch-swapping on T' and full data
        $T'' \leftarrow TBR(T', M)$;
        Check for better tree
        if $T''_{cost} < T_{cost}$ **then** $SSTree \leftarrow T''$;
    end
end
return $SSTree$

Goloboff et al., 2003 or POY, Varón et al., 2010) as a "base" method. DCM methods proceed in four sections:

1. Decompose the data set.

2. Solve subproblems.

3. Merge and reconcile subproblems.

4. Refine the merged tree.

The decomposition component was originally based on distance matrix analysis, but in Rec-I-DCM3 it is based on an initial tree. In this, it is similar to SS, recursively dividing the tree into smaller pieces that then can be analyzed easily. Unlike SS, these sectors overlap. The subproblems are analyzed as individual phylogenetic problems. Roshan et al. (2004) advocated sectors of $\frac{1}{4}$ to $\frac{1}{2}$ the original data set size. The trees resulting from the subproblems are then merged into a single tree via strict-consensus + random resolution. In the procedure,

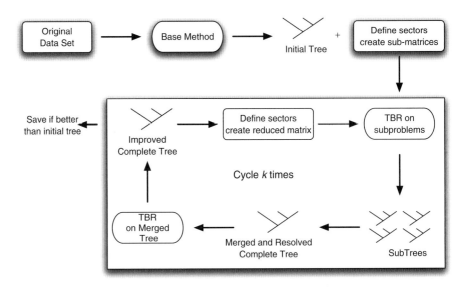

Figure 14.13: Rec-I-DCM3 refinement procedure of Roshan et al. (2004).

overlapping sectors are combined, often resulting in disagreement and irresolution. These polytomies are randomly resolved to create a binary tree, which is then refined via branch-swapping (*e.g.* TBR) (Fig. 14.13, Alg. 14.11). Rec-I-DCM3 has been shown to find improved solutions in reduced time for several sorts of problems (distances, ML, MP; Roshan, 2004).

The time complexity of SS and DCM will both be determined by the rearrangement procedure employed. If TBR, they will be linear in replicates and cubic in numbers of taxa.

The relative performance of these methods has been examined by Roshan et al. (2004) and Goloboff and Pol (2007). Goloboff and Pol (2007) point out a number of differences between SS and Rec-I-DCM3 that explain the performance advantages of combined search strategies (including SS), identifying the randomized polytomy resolution step in Rec-I-DCM3 as the likely source of its relative shortcoming. The motivation of this difference comes from the types of sectors defined by the two methods. SS defines non-overlapping sectors, whose relative placement is searched; Rec-I-DCM3 defines overlapping sectors whose arrangements are determined by reconciling subtrees with common leaf taxa. The searching for sector relationships in SS, and consensus-super tree like technique in Rec-I-DCM3 are the root cause of the difference in the behaviors of the methods.

14.7   Simulated Annealing

Simulated annealing is a refinement method designed to escape the problems of greedy algorithms (such as branch-swapping) by allowing tree modifications that (temporarily) increase the cost of the tree. The method is based on analogy

---

Algorithm 14.11: Rec-I-DCM3

Data: Input tree $T = (V, E)$
Data: Input data matrix M
Result: Return single Rec-I-DCM3 tree (*DCMTree*) if lower cost tree
        than input is found, $\varnothing$ if not
DCMTree $\leftarrow \varnothing$;
mincost $\leftarrow T_{cost}$;
foundBetter $\leftarrow true$;
while *foundBetter* = *true* **do**
    foundBetter $\leftarrow false$;
    Define set of data sectors in T by recursive centroid edge
    decomposition
    $S \leftarrow defineSectors(T, M)$;
    Analyze subproblems in S
    for $s_i \in S$ **do**
        Analyze subproblem s_i by favored method—*e.g.* RAS+TBR or
        exact if small
        $T_i^S \leftarrow methodTree(s_i)$;
    end
    Merge the T_i^S trees created from S via strict consensus merge and
    randomly resolve polytomies
    $T_S \leftarrow mergeTree(\{\bigcup_i T_i^S\})$;
    Perform TBR branch-swapping on T_S and full data;
    $T' \leftarrow TBR(T_S, M)$;
    Check for better tree;
    if $T'_{cost} < mincost$ **then**
        DCMTree $\leftarrow T'$;
        mincost $\leftarrow T'_{cost}$;
        foundBetter $\leftarrow true$;
    end
end
return *DCMTree*

to the annealing of metals, where more stable (*i.e.* lower energy) crystal configurations can be achieved by allowing the metal to anneal at moderate temperatures and cooling in a step-wise fashion (Chib and Greenberg, 1995). The basic procedure is a modification of the Metropolis–Hastings algorithm (Metropolis et al., 1953) proposed by Kirkpatrick et al. (1983) and Cerny (1985). The basic procedure involves establishing analogues to the parameters of metallurgical annealing. At each iteration, there are two states—s the current state and s' a second, candidate state. These states have energies E and E'. The probability of transitioning from state s to s' depends on the energy difference between the states (higher is worse) and the temperature, T (Eq. 14.1).

Nicholas Metropolis
(1915–1999)

$$P(s, s') = \min(1, e^{(E-E')/T}) \tag{14.1}$$

The temperature parameter defines how easily transitions between dissimilar states can take place. If the temperature value is high, radical changes may take place, if T is low, transitions to higher energies are rare. As T tends to 0, the algorithm becomes increasingly greedy, eventually only accepting lower energy (*i.e.* better) states. In general, for the algorithm to be able to transition between local optima, P must be > 0 when $e' > e$ although limiting to 0 when $T \to 0$. If $E' < E$, P must be $= 1$ (*i.e.* always accept a better solution). Lundy (1985) and Dress and Krüger (1987) first applied this approach to phylogenetic analysis. This formalism is used explicitly in the METRO module of PHYLIP (Felsenstein, 1993; Salter and Pearl, 2001), and Monte-Carlo-Markov-Chain refinements in Bayesian techniques (Yang and Rannala, 1997; Huelsenbeck and Ronquist, 2003) (Chapter 12).

Goloboff (1999a) modified this idea for parsimony in "Tree-Drifting" (DFT). DFT is an insertion into TBR swapping by altering the criterion for accepting a new tree T' over an existing T. In TBR, only if $T'_{cost} < T_{cost}$ is T' is accepted—a greedy technique. DFT changes this threshold rule to reflect the degree of cost difference between the trees and the relative degree of character support for T and T' (Eq. 14.2) with F the number of characters in favor of T' and C against.

$$\text{when } (T'_{cost} > T_{cost})$$

$$T' \text{ is accepted if } \left[\frac{F - C}{F} \leq \frac{RAND(0, 99)}{F - C + T'_{cost} - T_{cost}} \right] \tag{14.2}$$

$$F = \sum_{i}^{if(T^i_{cost} - T'^i_{cost}) > 0} T^i_{cost} - T'^i_{cost}$$

$$C = \sum_{i}^{if(T'^i_{cost} - T^i_{cost}) > 0} T'^i_{cost} - T^i_{cost}$$

$$RAND(x, y) = \text{a random integer} \in [x, y]$$

After a period of DFT (defined by number of tree changes), a TBR refinement is performed and the entire process repeated some number of times (Alg. 14.12). As with other refinements, the time complexity of DFT would be linear in the number of DFT replicates and cubic in number of taxa (due to

---

Algorithm 14.12: TreeDrift

Data: Input tree $T = (V, E)$ with leaves $L \subset V$
Data: Data matrix M of observations
Data: Frequency *period* of complete TBR branch-swapping
Data: Number *cycles* of complete DFT cycles
Result: Return Drift refined tree (*DFTTree*)
$DFTTree \leftarrow \varnothing$;
Number of repetitions
repetitions $\leftarrow 0$

Algorithm 14.12: (*Continued*)

for *repetitions* ≤ *cycles* **do**
    repetitions ← *repetitions* + 1
    numTreeChanges ← 0;
    Each edge in turn
    for $i = 0$ **to** $2|L| - 4$ **do**
        edge $e_i = \{u, v\}$;
        Delete edge and split tree
        $T_0 \leftarrow T \setminus e_i$; subtree without u
        $T_1 \leftarrow T \setminus e_i \setminus T_0$; remaining subtree from u
        Each edge in T_0
        for $e_j = \{u', v'\} \in E_0$ *and* $e_j \neq e_i$ **do**
            $v' \leftarrow$ new vertex between u_i and v_i of e_j;
            Each edge in T_1;
            for $e'' = \{u'', v''\} \in T_1$ **do**
                Define new 'root,' r for T_1
                $r \leftarrow$ new vertex between u'' and v'';
                Add back T_1 via edge $e' = \{r, v'\}$
                $e' \leftarrow \{r, v'\}$ in T_0;
                $T' \leftarrow T_0$;
                Determine tree cost
                $T'_{cost} \leftarrow treeCost(T, M, \sigma)$;
                T' better than T
                 if $T'_{cost} < T_{cost}$ **then**
                    $DFTTree \leftarrow T'$;
                    $T \leftarrow T'$;
                Check threshold to accept equal or worse tree
                else
                  $F = \sum_i^{if(T_{cost}^i - T'^i_{cost}) > 0} T_{cost}^i - T'^i_{cost}$;
                  $C = \sum_i^{if(T'^i_{cost} - T_{cost}^i) > 0} T'^i_{cost} - T_{cost}^i$;
                  Uniformly distributed random integer on [0,99]
                  $R \leftarrow RAND(0, 99)$;
                 if $\left[\frac{F-C}{F} \leq \frac{R}{F-C+T'_{cost}-T_{cost}} \right]$ **then**
                    $DFTTree \leftarrow T'$;
                    $numTreeChanges \leftarrow numTreeChanges + 1$;
                    Do full TBR after *period* tree changes
                    if *numTreeChanges* mod *period* = 0 **then**
                    $DFTTree \leftarrow TBR(DFTTree, M)$;
                 end
            end
        end
    end
  end
end
return *DFTTree*

TBR scaffold). DFT has been shown to be effective at improving locally optimal trees in large data sets, especially in combination with Sectorial Searches (Goloboff, 1999a; Giribet, 2007).

14.8   Genetic Algorithm

Genetic (sometimes referred to as "Genetical") algorithms (GA) simulate the evolutionary process to optimize complex functions (Holland, 1975; Goldberg et al., 1989; Sastry et al., 2005). The idea is to model the processes of mutation, recombination, and natural selection to identify solutions and escape local optima through random aspects of population dynamics. One of the key differences between GA and other approaches is that it operates on a collection of solutions simultaneously. One weakness is that it has no direct relationship to an exact solution (such as branch-swapping, which would, if sufficiently exhaustive, examine all possible trees).

There are six fundamental components to a GA procedure, which are repeatedly cycled (after *Initialization*) until stability or ennui sets in:

1. Initialization—An initial *population* of solutions (here trees) is generated.

2. Evaluation—Potential solutions are assigned an objective cost (fitness).

3. Selection—Those solutions with higher fitness (here lower cost) are preferentially replicated such that the more fit they are, the better represented they are in the population of solutions.

4. Recombination—Components of two or more candidate solutions are combined to create variation and potentially better solutions.

5. Mutation—Modifications are made to individual solutions.

6. Replacement—Those solutions created by selection, recombination, and mutation replace the existing generation in the population of solutions.

One of the tricks of GA is to balance mutation (which modifies existing trees) and selection. If the selection is too strong, the solution space will converge too quickly before finding new optimal solutions. If the mutation rate is too high, there will be little direction to the process and the search will approach randomization (as in SA when T is high). If the mutation rate is too low, there will be insufficient diversity to work with in finding new solutions.

Moilanen (1999, 2001) introduced GA to parsimony-based tree searching with PARSIGAL (although Matsuda, 1996, had used GA for likelihood calculations on protein sequences) explicitly to escape the local minima in which trajectory searches often found themselves (in multiple sequence alignment, SAGA Notredame and Higgins, 1996, is based on GA). Moilanen's GA did not emphasize mutation, other than through perturbed distance matrix tree reconstruction in the initialization step, and added limited branch-swapping to improve the post-recombination trees (Fig. 14.14, Alg. 14.13). The time complexity of

---

Algorithm 14.13: GeneticAlgorithmTreeSearch

Data: Input tree $T = (V, E)$

Data: Data matrix M of observations

Data: Element character cost matrix σ of pairwise distances between all character states

Data: Number *generations* of complete GA cycles

Data: Number *popSize* of trees in each generation

Result: Set of GA refined trees (*GATrees*)

$GATrees \leftarrow \varnothing$;

$minCost \leftarrow \infty$;

Create initial population of trees

for $i = 0$ **to** $i < popSize$ **do**

    $M' \leftarrow perturbMatrix(M)$;

    Generate distance-based tree from perturbed matrix

    $T_0^i \leftarrow makeTree(M')$;

    Set initial minimum tree cost based on original matrix

    $(T_0^i)_{cost} \leftarrow getCost(T_0^i, M)$;

    if $(T_0^i)_{cost} < minCost$ **then** $minCost \leftarrow (T_0^i)_{cost}$;

end

Number of repetitions

for $j = 0$ **to** $j < generations$ **do**

    $numNextGen \leftarrow 0$;

    for $i = 0$ *to* $i < popSize$ **do**

        Perform local (SPR or NNI) refinement

        $T' \leftarrow localRefine(T_{j-1}^i)$;

        Found lower cost tree than initial population

        if $T'_{cost} < minCost$ **then**

            $minCost \leftarrow T'_{cost}$;

            $T_j^{numNextGen} \leftarrow T_j^{numNextGen} \cup T'$;

            $numNextGen \leftarrow numNextGen + 1$;

        end

    end

    Perform tree recombination

    while $numNextGen < popSize$ **do**

        Choose first parent tree randomly based on cost

        $P1 \leftarrow randFitChoice(T_{j-1})$;

        Choose second parent tree randomly based on cost.

        $P2 \leftarrow randFitChoice(T_{j-1})$;

        Recombine subtrees from two parents and produce

        $(T_j^{numNextGen}, T_j^{numNextGen+1}) \leftarrow$ recombine (P1, P2) two offspring trees placed in population;

        $numNextGen \leftarrow numNextGen + 2$;

    end

    Determine minimum cost over trees in generation j

    $minCost \leftarrow \min_i \left[(T_j^i)_{cost} \right]$;

end

return *GATrees*

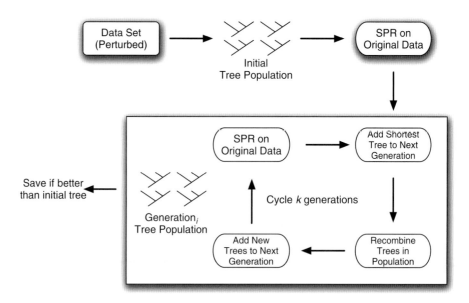

Figure 14.14: Genetic Algorithm tree search procedure of Moilanen (1999).

Moilanen's complete GA strategy will depend on the number of generations, g, the population size, p, and the complexity of the local swapping operations (SPR) on n taxa, for an overall time complexity of $O(gpn^2)$.

Goloboff (1999a) focused on the recombination step of GA (Fig. 14.15) in defining his Tree-Fusing (TF) procedure. Goloboff modified Moilanen's approach to enhance the capabilities of tree recombination to attack the problem of composite optima in large trees. Goloboff added SPR rearrangement to the recombination step (as opposed to after the trees are constructed) to optimize the placement of the exchanged subtrees. TF has been extremely effective at lowering tree costs—even when presented with clearly suboptimal inputs, TF is able to improve the result substantially. This approach has been used in large sensitivity analysis (Wheeler, 1995) studies. Combining the results of different parameter runs with TF and subsequent TBR (SATF) simultaneously enhanced the quality of results across parameter space (Wheeler et al., 2004; Giribet, 2007).

14.9   Synthesis and Stopping

As observed by Nixon (1999) and Goloboff (1999a), implemented in TNT (Goloboff et al., 2003) and POY4 ("search" command; Varón et al., 2010), and reviewed by Giribet (2007), no one search strategy is likely to be effective for all data sets, and combinations of procedures have the best opportunity to result in satisfactory results. Goloboff (1999a) described and tested 11 different strategies.

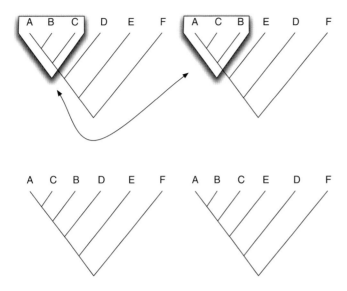

Figure 14.15: Tree recombination (Tree-Fusing) procedure of Moilanen (1999) and Goloboff (1999a).

A remaining issue is how much is enough? When can an investigator feel confident that sufficient effort has been expended to ensure a stable result? Two notions of completion are stability in optimality value, and stability in tree topology (usually strict consensus of resulting trees). Goloboff (2002) and Giribet (2001) employed these criteria in "driven" searches (Goloboff et al., 2003). The motivation is to continue to perform replicates of combined search strategies until best cost solutions have been found in repeated, randomized "hits." Multiple hits on a given tree cost, with additional hits not altering the strict consensus of their results, is a potential indicator of robust results (perhaps by progressively doubling the heuristic intensity until results no longer change). This is, of course, no guarantee of a minimum cost solution (the problem is NP–hard after all), but it is a useful stopping rule.

14.10   Empirical Examples

Frost et al. (2006) performed an analysis of 522 amphibian and outgroup taxa based on eight molecular loci and morphology (Fig. 14.16). In order to deal effectively with such a large tree, they employed a diversity of heuristic tree search strategies (Fig. 14.17) including random addition sequence Wagner builds with TBR refinement (RAS+TBR), Genetic Algorithm (Tree-fusing), Simulated Annealing (Tree-Drifting), and other Tree-Alignment (Chapter 10) heuristics.

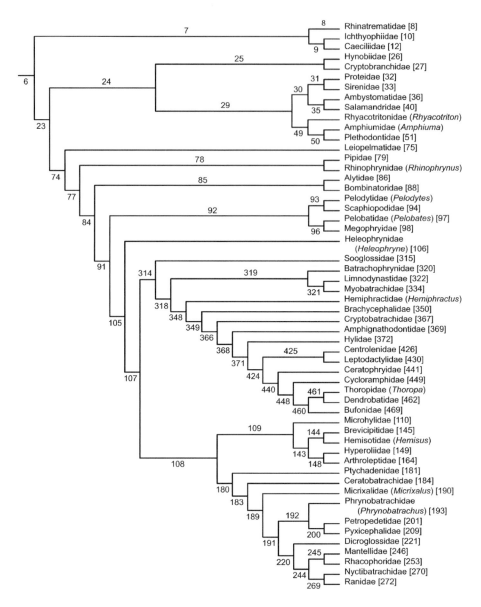

Figure 14.16: Familial level abstract of the 522 taxon amphibian analysis of Frost et al. (2006). The edge numbers are links to branch information in the paper.

Summary of Tree-Searching Methods Combined in Overall Search Strategy
See the text for more detailed explanations and references. Different runs combined multiple procedures, and all runs included SPR and/or TBR refinement.

Searching method	Description of procedure
RAS	Random addition sequence Wagner builds
Constrained RAS	As above, but constrained to agree with an input group inclusion matrix derived from the consensus of topologies within 100–150 steps of present optimum
Subset RAS	Separate analysis of subsets of 10–20 taxa; resulting arrangements used to define starting trees for further analysis of complete data set
Tree drifting	Tree drifting as programmed in POY, using TBR swapping; control factor = 2 (default)
Ratcheting (fragment)	Ratcheting as programmed in POY, with 15–35% of DNA fragments selected randomly and weighted 2–8 times, saving 1 minimum-length tree per replicate
Ratcheting (indel, tv, ts)	Ratcheting approximated by applying relative indel-transversion-transition weights of 311, 131, and 113, saving all minimum length trees
Constrained ratcheting (fragment)	As above, but beginning with the current optimum input as a starting tree and constrained to agree with an input group inclusion matrix derived from the consensus of topologies within 100–150 steps of present optimum
Tree fusing	Standard tree fusing followed by TBR branch swapping, with the maximum number of fusing pairs left unconstrained
Manual rearrangement	Manual movement of branches of current optimum
Ratcheting (original) of final implied alignment	Parsimony ratchet of fixed matrix, as implemented in Winclada

Figure 14.17: Analytical strategies of Frost et al. (2006).

Genes used in the analysis, indicating the number of fragments in which every gene was split for alignment

Gene	Fragments	Taxa	Characters	Scope	Type	Genome
LSU rRNA	11	11700–1267	312–115	Global	DNA	Nuclear
MatK	1	11855	792	Embryophyta	DNA	Plastid
NdhF	1	4864	1209	Embryophyta	DNA	Plastid
RbcL	1	13043	13 043	Embryophyta	DNA	Plastid
COXI	1	7310	1296	Metazoa	PROT	Mitoch
COXII	1	8315	437	Metazoa	PROT	Mitoch
COXIII	1	2309	272	Metazoa	PROT	Mitoch
CytB	1	13766	337	Chordata	PROT	Mitoch
NDI	1	4123	349	Metazoa	PROT	Mitoch
SSU rRNA	6	20462–19336	293–26	Global	DNA	Nuclear
SSU rRNA	1	1314	464	Hexapoda	DNA	Mitoch
RNAPII	2	869–333	515–203	Fungi/Global	PROT	Nuclear
LSU rRNA	1	752	314	Ascomycota	DNA	Mitoch

Figure 14.18: Molecular data table of Goloboff et al. (2009).

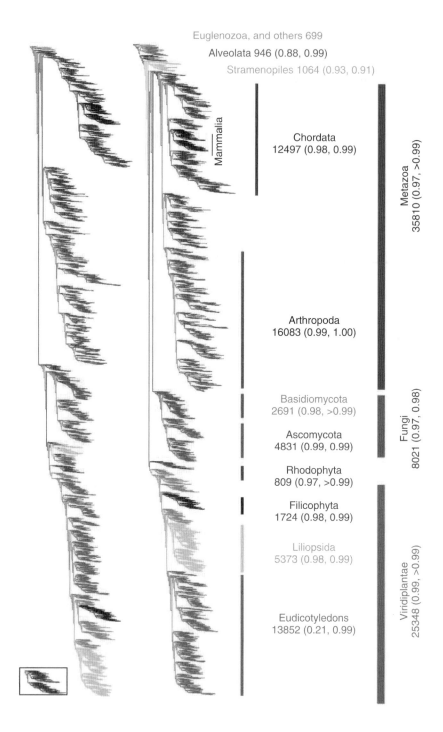

Figure 14.19: Combined eukaryote analysis tree at 730,435 steps of Goloboff et al. (2009). The values beneath the taxon names are the number of taxa included and % placement in agreement with GenBank taxonomy. See Plate 14.19 for the color Figure.

Goloboff et al. (2009), in the largest empirical analysis to date, created a data set of 73,060 eukaryotic taxa for 17 loci and morphology (Fig. 14.18). The analysis relied heavily on Sectorial Searches and Genetical Algorithm (in the guise of Tree-fusing) after many RAS+TBR starting points for trees based on individual loci. The final result was a tree of length 730,435 parsimony steps (Fig. 14.19).

In both the Frost et al. (2006) and Goloboff et al. (2009) cases, it is very unlikely that any one of these heuristic approaches would have yielded satisfactory results on their own. Considerable effort was expended in combining multiple strategies until stable (if not necessarily optimal) results were obtained.

14.11 Exercises

1. When we search for a tree, how good is good enough?

2. Should we strive for exact solutions at the cost of sampling?

3. Write down the NNI neighborhood for the tree (A (B (C (D (E, F))))).

4. Write down the SPR neighborhood for the tree (A (B (C (D (E, F))))).

5. Write down those trees in the TBR neighborhood for the tree (A (B (C (D (E, F))))) that are not in its SPR neighborhood.

6. If we double the number of taxa in an analysis, how much longer will an NNI search take? SPR? TBR?

7. As we raise the temperature in a simulated annealing search, what happens to the acceptance of suboptimal trees?

8. If Pablo Goloboff were a tree, what sort of tree would he be?

Chapter 15

Support

Support means different things to different people. In general, "support" covers a variety of indices designed to measure confidence in a vertex or clade. There is a diversity of flavors of support specific to optimality criteria (parsimony, likelihood, posterior probability) and the aspect that is desired to be quantified. There can be no single "best" support value given that the goals of different investigators and support indices vary.

The discussion here covers the use and measure of support in different contexts. Although there is a broad variety of support objectives and indices, there are organizing themes that run through the topic and serve to organize the principles of support.

The earliest notions of support were derived from edge weights (branch lengths as number of synapomorphies "supporting" a clade). These informal ideas were logical in that the edge weight quantified the amount of change (distance, parsimony changes or likelihood), but was thought inadequate since convergent and parallel change (homoplasy) would be conflated with unique historical change (synapomorphy). This led to resampling and optimality-based indices.

15.1 Resampling Measures

Resampling is a technique in general use in statistics to replace an unknown distribution with a known one based on an observed sample. This is done, in general, in two ways: bootstrapping and jackknifing. In short, bootstrap methods sample a set of observations to create a new sample of equal size with replacement, while jackknife methods sample without replacement, effectively deleting some fraction of the original data.

In systematics, two entities present themselves for resampling—taxa and characters. In principle, elements in either set could be used as the basis for jackknife or bootstrap analysis. If we are concerned with the evidentiary basis for group support, it is logical to focus attention on characters. Given that any

Systematics: A Course of Lectures, First Edition. Ward C. Wheeler.

modified taxon set would create trees different from the one whose support is desired, taxon resampling has not played a large role in support indices or descriptors.

15.1.1 Bootstrap

The bootstrap was first described by Efron (1979) (summarized in Efron and Tibshirani, 1993) to estimate a variety of parameters (mean, variance *etc.*) of unknown distributions. The idea is to replace F, an unknown distribution, with an approximation $\hat{F}$. If X^* is a sample drawn from $\hat{F}$, and X from F, aspects of F can be approximated by X^* and $\hat{F}$.

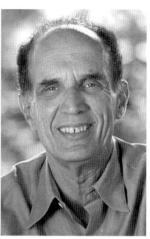

Bradley Efron

The central aspect of the bootstrap is the choice of $\hat{F}$. When the observed data (sample distribution function) are used, the method is a *non-parametric bootstrap*. When $\hat{F}$ is chosen from a parametric family of distributions (such as χ^2 or Gaussian) the method is a *parametric bootstrap*. Simulated resampling is a staple of both techniques. In the non-parametric case, the sample distribution function (observed data) are used (with replacement) to create simulated samples. The random sampling is repeated some large number of times and features of interest calculated from the simulated samples. When parametric bootstrapping is performed, the sample is used to estimate the parameters of a specific distribution (*e.g.* μ and σ of a normal distribution) and the resampling simulations are drawn from that parameterized distribution. Clearly, parametric bootstrapping will have more power, but this relies on knowledge of the underlying family of distributions. Furthermore, for $\hat{F}$ to be a good approximation for F, the observations must be independent and identically distributed (i.i.d), meaning there can be no dependence among observations (here characters), and they must follow the same distributional model. In systematics, such information is, in general, unknown, hence non-parametric bootstrapping is the usual method of choice.

Felsenstein (1985) first employed the bootstrap on trees in attempting to create confidence intervals on vertices. Felsenstein wanted to be able to establish these vertices (or clades or groups) as statistically significant if their bootstrap values were ≥ 0.95. The procedure advances in five steps:

1. Determine an optimal tree (T_o) by a search procedure on observed data (D) under an optimality criterion.

2. Resample the observed data (columns of aligned matrix) with replacement, yielding a new data set (D') with the same number of characters as the original. Characters will be present zero, one, two or more times.

3. Determine an optimal tree (T'_o) from D' using the same search procedure and optimality criterion as in step 1.

4. Repeat steps 2 and 3 k times adding the generated trees to the set of bootstrap trees (T_B).

5. Construct the tree of all nodes with frequency ≥ 0.50 in T_B (majority-rule consensus).

Taxon	Characters							
A	0	0	0	0	0	0	0	0
B	1	0	0	0	0	0	0	0
C	1	1	0	0	0	0	0	0
D	1	1	1	0	0	0	0	0
E	1	1	1	1	0	0	0	0
F	1	1	1	1	1	0	0	0
G	1	1	1	1	1	1	0	0
H	1	1	1	1	1	1	1	0
I	1	1	1	1	1	1	1	1
J	1	1	1	1	1	1	1	1

Table 15.1: Simple data set of 10 taxa (A-J) and 8 binary characters.

This consensus is the *bootstrap tree*, and the node frequencies are the bootstrap support values of the clades on the tree. In the simple case of n characters, r of which support a vertex V (without disagreeing characters), the bootstrap support of V (V_{Bt}) will be:

$$V_{Bt} = 1 - \left(1 - \frac{r}{n}\right)^n \tag{15.1}$$

Consider the data in Table 15.1. A single uncontroverted binary character supports each node; the most parsimonious tree is the expected pectinate shape with a cost of eight steps (Fig. 15.1). The bootstrap support values are shown for each vertex. If we add 80 uninformative characters (autapomorphies), the bootstrap values decrease to the expected value of 0.634 (on average) (Fig. 15.2).

15.1.2   Criticisms of the Bootstrap

Bootstrap as Confidence Interval

Hillis and Bull (1993) showed that in simulations, bootstrap values did not reflect well either the repeatability or accuracy of the reconstruction. As far as repeatability is concerned, they found the method largely useless due to the high variance in the bootstrap results. Regarding accuracy (in a simulation context), bootstrap values could either grossly underestimate or overestimate the actual values. The non-independence of vertices in trees may contribute to this (as well as violation of independent and identical distribution and random sampling assumptions in empirical comparative data). Furthermore, the bootstrap tree need not agree with the tree based on the original observations (even in principle), which can lead to suboptimal, refuted groups seeming to be better supported than optimal, corroborated groups.

Even though the validity of bootstrap values as statistical tests may have been refuted both in theory and practice, as a means of assessing support, bootstrap values remain a popular measure of a form of support.

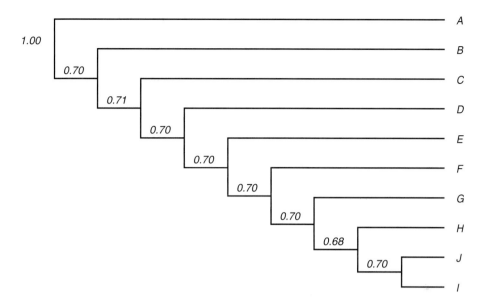

Figure 15.1: Most parsimonious tree based on the data in Table 15.1 with boot-strap frequencies shown (1000 replicates, 10 Wagner builds and TBR for each replicate using Varón et al. (2008, 2010)).

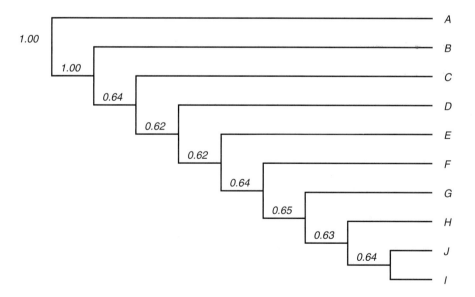

Figure 15.2: Most parsimonious tree based on the data in Table 15.1 and 80 uninformative characters with bootstrap frequencies shown (1000 replicates, 10 Wagner builds and TBR for each replicate using Varón et al. (2008, 2010)).

Effects of Uninformative Characters

From Equation 15.1, it can be seen that the support for a group will depend on the number of characters whether they are informative or not. Harshman (1994) pointed out that the bootstrap values could, in principle, depend on the number of unique and invariant characters in the data set. In fact, for a given number of informative r, uninformative features would increase n (Eq. 15.1) thereby decreasing the bootstrap support. Carpenter (1996) showed this effect to occur in real data as well as in theory.

15.1.3   Jackknife

John Tukey
(1915–2000)

The jackknife was originally defined (Quenouille, 1949, 1956) as a method that recalculated aspects of an empirical distribution by sequentially removing a single observation, the idea being that the variation in a sample-based parameter estimation could be reliably assessed in this manner. Later, Tukey (1958) generalized this idea and encouraged broader use by deleting one-half of the empirical sample. Farris et al. (1996) brought this to systematics by noting that the limit of Equation 15.1 as $n \to \infty$ is $1 - e^{-r}$, leading to the asymptotic convergence of the bootstrap and jackknife. If a group is supported by one uncontroverted synapomorphy, among an unboundedly large number of characters, the support for that group under the bootstrap and jackknife procedures will be identical if the delete percentage is e^{-1}. Such a resampling procedure under parsimony was named the "Parsimony Jackknife."

Diana Lipscomb

The procedure was originally described and used as a way to generate phylogenetic trees based on jackknife support (Lipscomb et al., 1998; Little and Farris, 2003). As opposed to optimality-based procedures, the method was rapid, generating a tree with support values in a fraction of the time of a conventional search. As with other non-optimality-based approaches, however, this was eventually abandoned in favor of its use as a support measure.

The jackknife proceeds in five steps:

1. Determine an optimal tree (T_o) by a search procedure on observed data (D) under an optimality criterion.

2. Resample the observed data (columns of aligned matrix) generating a new data set (D') where each character has the probability e^{-1} of deletion. This data set will have, on average, $1 - e^{-1}$ characters.

3. Determine an optimal tree (T'_o) from D' using the same search procedure and optimality criterion as in step 1.

4. Repeat steps 2 and 3 k times adding the generated trees to the set of jackknife trees (T_J).

5. Construct the majority-rule consensus tree of all nodes with frequency ≥ 0.50 in T_J.

As with the bootstrap, the jackknife tree need not agree completely (in unresolved or resolved groups) with the original, unsampled tree. Figure 15.3 shows

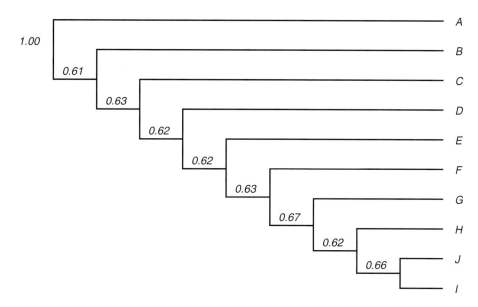

Figure 15.3: Most parsimonious tree based on the data in Table 15.1 with jack-knife frequencies shown (1000 replicates, 10 Wagner builds and TBR for each replicate using Varón et al. (2008, 2010)).

jackknife frequencies of the data in Table 15.1 at their expected values of approximately $1 - e^{-1}$ or 0.632.

15.1.4   Resampling and Dynamic Homology Characters

Resampling in the context of entirely linked, dependent observations is mean-ingless. The only way to employ such methods (Wheeler et al., 2006a; Varón et al., 2008) would be to convert the dynamic characters to a static set (presum-ably tree-based). The sampling would be entirely conditioned on the underlying tree, rendering the resulting values inflated for the source tree and unclear for any others. Guide-tree based MSA heuristics (Chapter 8) suffer this same prob-lem, if not to the same degree. If loci or entire genetic regions are treated as characters, resampling can take place at the level of loci or transcriptome (*e.g.* EST analysis such as Dunn et al., 2008). This really just bumps the problem up a level. If entire genome data were considered, all loci would be compo-nents of a larger dynamic homology problem and resampling, once more, loses applicability.

15.2   Optimality-Based Measures

As opposed to resampling measures, optimality-based measures are calculated based on the cost (length, parsimony score, likelihood, posterior probability)

Kåre Bremer

of trees. Usually, the comparison is between the best tree found and other, suboptimal trees. The indices commonly used by different optimality criteria are closely related, even interconvertible in certain circumstances.

15.2.1   Parsimony

The most commonly used optimality-based measure of vertex support (S) under this optimality criterion is the difference between the cost of the minimal cost tree (T^o) without a vertex (v–group or clade) and that with (Eq. 15.2).

$$S_v^{Bremer} = T^o_{v \notin V} - T^o_{v \in V} \qquad (15.2)$$

This value is determined for each vertex on the tree and is usually referred to as Bremer (1988) support.

This form of support is difficult to calculate. In principle, the lowest cost tree overall would be compared to all other trees in order to identify the minimum cost tree without a particular group. This is obviously an NP–hard problem. In actual usage, heuristic searches are performed (*e.g.* based on branch-swapping, constrained searches, or trees visited during the optimal search) and the support values generated are upper bounds. Figure 15.4 shows Bremer values of the

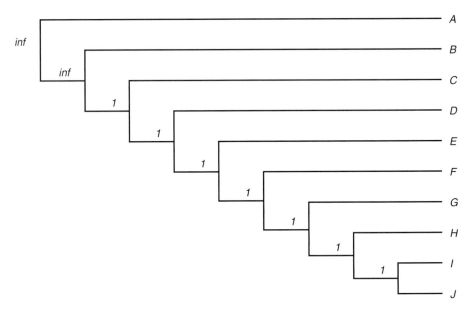

Figure 15.4: Most parsimonious tree based on the data in Table 15.1 with Bremer values shown (10 Wagner builds and TBR for each constrained vertex using Varón et al. (2008, 2010)). The infinity values are for edges not broken in heuristic Bremer search.

data in Table 15.1 at their expected values of 1 (a single synapomorphy at each vertex).

Several modifications have been made to Bremer values to explore more fully particular aspects of support desired for different purposes. One of the perceived shortcomings of this form of support is that it measures only overall support and makes no allowance for the balance between evidence in favor of and against a particular clade or vertex. As Goloboff and Farris (2001) point out, two groups, one with 100 character changes in favor and 95 against, and a second with five uncontroverted synapomorphies, would have identical Bremer values of five, yet our intuition might suggest that these are very different situations. Some have responded to this by employing resampling methods (such as the jackknife above) to evaluate such situations. Goloboff and Farris (2001) suggested the Relative Fit Difference (RFD) to deal with this issue.

The RFD examines the number of characters that favor a group (F) and compares that to the number that support some alternative (C). The standard Bremer value is their difference, $F - C$. The RFD normalized this value by dividing by F (Eq. 15.3), yielding an index (on the optimal tree $F \geq C$) varying from 0 (ambivalent support) to 1 (uncontroverted).

$$RFD_v = \frac{F - C}{F} \qquad (15.3)$$

One shortcoming of this measure (and with resampling as well) is that its extension to dynamic character types is unclear. If there were several dynamic characters (*e.g.* multiple loci), F could be defined as the sum of the costs of all loci that had a lower cost when a vertex was present than when it was not, and C defined as its complement. With a completely dynamic system such as a total genome, such distinctions would be impossible to make since there would be only one character—the genome. Perhaps the difference between costs of a character on the optimal and other trees could be used in place of $F - C$, but this would be time consuming since the determination of F and C would require NP–hard optimizations.

Grant and Kluge (2007), out of a desire similar to that of Goloboff and Farris (2001), suggested a normalization of Bremer support based on the difference between the most costly tree (a complete bush—but the "worst" binary tree might be a better comparison) and the least (the optimal tree). This they termed the Relative Explanatory Power (REP) value. Unlike the RFD, the normalization is not vertex specific; each Bremer value is divided by the same factor. In this way, REP values, like RFD, can be compared between data sets, but unlike RFD values have no effect on vertex support within a given tree.

Another modified use of Bremer values was proposed by Baker and DeSalle (1997) in cases of multiple sources of information in a "total evidence" (Kluge, 1989) analysis. Standard Bremer values would describe the support based on the entire data set. Baker and DeSalle (1997) calculated the individual contributions of data sets by determining the Bremer supports of the individual data sources (partitions) for each vertex on the favored tree. In this way, those nodes that

| Node | Bremer | Gene partitions | | | | | | | |
no.[a]	support	hb	wg	ACHE	ADH	COIII	16S	ND1	COII
1	21	4	4	0	2	8	1	1	1
2	8	4	2	2	0	1	0	0	−1
3	15	5	1	2	2	4	−3	1	3
4	13	−4	0	2	2	3	4	2	4
5	14	1	3.5	0	1.5	3.5	2	2	0.5
6	34	10	4	2.5	−2	1.5	8.5	4	5.5
7	4	3	2	0	1	0	0	0	−2
8	5	2	1	−1	−2	3	1	−2	0
9	11	6	2	−1	1	3	−1	1	0
10	5	1	2.25	−1	−0.25	3.5	−0.75	0	0.25
11	3	−1	3	0	2	0	−2	0	1
12	30	16.5	4	2.5	2.5	0	3	1	0.5
Total	163	47.5	28.75	8	9.75	30.5	15.75	10	12.75

Figure 15.5: Partitioned Bremer support for a series of nodes derived from a collection of molecular loci (Baker and DeSalle, 1997).

Morris Goodman
(1925–2010)

had consensual support could be identified and distinguished from those whose support was based on a minority of data sources (Fig. 15.5).

A Note on Nomenclature

The notion of support embodied in Bremer support values has been described several times and given different names. Faith (1991) used "length-difference," and Donoghue et al. (1992) suggested "decay index." Since Bremer (1988) was published earlier, that name has stuck, but recently Grant and Kluge (2008b) have pointed out that Goodman et al. (1982) used the same concept as that of Bremer in their discussion of primate relationships. This they called the "Strength of Grouping" or SOG value. Based on this seeming priority, Grant and Kluge (2008b) suggest the index be referred to as Goodman–Bremer.

15.2.2   Likelihood

In the same way that optimal tree costs are compared by their difference in parsimony, optimal tree likelihoods are compared by their ratio in likelihood methods. The support for a vertex, then (S_v^{LR}), is the ratio of the likelihoods of the maximum likelihood tree (T^v) with a vertex divided by that without (Eq. 15.4).

$$S_v^{LR} = \frac{T_{v \in V}^v}{T_{v \notin V}^v} \tag{15.4}$$

Figure 15.6 shows the likelihood ratio values of the data in Table 15.1 at their expected values of 0.693 (under No-Common-Mechanism model; Tuffley and Steel, 1997).

Furthermore, as mentioned earlier (Chapter 11), twice the log-transformed likelihood ratio is distributed approximately as χ^2 with a single degree of freedom (Eq. 15.5).

$$2 \cdot \left(\log T^o_{v \in V} - \log T^o_{v \notin V} \right) \approx \chi^2_1 \qquad (15.5)$$

This implies that a vertex will be "significantly" supported if the log likelihoods differ by 1.903 or likelihoods by a factor of 6.826. If all $n - 2$ vertices were to be tested simultaneously, however, a much stricter criterion would be required due to multiple testing issues (DeGroot and Schervish, 2006; Kishino and Hasegawa, 1989; Shimodaira and Hasegawa, 1999). Along these lines, Lee and Hugall (2003) used resampling measures to assign significance values to Partitioned Bremer Support values (Baker and DeSalle, 1997).

Another form of test, "interior branch test" (Felsenstein, 1981), differs from the likelihood ratio test (LRT) of tree optimality by examining whether a particular branch length ($\mu \cdot t$) is greater than zero. This is accomplished by determining the likelihood of the best tree with the added constraint that that particular edge has zero length. The same LRT test can be performed on the overall tree likelihoods with the same critical values and multiple testing caveats. At first, this was suggested to test whether a branch were "true" or not. Later (Yang, 1994b), it became clear that with sufficient data, every edge on a simulated "wrong tree" could have significantly non-zero length.

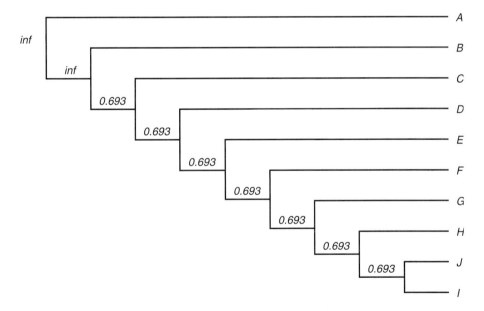

Figure 15.6: Maximum likelihood tree ($-\log \text{lik} = 11.09$) based on the data in Table 15.1 under No-Common-Mechanism model (Tuffley and Steel, 1997) with likelihood ratio values shown (10 Wagner builds and TBR for each constrained vertex using Varón et al. (2008, 2010)). The infinity values are for edges not broken in heuristic search.

15.2.3   Bayesian Posterior Probability

Bayesian support measures are, not surprisingly, very similar to those of likelihood. Three sorts of support are commonly generated in Bayesian analysis: "credibility" intervals, Bayes factors, and vertex or clade posterior probabilities.

Credibility

Credibility intervals are the Bayesian version of confidence intervals. They represent a central fraction (frequently 95%) of the posterior probability distribution. The idea for this measure would be that some aspect of a tree, such as a vertex, would or would not be included in those trees which contain 95% (for example) of the posterior probability. This is analogous to the confidence interval interpretation of bootstrap values (Felsenstein, 1981).

The process of such a test would be first to construct the credibility interval, or in the case of systematics, the minimal set of highest posterior probability trees (MAP) containing the desired fraction of credibility. If trees with a vertex are in that interval, they are supported; if not, not[1].

There is a significant difficulty with such an approach. The full posterior probability of hypotheses (trees) are required. This includes the denominator term of Equation 12.1, $\sum_{T_j \in \tau} pT_j \cdot p(D|T_j, \theta)$. Typically, this is not calculated due to the difficulty of the integration over all possible trees. A second problem can occur in standard statistical problems when the optimal set is multi-modal, but this is not a problem with trees since trees do not comprise an ordered set.

Bayes Factors and MAP measures

A more common index for Bayesian support is the Bayes factor (Kass and Raftery, 1995; Lavine and Schervish, 1999). The Bayes factor (B) measures the ratio of posterior probability to prior probability for a pair of hypotheses (Eq. 15.6).

$$B(T_{ij}) = \frac{p(T_i|D)/p(T_j)}{p(T_i|D)/p(T_j)} \tag{15.6}$$

This can be applied in several contexts. First, whether an entire hypothesis (tree) is superior to another, or second, to examine the effects of the presence or absence of a group or vertex (Eq. 15.7).

$$B(T_v) = \frac{p(T_{v \in V}|D)/p(T_{v \in V})}{p(T_{v \notin V}|D)/p(T_{v \notin V})} \tag{15.7}$$

If the priors are flat ($p(T_i) = p(T_j), \forall i, j$), the Bayes factor is identical to the likelihood ratio. A comparison can also be made that is analogous to the interior

---

[1]Such intervals need not be unique and may not even include the MAP tree if its posterior probability is less than 0.05.

branch test by comparing posterior probabilities where an edge weight (branch length) is constrained to zero. One of the issues with this test (as with many Bayesian tests) is that the prior distribution for branch lengths is unclear. Various distributions can be used (*e.g.* uniform, exponential *etc.*), but it is not obvious (at least to me) which, if any, are reasonable or model ignorance appropriately (see Chapter 12).

Bayes factors also contain an element of subjectivity when it comes to interpretation. Commonly, some hierarchy of comfort with the strength of evidence is used such as:

- $B(T_v) \geq 10^0$: T_v is supported.

- $10^0 < B(T_v) \leq 10^{1/2}$: T_v is minimally supported.

- $10^{1/2} < B(T_v) \leq 10^1$: T_v is substantially supported.

- $10^1 < B(T_v) \leq 10^2$: T_v is strongly supported.

- $10^2 < B(T_v)$: T_v is decisively supported.

Clade Posteriors

These values were first used by Mau et al. (1999) and Larget and Simon (1999) to summarize the posterior probability of individual vertices. They are created from the stationary pool of trees, whose frequency is determined by their overall posterior probability (Sect. 12.3.6). These are the support measures that are usually reported as Bayesian support in commonly employed implementations (Huelsenbeck and Ronquist, 2003). Clade posteriors are summaries over multiple trees, and their values exhibited by no single tree. As support measures, clade posteriors are more akin to resampling measures due to their summary nature and are discussed in that context below. As with resampling measures, there is no necessity that a tree constructed of those vertices with ≥ 0.50 posterior probability will agree with the MAP tree (Wheeler and Pickett, 2008), and "wrong" clades may achieve near 1.0 support (Goloboff, 2005; Yang, 2006; Chapter 12).

15.2.4   Strengths of Optimality-Based Support

One of the main benefits of optimality-based support is that its meaning is clear. The support value for a given vertex is a function of the best (optimal) tree with that node, and the best tree without. As such, it is related to the overall objective of a tree search—the best tree by whatever criterion—and only those vertices in the best tree are evaluated. Resampling measures (and clade posteriors) may identify nodes not found in the best tree as more supported than those that are. This is due in part to the summary aspect of integrating support over a series of trees which may differ in many fundamental respects, but agree in the presence of a given vertex. Treating vertices as entities independent of their source trees seems fundamentally at odds with the hierarchical nature of phylogenetic trees.

15.3   Parameter-Based Measures

Although often discussed in the context of support, parameter-based support (*sensu* Wheeler, 1995; as opposed to a distributional parameter) is more of a statement of robustness of phylogenetic results to variations in those parameters (Wheeler, 1995). As summaries of vertex presence over multiple trees, they are akin to resampling measures and have no direct relationship to any of the component "best" trees. Parameter-based trees are also like resampling (or clade-posterior) trees in that they need not agree completely with the "best" tree by any particular measure. These depictions convey the breadth of parameter choices under which particular groups are supported—usually those on an optimal tree (Fig. 15.7).

15.4   Comparison of Support Measures— Optimal and Average

As mentioned above, there is an often bewildering variety of things called "support." One immediate question is how do these descriptors relate to each other? They all measure something, but what? Given the discussions above, we can define the *optimal* support of a vertex (S_v) as a function of the costs of the optimal tree with a vertex ($T^o_{v \in V}$) and the optimal tree without ($T^o_{v \notin V}$) (Eq. 15.8) (Wheeler, 2010).

$$S_v = f\left(T^o_{v \in V}, T^o_{v \notin V}\right) \tag{15.8}$$

For parsimony this function is their difference (Eq. 15.9),

$$S_v = T^o_{v \notin V} - T^o_{v \in V} \tag{15.9}$$

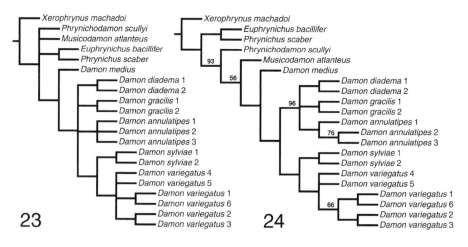

Figure 15.7: Strict (left) and majority-rule (right) consensus trees over parameter space (Prendini et al., 2005).

for likelihood their ratio (Eq. 15.10),

$$S_v = \frac{T^o_{v \in V}}{T^o_{v \notin V}} \tag{15.10}$$

and for Bayes a prior normalized ratio (with $p()$ denoting prior probability; Eq. 15.11).

$$S_v = \frac{T^o_{v \in V}/p(T^o_{v \in V})}{T^o_{v \notin V}/p(T^o_{v \notin V})} \tag{15.11}$$

Furthermore, if we examine the Bremer support of parsimony and the likelihood ratio of maximum likelihood under the No-Common-Mechanism (NCM) model (Tuffley and Steel, 1997), we can identify further links.

Taking the log of the likelihood ratio of Equation 15.10 (Eq. 15.12),

$$\log S_v = \log T^o_{v \in V} - \log T^o_{v \notin V} \tag{15.12}$$

and recall that under NCM the likelihood for a given tree with n characters with r states and l_i changes in character i is:

$$T_c = \prod_i^n r^{-(l_i+1)}$$

the log of which would then be:

$$\log T_c = \sum_i^n -(l_i + 1) \log r$$

rearranged becomes:

$$\log T_c = n - \log r \sum_i^n l_i$$

noting that $\sum_i^n l_i = l^p$, the parsimony length, and substituting back into Equation 15.12:

$$\log S_v = \log T^o_{v \in V} - \log T^o_{v \notin V} = \log r \cdot \left(l^p_{v \notin V} - l^p_{v \in V} \right)$$

Hence, the log likelihood ratio is equal to the Bremer support multiplied by the log of the number of states in the (non-additive) characters. With uniform topological priors, the Bayes factor will show the same equivalence under NCM. Bremer support, likelihood ratio, and Bayes factor, then, are closely related as optimal support measures.

Clade posterior, bootstrap, and jackknife supports are also related, but as *average* or expected support measures. Consider a distribution of trees $p(T) = g(T)$ (*e.g.* NCM with $p(T) = r^{l^p+1}$). We can define a value v_T for each vertex on each tree (Eq. 15.13).

$$v_T = \begin{cases} 0 & v \notin V \\ 1 & v \in V \end{cases} \tag{15.13}$$

The expectation of a vertex is the product of the value and distribution of v over all trees (Eq. 15.14).

$$E(v) = \sum_{T \in \Omega_T} v_T \cdot g(T) \tag{15.14}$$

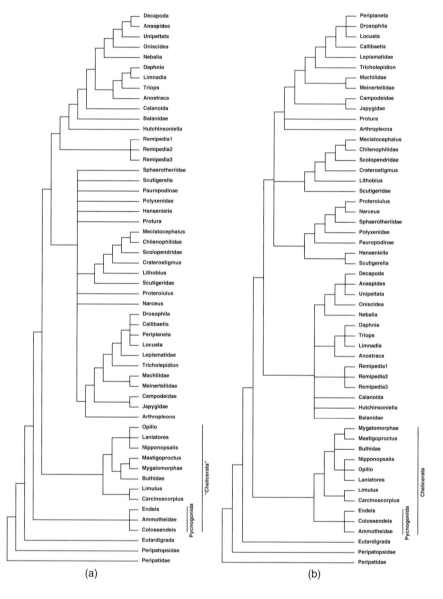

Figure 15.8: Parsimony (left) and likelihood (right) (under NCM (Tuffley and Steel, 1997)) analysis of Giribet et al. (2001) arthropod anatomical data (Wheeler and Pickett, 2008).

For Bayesian methods, $g(T)$ is the posterior probability of a tree given data $(T|D)$. Resampling methods yield a distribution of pseudoreplicate data sets, each of which has a single strict consensus tree (Eq. 15.15).

$$
\begin{aligned}
D' &\leftarrow RAND(D) \\
T' &= f(D') \\
g(T_i) &= \frac{1}{n} \sum_{i}^{replicates} 1 \text{ if } T_i = T'
\end{aligned}
\qquad (15.15)
$$

Resampling methods and vertex posterior probabilities are measuring an expected or average level of support over all trees. This is still support, but of a different type than optimal[2]. Wheeler and Pickett (2008) demonstrated this in their analysis of arthropod anatomical data (Fig. 15.8). The branch supporting chelicerate monophyly had a clade posterior probability of 0.6 (Huelsenbeck and Ronquist, 2003), while the alternate paraphyly in the parsimony analysis had a Bremer support of 2, a jackknife support of 0.57, and a log likelihood ratio/Bayes factor (under uniform tree priors) of 0.25. These are each support measures, but support alternate groupings of the same taxa under the same data and models.

15.5   Which to Choose?

There is no all-satisfactory measure of support since there is no consensual notion of support. Alternate measures based on resampling, relative optimality, and statistical distributions are all support measures. The key requirement is to be able to relate the measure to the entity being measured. When we are concerned with optimal trees, optimality-based measures are most closely linked.

15.6   Exercises

1. Consider the data of Dunn et al. (2008) treating gaps as character states.

   Determine a reasonably good tree using parsimony.

   Estimate the bootstrap tree using parsimony.

   Estimate the bootstrap support values for your "good" tree using parsimony.

   Estimate the delete one-half jackknife tree using parsimony.

   Estimate the delete one-half jackknife support values for your "good" tree using parsimony.

   Estimate the delete e^{-1} jackknife tree using parsimony.

---

[2]Clade posterior probabilities as support are also burdened with unavoidable clade size effects (Sect. 12.3.6).

Estimate the delete e^{-1} jackknife support values for your "good" tree using parsimony.

Estimate the Bayesian clade posterior tree using No-Common-Mechanism.

Estimate the Goodman–Bremer support for your best tree.

2. Repeat the above treating gaps as missing data.

3. What general conclusions can you draw from the previous two exercises?

4. If the number of character states were given as 21, when gaps were treated as states, and 20 when gaps were treated as missing data, what would the likelihood ratios be for the nodes on the two best trees under No-Common-Mechanism using a Neyman model?

5. If the prior distribution on trees were taken to be uniform, and the universe of topologies and their likelihoods determined by the Goodman–Bremer results above (*i.e.* the $n - 3$ edges with support values represent the entire universe of trees), what level of Bayesian credibility is represented by the best tree(s) under No-Common-Mechanism with gaps treated as states and missing data? How many trees are required to achieve 95% credibility? How might you summarize this result?

Chapter 16

Consensus, Congruence, and Supertrees

The methods discussed here take as input a set of trees and output a single tree. As commonly used, consensus methods operate on tree sets where each tree has the same leaf set as opposed to supertree methods, which operate on trees with different leaf sets. In reality, such consensus methods are a subset of supertree methods. In all cases, the primary goal is to represent areas of agreement and disagreement among the input trees and produce a summary statement in the form of a tree. None of the methods discussed below take account of the underlying character data directly, but operate solely on the topologies of their derived trees.

In general, consensus and supertree methods operate on sets of leaf taxa defined by splits or rooted clades, looking for "common" information. As a result, there are many shared operations among the procedures as well as subtle and not so subtle differences.

16.1 Consensus Tree Methods

16.1.1 Motivations

Originally, consensus methods were derived to identify common information between rival classifications (Adams, 1972). The notion was that inputs would be rooted trees with identical leaf sets and the consensus should represent generally shared information. The definition of shared has never been consensual, hence a variety of techniques have arisen to solve this problem.

16.1.2 Adams I and II

Adams (1972) defined two consensus procedures to operate on rooted trees. For convenience we can refer to these as Adams I and Adams II. Adams I

Systematics: A Course of Lectures, First Edition. Ward C. Wheeler.
© 2012 Ward C. Wheeler. Published 2012 by Blackwell Publishing Ltd.

operates on fully labeled trees (internal vertices as well as leaves). Unlike other methods, Adams I is ancestor-based, not descendant-based. If a labeled node M has an ancestor L shared on all input trees, the edge $L \to M$ will exist on the consensus even if $M \to (A, B)$ on one tree and $M \to (C, D)$ on another. The method proceeds as follows:

1. Define the vertex set of each input tree T_i as V_{T_i}

2. Identify one of the input trees as T_0

3. Begin at root

4. Choose vertex X on T_0 that has no ancestor or a known ancestor (determined by a previous iteration)

   Define $Y = X \cap_i V_{T_i}$

   If $Y = \emptyset$, discard X

   Else, if X has no ancestors in Y, then X is a root of the consensus (which could be a forest)

   Else, define Z to be the set of vertices ancestral to X on all trees. Connect X to the closest element in Z

5. Return to step 4 until all vertices are connected to an ancestor

The algorithm can be followed using Adams' example (Fig. 16.1). The Adams I consensus is rarely, if ever, employed. This is largely due to the fact that equivalent labeled ancestors (with potentially different descendants) are unknowable.

   The Adams II consensus operates on rooted trees with unlabeled internal nodes. In order to construct the consensus, the method examines each progressive split from the root to the leaves, joining the non-empty intersections (least upper bound or LUB) of the splits in the input trees as sister taxa (Fig. 16.2).

1. Begin at the root vertex of the input trees

2. Define the sets of leaf taxa $T_{i,j}$ for each tree i and groups j descendent from the vertex (two if binary, more otherwise) after removing those leaf taxa that have been placed on the consensus previously

3. Determine the LUB (Fig. 16.2) of each group (j) on each tree (i) with each on the others

4. For each LUB

   If the LUB is empty, ignore

   Else, if the LUB had ≤ 2 leaves, place that LUB in the consensus as a descendant of the vertex

   Else, the LUB is a vertex for further analysis. Go to step 2

The algorithm can be followed using the example above (Fig. 16.3). Adams II trees are also not in broad use in systematic analysis since (as in the example) groups not present in *any* of the input trees can be part of the consensus. This

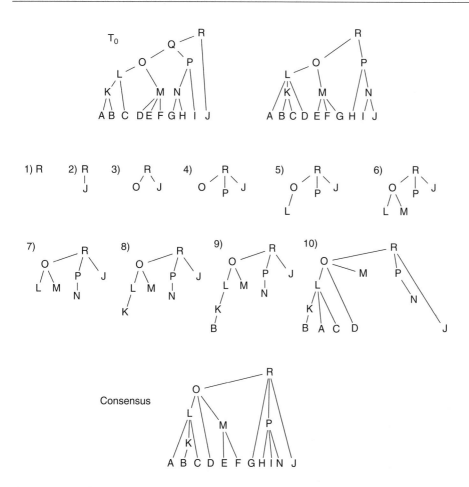

Figure 16.1: Steps in the Adams (1972) I consensus. Input trees (left as T_0) above, and the consensus below. Note that ancestor (non-leaf vertex) N becomes a leaf in the consensus and group (H, I, N) is not present in either of the input trees.

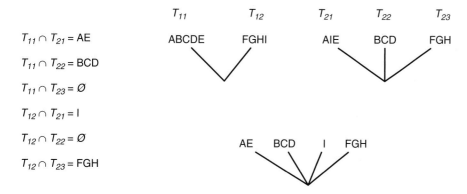

$T_{11} \cap T_{21} = AE$

$T_{11} \cap T_{22} = BCD$

$T_{11} \cap T_{23} = \emptyset$

$T_{12} \cap T_{21} = I$

$T_{12} \cap T_{22} = \emptyset$

$T_{12} \cap T_{23} = FGH$

Figure 16.2: LUB of the Adams (1972) II consensus.

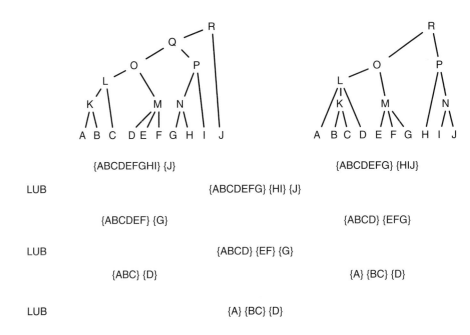

{ABCDEFGHI} {J}   {ABCDEFG} {HIJ}

LUB   {ABCDEFG} {HI} {J}

{ABCDEF} {G}   {ABCD} {EFG}

LUB   {ABCD} {EF} {G}

{ABC} {D}   {A} {BC} {D}

LUB   {A} {BC} {D}

Figure 16.3: Steps in the Adams (1972) II consensus. Input trees above, and the consensus below. Note that group (*HI*) is paraphyletic in both input trees, yet monophyletic in the consensus.

feature can be useful, however, in identifying "wild card" taxa that have widely differing positions on input trees with an underlying stable structure (*e.g.* as can happen with large amounts of missing data[1]).

16.1.3   Gareth Nelson

Many of the ideas employed by consensus tree techniques can be traced to Nelson (1979). Although not an easy paper, and often misunderstood and misinterpreted, this work has all the basic elements of consensus methods. Nelson

Gareth Nelson

---

[1] At least Darrel Frost says so.

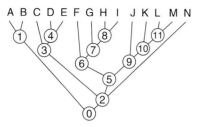

Phylogram

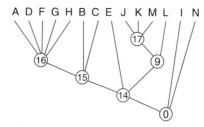

Phenogram: Adults

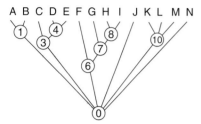

Phenogram : Larvae

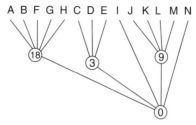

Phenogram : Males

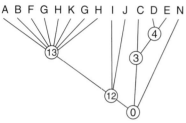

Phenogram : Pupae

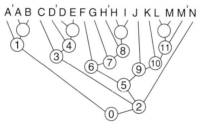

Phylogram : Alternative

Figure 16.4: Michener (1977) analysis of allopadine bees based on multiple data sources (Nelson, 1979).

began with a study of bees by Michener (1977) based on multiple sources of morphological data (Fig. 16.4).

Nelson defined objects, relations, and operations to formalize consensus. First off, he defined a *component* as a subtree or clade with ≥ 2 members (vertex with ≥ 2 descendent leaves). He then defined three relationships possible among components A and B on alternate trees: *exclusion*, *inclusion*, and *replication* (Eq. 16.1).

$$
\begin{aligned}
\text{exclusion:} \quad & A \cap B = \emptyset \\
\text{inclusion:} \quad & A \cap B \in \{A, B\} \\
\text{replication:} \quad & A = B
\end{aligned}
\tag{16.1}
$$

Two operations were also defined (Eq. 16.2).

$$
\begin{aligned}
\text{combinability:} \quad & A \cap B \in \{A, B, \emptyset\} \\
\text{non-combinability:} \quad & A \cap B \notin \{A, B, \emptyset\}
\end{aligned}
\tag{16.2}
$$

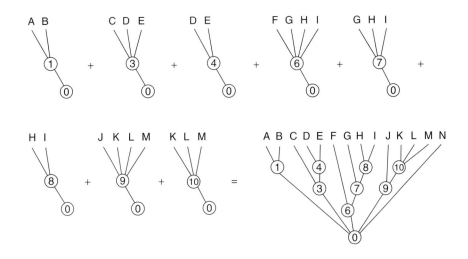

Figure 16.5: Nelson I consensus based only on replicated components (Nelson, 1979).

Girded with these definitions, Nelson went on to define two consensus operations. The first was based on replication alone. We can call this the Nelson I consensus. The Nelson I consensus of n input trees contains only those identical (replicated) components (subtrees) present in ≥ 2 input trees (Fig. 16.5). The second, Nelson II, for lack of a better name, was based on both replicated and combinable components (Fig. 16.6) giving a "General Cladogram."

Confusion has surrounded "Nelson" consensus due to Nelson's definitions of repeatability in terms of the number of input trees. When only two input trees are considered, Nelson I is equivalent to the "Strict" consensus (below) and Nelson II "Semi-Strict" (also below). When applied to more than two input trees, the methodology is unique and possibly contradictory (there can be conflicting replicated elements). Nelson consensus has been said to be strict,

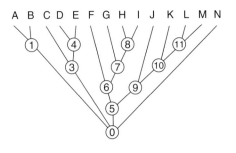

Figure 16.6: Nelson II consensus ("General Cladogram") based on replicated and combinable components (Nelson, 1979).

semi-strict, majority-rule (if $n = 3$), and none of these, due to this factor. The procedure, like the author, is brilliant, but often confusing. Nonetheless, the elements and operations Nelson defined are carried through much of the later consensus literature.

16.1.4   Majority Rule

Margush and McMorris (1981) made more precise the notion of comparison among multiple trees (consensus n-trees) and advocated a *majority-rule* approach. In essence, if components conflicted among trees, those with more than one-half representation would be present in the consensus tree. They showed that a tree must exist for such groups. After presenting a clarified algorithm for Adams II consensus, Margush and McMorris (1981) defined the majority-rule consensus based on groups of leaf taxa, but it works equally well for splits. In both cases, those groups components or splits present in more than one-half of the input trees are present in the consensus (Fig. 16.7). Majority-rule consensus is used often as a summary statistic in Bayesian analysis of clade posterior probabilities as well as general summaries of cladogram variation. A benefit of this form of consensus over more restrictive forms (such as strict, below), is that the resulting consensus cladograms are, in general, more resolved. A drawback is that there may be nearly as many groups that conflict as agree with the consensus. Various groups may be universally supported, others only marginally so.

16.1.5   Strict

Schuh and Polhemus (1980) (citing Nelson, 1979 in their study of the Leptopodomorpha) restricted themselves to universally replicated (*sensu* Nelson) elements in describing *Strict* consensus[2]. In a rooted context, the consensus tree

Randall T. Schuh

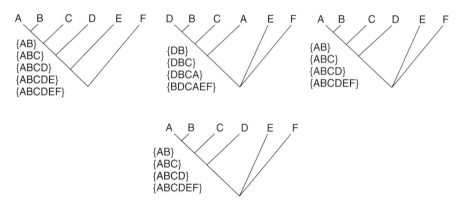

Figure 16.7: Majority-rule consensus (below) based on frequently replicated components in input trees (above) (Margush and McMorris, 1981).

---

[2]The name, however, comes from Sokal and Rohlf (1981).

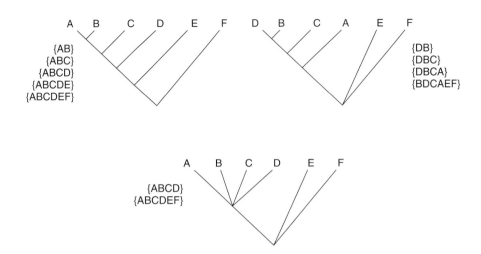

Figure 16.8: Strict consensus (below) based on universally replicated components in input trees (above) (Schuh and Polhemus, 1980).

is constructed from those sets of leaf taxa common to all input trees. When dealing with unrooted input trees, the consensus is based on splits that are common to all inputs (Fig. 16.8). A strength of strict consensus is that its meaning is absolutely clear. Any group (or split if unrooted) in the consensus is present in *all* the input trees. A potential drawback is that a shift in a single taxon from one side of a tree to another (in an otherwise stable tree of many taxa) will result in a completely unresolved consensus, signifying correctly that there are no shared groups.

16.1.6 Semi-Strict/Combinable Components

The *semi-strict* consensus (Bremer, 1990) attempts to strike a middle course between the strict and majority-rule methods. An extension of the Nelson II method, both universally replicated and universally combinable are included in the consensus cladogram (Fig. 16.9). Semi-strict trees are often used in biogeography and in summarizing competing cladograms where it is thought that absence of information should play no role in the final result. When the unresolved members of the combinable component set are due to character conflict, the method is less favored.

Kåre Bremer

16.1.7 Minimally Pruned

In cases where "wild-card" taxa present themselves due to missing data or extreme autapomorphy, it can be appealing to use *common/minimally pruned* consensus (Gordon, 1980). The motivation comes from the observation that there can be a small number of taxa that are placed seemingly randomly on a backbone of well-structured taxa—the most extreme case being a taxon with no

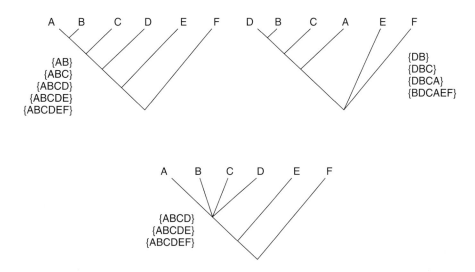

Figure 16.9: Semi-strict consensus (below) based on universally replicated and combinable components in input trees (above) (Bremer, 1990).

data at all added to a perfect phylogenetic data set. There will be $2n - 5$ placements of the all-missing taxon and no shared groups or splits among them. If that single taxon is pruned, however, absolute agreement is revealed (Fig. 16.10). Although intuitively appealing, the determination of such a minimally pruned tree is NP–hard and may be non-unique (multiple equally resolved prunings) severely limiting its utility[3].

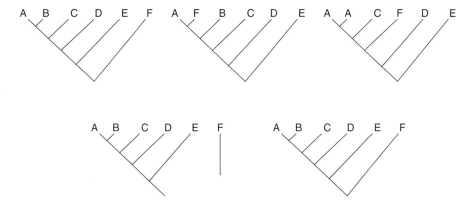

Figure 16.10: Common-pruned consensus (below) based on removing taxon F in the input trees (above) and placing it at the basal-most position on the tree consistent with the input trees (Gordon, 1980).

---

[3]An alternative to this would be only those elements and taxa in common, the *maximum agreement subtree* MAST (Finden and Gordon, 1985).

16.1.8   When to Use What?

Not all instances of tree consensus are equal, requiring different applications of consensus techniques. As pointed out by Nixon and Carpenter (1996a), when faced with multiple equally optimal trees from a single analysis, only strict consensus appropriately summarizes the implications of the data. Only universally supported clades are unambiguously reported by a strict consensus approach. Semi-strict, combinable component consensus trees may still contain groups that are not required to make the input trees of minimal cost. In principle, the situation can occur where a group resolved (but not incompatible with others) in a single output tree among k equally optimal solutions would be present in the single output consensus. When summarizing results from multiple analyses, however, (especially in biogeography where absence of information is often thought not to be informative) semi-strict trees have a reasonable place.

James Carpenter

Majority-rule consensus is frequently used to create summary trees for resampling measures (bootstrapping, jackknifing) and Bayesian analysis. Parameter sensitivity variation can also be presented in this fashion. As long as the distinctions between optimal and support-based trees are clear, these objects can be useful summaries of support.

Those methods (*e.g.* Adams I and II) that produce groups not seen in any of the input trees are never appropriate summaries of input trees. This seems a basic point that is followed with near absolute agreement for consensus within an analysis. Unfortunately, this clarity of reason has not fully carried over to all areas of systematics and still lives in several commonly used techniques of supertree analysis.

16.2   Supertrees

16.2.1   Overview

Supertree methods, like those of consensus, take a set of trees as input and produce a single tree as output. Unlike consensus techniques, supertree inputs can vary in leaf complement, hence, are a more general approach. While consensus approaches grew out of the desire to summarize the results of single analyses, supertrees were initially conceived as ways to combine multiple analyses into a single result without recourse to data analysis and combination. Specifically, supertrees had the ability to place taxa that had never appeared together on a single tree in any previous analysis. Supertrees, then, stand in contrast to "supermatrix" approaches of data combination and subsequent analysis. The speed and avoidance of laborious data examination and time consuming searches on large data sets recommends the method to many. The divorce of trees from data-based hypothesis testing has equally discouraged many. This contrast will be explored after the basic methods of supertree analysis are presented.

16.2.2   The Impossibility of the Reasonable

Michael Steel

Steel et al. (2000) enumerated a set of reasonable behaviors that should be possessed by supertree methods. They concern the relationships among the input trees and the (single) output supertree. The authors showed that no supertree method could simultaneously meet many of these requirements a salutary, and perhaps depressing, result.

The first three properties Steel et al. (2000) discussed were:

1. Order of input trees should be irrelevant

2. If leaves are renamed, the result (output supertree) is unchanged other than the names

3. The output tree displays (contains the relationships of) the input trees when they are compatible

When dealing with unrooted trees, these properties (P1-P3) cannot be realized simultaneously. Steel et al. (2000) considered the input trees 12|45, 34|16, and 56|23 and their two parent trees (Fig. 16.11). The two parent trees of Figure 16.11 are the only two candidates for a supertree solution that satisfy P3. If the names of taxa 2 and 6 and 3 and 5 are exchanged, the input trees are simply permuted (Fig. 16.12). Though the renaming simply permutes the input trees, the same cannot be said of the output trees. The output trees are interchanged, hence, if either had been chosen as the output supertree, it would have been changed by the renaming—violating P3.

This result holds for unrooted trees, but P1-P3 can be realized for rooted trees (Semple and Steel, 2000) via a graph method after Aho et al. (1981) (BUILD below). Furthermore, as can the two additional properties:

- P4 Each leaf that occurs in at least one input tree is present in the output.

- P5 The output tree can be determined in polynomial time.

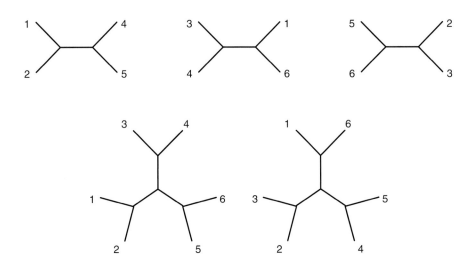

Figure 16.11: Three input trees (above) displayed by the two candidate supertrees (below) (Steel et al., 2000).

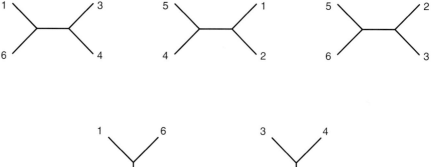

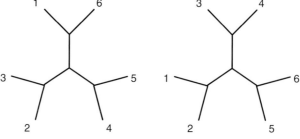

Figure 16.12: The trees of Figure 16.11 with taxa 2 and 6 and 3 and 5 exchanged (Steel et al., 2000).

By the same logic of the P1-P3 case above, the authors showed that the condition:

- P6 If all input trees contain $ab|cd$, then the output tree must contain $ab|cd$ cannot be realized simultaneously with P1 and P2. This result also applies to rooted trees.

The two conditions for the rooted version are:

- P6' If all input trees contain $ab|c$, then the output tree must contain $ab|c$.

- P7 If at least one input tree contains $ab|c$ and no other trees contain $ac|b$ or $bc|a$, then the output tree must contain $ab|c$.

P1–P5 and P6' can be achieved (*e.g.* Adams II; Chapter 15), but not P7. This was shown by the example of four trees of five leaves each, each containing a single resolved group (ab, bc, cd, de) (Fig. 16.13). A supertree result satisfying P7 would have to simultaneously contain $ab|e$, $bc|e$, $cd|a$, and $de|a$, which cannot occur.

Figure 16.13: Five taxon–four tree case of Steel et al. (2000) showing the impossibility of satisfying condition P7.

16.2.3   Graph-Based Methods

One of the first challenges to graph-based methods is to determine the compatibility of input trees. If they are compatible (there are no contradictory splits among the input trees) the trees can be combined directly. This problem was shown to be NP–hard by Steel (1992) in the case of unrooted trees. For rooted trees, this determination can be made in polynomial time by the BUILD algorithm of Aho et al. (1981), with time complexity $O(n^2 \log n)$.

BUILD Algorithm

The BUILD algorithm is based on identifying triples, $ab|c$ in trees. At each step, BUILD creates a graph where the vertices are the total leaf set (S) of all input trees (T), and the edges connect vertices a and b that have an $ab|c$ relationship on at least one tree. If the resulting graph has more than one connected component, recursively apply the graph step to each component until either there are fewer than 3 vertices (which are then added to the output tree) or a single component of three or more vertices is found signifying that the input trees are not compatible.

1. Set S to leaf set from T (set of input trees).

2. If $|S| < 3$, connect to output tree C.

3. Else, construct a graph $[R, S]$ of s via $ab|c$ edges (creating an edge for each derived pair in a triplet).

4. Let $S_0, \ldots, S_{k-1}$ denote connected components of $[R, S]$. If $k = 1$, then STOP and output INCOMPATIBLE.

5. Else, for each $i \in 0, \ldots, k - 1$ BUILD (T_i) from pruned T.

If the algorithm completes, the input trees are compatible (Fig. 16.14).

MinCut Algorithm

Semple and Steel (2000) modified the BUILD algorithm to produce a tree even if the algorithm reaches a non-trivial single connected component (BUILD step 4 above). This is accomplished by noting not only that vertices a and b share an $ab|c$ relationship on at least one tree, but by enumerating how many input trees display that triple. If the input trees are incompatible, when step 4 (above) is reached, the maximum "weight" (in terms of triplets) edge is contracted and the process continues (Fig. 16.15).

1. Set S to leaf set from T (set of input trees).

2. If $|S| < 3$, connect to output tree C.

3. Else, construct a graph $[R, S]$ of s via $ab|c$ edges.

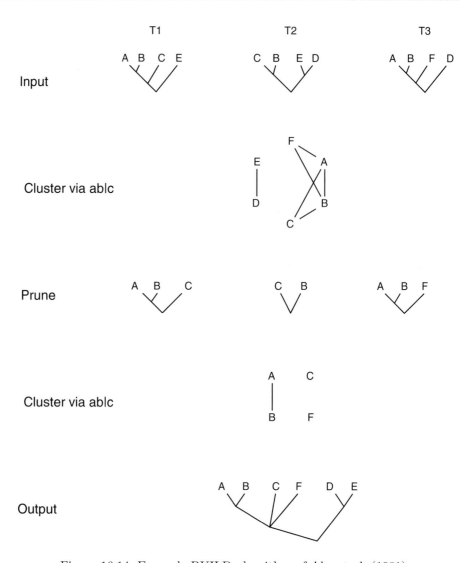

Figure 16.14: Example BUILD algorithm of Aho et al. (1981).

4. Let $S_0, \ldots, S_{k-1}$ denote connected components of $[R, S]$. If $k = 1$, then create $S \setminus E^{max}$ from $ab|c$ edge weights. Contract E^{max} and cut all minimum weight edges.

5. For each $i \in 0, \ldots, k-1$, MinCut (T_i) from pruned T.

In terms of the reasonable conditions enumerated above, MinCut realizes P1–P5 and P6'. The method is not, however, perfect. Like Adams II consensus trees (above), MinCut can produce groups not present (and not implied) in any of the input trees, even when the leaf sets are identical (Fig. 16.16).

Input

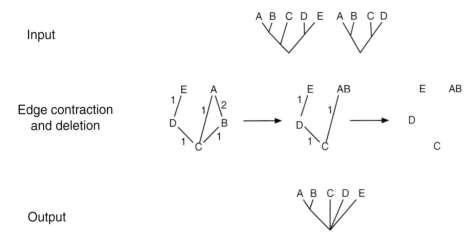

Edge contraction
and deletion

Output

Figure 16.15: Example MinCut algorithm of Semple and Steel (2000).

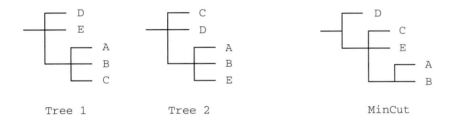

Tree 1   Tree 2   MinCut

Figure 16.16: Example of MinCut algorithm showing two input trees and Min-Cut output displaying two groups (AB and ABCE) neither supported nor implied by the input trees (Goloboff and Pol, 2002).

16.2.4   Strict Consensus Supertree

An alternate treatment of incompatible trees came from Gordon (1986). Gordon generalized the notion of strict consensus to overlapping, non-identical leaf sets. The method is straightforward and clear. Groups contradicted among inputs are not components of the output supertree (Fig. 16.17).

16.2.5   MR-Based

The most commonly used supertree methods are based on matrix representations of trees. These derive from Farris (1973b) group-inclusion characters where the descendent leaves of each vertex (subtree) on a rooted tree are assigned "1" and all others "0" (this can also be done on splits in an unrooted context) (Fig. 16.18).

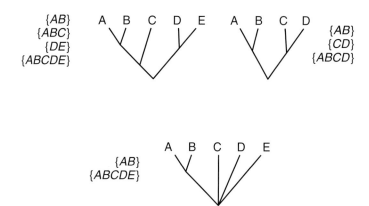

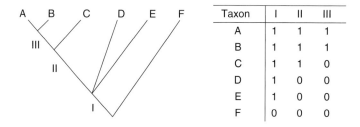

Figure 16.17: Example Strict Supertree of Gordon (1986).

Taxon	I	II	III
A	1	1	1
B	1	1	1
C	1	1	0
D	1	0	0
E	1	0	0
F	0	0	0

Figure 16.18: Matrix representation after Farris (1973b).

Matrix Representation of Parsimony

Matrix Representation of Parsimony (Baum, 1992) (MRP), or its variations, is the most commonly used supertree method. The central idea is to create a series of characters, one for each vertex on each input tree as above, and treat this matrix as a standard parsimony data set with binary characters and missing data (Fig. 16.19).

Although popular, MRP is not without its critics. As pointed out by empirical (*e.g.* Gatesy et al., 2004) and theoretical analysis (*e.g.* Goloboff and Pol, 2002; Goloboff, 2005), and even acknowledged by its most devoted adherents (*e.g.* Bininda-Emonds et al., 2005), groups not present in *any* of the inputs can be produced (Fig. 16.20) as well as groups contradicted by a majority of inputs (Fig. 16.21). Furthermore, the time complexity of MRP is the same as that for the original tree search, only with a new character matrix.

To me, these behaviors are so seriously pathological as to be disqualifying.

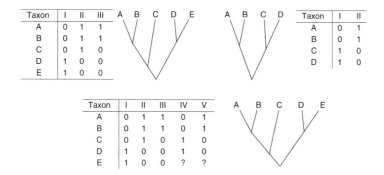

Figure 16.19: Matrix Representation of Parsimony (Baum, 1992). The two upper trees and their matrix representations are combined into the lower matrix that then produces the lower tree after parsimony analysis.

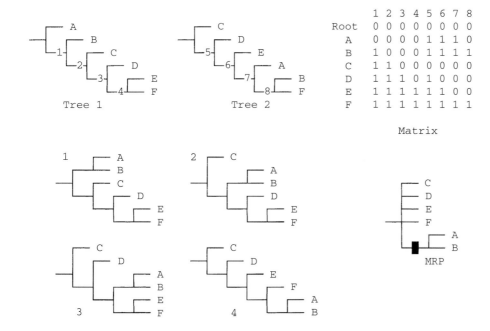

Figure 16.20: Example of MRP algorithm showing two input trees (T1 and T2), their MRP matrix and resulting supertrees (1–4 and strict consensus for MRP) displaying group AB that is not present in either of the input trees (Goloboff and Pol, 2002).

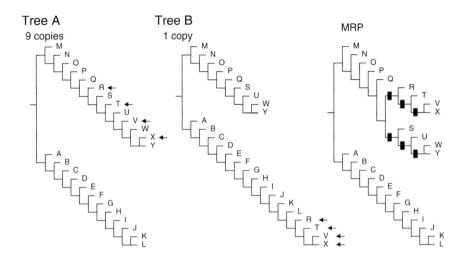

Figure 16.21: Example of MRP algorithm showing two input topologies, and resulting supertree displaying groups (RTVX) and (SUWY) contradicted by 9 of 10 input trees (Goloboff, 2005).

Variations

There are many variations on the MRP theme (14 alternates reviewed in Wilkinson et al., 2005; Fig. 16.22). Each has adherents and detractors, and the justifications are generally *ad hoc*. The problems of novel and contradicted groups are either likely to or known to occur in all these methods.

- Standard MRP (Baum, 1992).

- Irreversible MRP (Bininda-Emonds and Bryant, 1998).

- Compatibility MRP (Purvis, 1995b).

- Sister-Group MRP (Purvis, 1995a).

- Three-Taxon-Analysis (Nelson and Ladiges, 1994).

- Quartet MRP (Wilkinson et al., 2001).

- Min-Flip (Chen et al., 2003).

16.2.6 Distance-Based Method

Lapointe and Cucumel (1997) took a different, distance-based, approach to the supertree problem. These authors took each phylogenetic matrix and converted

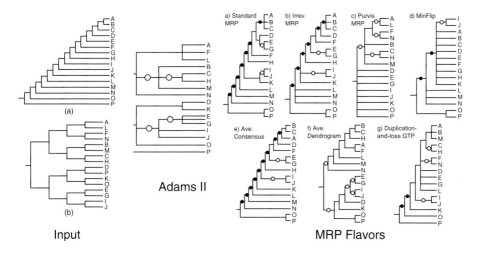

Figure 16.22: Example behavior of a variety of supertree methods (Wilkinson et al., 2005).

it into a distance. An overall average distance ($\bar{d}$) among $|T|$ datasets was then created from these values (Eq. 16.3).

$$\bar{d}_{ij} = \frac{1}{|T|} \sum_{k=0}^{|T|-1} d_{ij}^k \qquad (16.3)$$

The supertree itself is contructed via standard distance techniques such as minimum evolution (Chapter 9; Fig. 16.23).

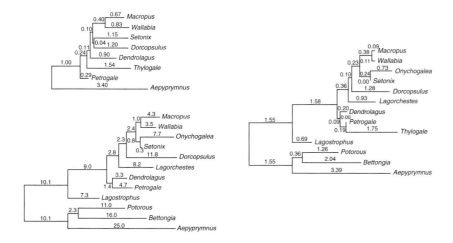

Figure 16.23: Distance supertree (Lapointe and Cucumel, 1997) based on marsupial data.

16.2.7   Supertrees or Supermatrices?

Supertrees have been proposed as a solution to the problems presented by large, combined data sets. As such, they are to be compared to total-evidence based "supermatrix" analyses (Fig. 16.24). Given that analyses can now be performed on commodity hardware that contain tens of thousands of taxa (Goloboff et al., 2009) and multiple genetic and morphological data sources, computational effort can no longer be viewed as a justification[4].

A second rationale is that supertrees are agnostic as to the epistemological basis of the input trees. The source trees may be based on distance, parsimony, likelihood, or posterior probability—or any combination. This is true, but can only be viewed as a strength if the investigator has no rationale or preference for optimality criteria (*e.g.* would alphabetical order be an acceptable criterion?). A further implication of this is that supertrees can only be tested against each other in their ability to minimize an overall tree difference cost. That this is a legitimate form of hypothesis testing is yet to be justified.

Perhaps the most important distinction between supermatrix and supertree approaches is the use of observational data (from whatever source). Supermatrix analysis evaluates hypotheses based on the totality of data. Hypotheses are

	Pro	Con
Super-Matrix	Evidence-based Differential support Transformational optimality Hypothesis testing via observation = homology	Hard work Missing/Inapplicable data Need to choose a criterion Higher time complexity
Super-Tree	No data work Lower time complexity Combine results from different optimalities	Disconnected from data Include groups found in no trees Lack groups found in all trees Hypothesis testing via concordance (at best) Data quality, non-independence untested

Figure 16.24: Pros and cons of supertree and supermatrix analyses.

---

[4]Furthermore, the simulation study of Kupczok et al. (2010) reported superior results for supermatrix analysis with respect to supertree methods in recovering the simulated "true" tree.

erected and competed via their ability to explain variation in nature. Supertrees are completely divorced from this idea. Whether a tree is supported by one character or an entire genome, by decisive or ambiguous data, is completely irrelevant to the process. Supertrees make no allowance for differential support or underlying patterns. In doing so, they avoid the hard work in creating large coherent data matrices, but pay a price in terms of interpretive opacity and data free results.

The central question is would anyone ever favor a supertree result over one based on a supermatrix[5]? Until this question is resolved, it is unclear whether supertrees are necessary or desirable.

16.3 Exercises

For the first seven exercises below, consider the tree set of Figure 16.25. For the final two exercises below, consider the tree set of Figure 16.26.

1. Construct the Adams I consensus tree.

2. Construct the Adams II consensus tree.

3. Construct the Nelson I consensus tree.

4. Construct the Nelson II consensus tree.

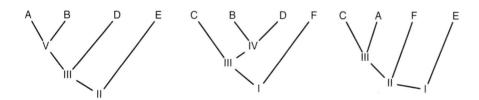

Figure 16.25: Input trees for consensus Exercises 1 to 7.

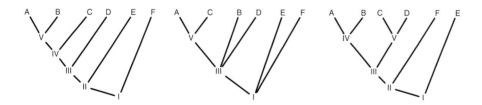

Figure 16.26: Input trees for supertree Exercises 8 and 9.

5. Construct the Majority Rule consensus tree.

6. Construct the Strict consensus tree.

7. Construct the Semi-Strict consensus tree.

8. Construct the MinCut supertree.

9. Construct the MRP supertree.

10. What is your position in the supermatrix–supertree debate?

Part V

Applications

Chapter 17

Clocks and Rates

Emil Zuckerkandl

Linus Pauling
(1901–1994)

Since Zuckerkandl and Pauling (1962, 1965), the dating of phylogenies with molecular data has been a siren call. At first, the "molecular clock" was embraced (by some) and used in simple linear dating based on the regression of molecular distance and one or more fixed dates based on fossils or vicariant events. More recently, this has become more sophisticated with the use of multiple rates and sequence change models. The goal is the same, and perhaps just as elusive.

17.1 The Molecular Clock

Zuckerkandl and Pauling (1962) examined vertebrate hemoglobin protein sequences and noted the approximate constancy of changes in lineages dated with fossils from the presumed root to observed extant taxa. From this, they proposed an extrapolation to all molecules and lineages, stating that such molecular analysis could date events about which we have no fossil information.

> It is possible to evaluate very roughly and tentatively the time that has elapsed since any of the hemoglobin chains present in a given species and controlled by non-allelic genes diverged from a common chain ancestor. ... From paleontological evidence it may be estimated that the common ancestor of man and horse lived in the Cretaceous or possibly the Jurassic period, say between 100 and 160 million years ago. ... the presence of 18 differences between human and horse α-chains would indicate that each chain had 9 evolutionary effective mutations in 100 to 160 millions of years. This yields a figure of 11 to 18 million years per amino acid substitution in a chain of about 150 amino acids, with a medium figure of 14.5 million years. (Zuckerkandl and Pauling, 1962)[1]

[1]More can be found on the development and controversies over attribution in Morgan (1998).

Zuckerkandl and Pauling did not go into great detail in their analysis, and later authors expanded upon this idea to both general molecular calibration and "neutral" molecular evolution (Kimura, 1983).

17.2 Dating

Before immersing ourselves in rate models and tests, it is worth returning to the nature of the dating process used to calibrate molecular time analyses.

First, as Hennig (1966) described, if a fossil date is known for a taxon, its sister taxon *must* have been present at that time as well. This is due to the simple logic of lineage splitting. When a lineage splits, two (sister) taxa are created at the same time, hence have the same age. Furthermore, if two sister lineages have been dated, the older date applies to both lineages. All age scenarios are compatible with all trees. The sister-taxon relationships, and hence minimum taxon age dates, may change, but dating scenarios do not discriminate among trees. This idea has been developed further into the *ghost* taxon concept (Norell, 1987, 1992) for estimating paleo-diversity.

Second, fossil dates yield *minimum* ages. Fossilization is not a high probability event, and the chance that this occurred immediately following a splitting event is impossibly small. Furthermore, the fact that the fossil specimen is recognizable as a member of a particular lineage implies that enough time has intervened to allow for the acquisition of apomorphies. Again, unlikely in the immediate aftermath of a splitting event.

Third, ages are not absolute, precise values. Many dates are relative (stratum-based), or derived from radiometric dating above and below (not exactly at) the specimen. Additionally, the dating methods themselves are not without measurement error. These factors are discussed in greater detail below.

Fourth, vicariance-based dates have the same sort of imprecision of radiometric dates, compounded by uncertainty in timing and sequence of land-mass splitting and union. An additional factor is the confounding effect of dispersal. Vicariance is not the only means to account for taxonomic distributions, and dispersal may have occurred well after dramatic events (*e.g.* primates found in South America and Africa are well supported to have a more recent shared history than the land masses). Furthermore, lineage splitting could have occurred well before that of the land mass. In summary, lineages could have split before, during, or after tectonic change and not necessarily in sync.

17.3 Testing Clocks

17.3.1 Langley–Fitch

The first test of clock like behavior in molecular data was that of Langley and Fitch (1974). This method treats amino-acids (or other sequence data) as non-additive (Fitch, 1971) characters. Parsimonious character changes are assigned to each of the edges of the (given or reconstructed) tree. These values are used to estimate the edge parameters under a (now standard) Poisson process likelihood

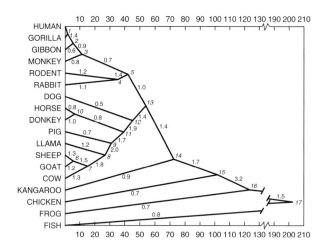

Figure 17.1: Langley and Fitch (1974) "observed" and corrected α and β hemoglobins, cytochrome c, and fibrinopeptide A edge lengths.

	Uncorrected			Corrected		
	χ^2	(df)	P<	χ^2	(df)	P<
Among proteins within legs (relative rates)	102.7	(62)	10^{-3}	102.7	(62)	10^{-3}
Among legs over proteins (total rates)	63.0	(26)	10^{-4}	48.7	(24)	0.002
Total	165.7	(88)	10^{-5}	151.4	(86)	10^{-4}

Figure 17.2: Langley and Fitch (1974) test of clocklike behavior of α and β hemoglobins, cytochrome c, and fibrinopeptide A.

model (Fig. 17.1). The clock test in the model is based on the proportionality of probabilistic change to "observed" (actually inferred) parsimony changes. The proportionality of change over the edges of the tree is tested against a χ^2 distribution with $2n - 3$ degrees of freedom for n taxa (Fig. 17.2).

With this method, Fitch and Langley rejected clocklike behavior among the vertebrate lineages and among proteins. They did, however, say that the implied molecular dates correlated well with paleontological dating.

17.3.2   Farris

Farris (1981) tested (and attacked) the clock idea in a more direct and simple manner. If there were a molecular clock, at least in the strict sense, the distance data would be ultrametric (or at least additive; Sect. 9.3). Furthermore, if this were the case, then UPGMA (Chapter 9) would be the correct and efficient means to reconstruct phylogenetic trees. Farris then pointed out that ultrametric non-trivial data sets are exceedingly rare (if they exist at all—I for one do not know of any). Furthermore, corrections to real distance data to account for unobserved change do not restore ultrametricity or additivity (and may even increase their violation; see Sect. 9.2.1). More hopeful investigators (*e.g.* Felsenstein, 2004) held out that even though Farris was correct, "minor" violation of a clock was not enough to falsify the idea.

17.3.3 Felsenstein

Felsenstein (1984) proposed an alternate test of clocks based on the least-squares fitting of distance data to a tree (see Chapter 9). This method employs a common strategy of testing alternate tree fit scenarios, one where the clock is enforced via leaf to root distances being equal (S_1), and the second where this constraint is absent (S_0). The unconstrained tree cost cannot be worse than the constrained, and the test is to see whether there is a significant difference between the two. In the case of least-squares estimated distances, the test statistic is the difference in sum of squares between the two trees divided by that of the unconstrained, $\frac{S_1 - S_0}{S_0}$. This would be distributed as F with $(n-1, n^2 - 3n + 3)$ degrees of freedom (or $\frac{(n-2)(n-3)}{2}$ in the denominator if distances are averaged between $i \to j$ and $i \leftarrow j$).

An analogous test for likelihood was proposed by Felsenstein (1981). Consider the tree of Figure 17.3. A clock (not absolute in this case, but for sister taxa) would add the following constraints:

$$
\begin{aligned}
V_1 &= V_2 \\
V_4 &= V_5 \\
V_3 &= V_4 + V_7, V_5 + V_7 \\
V_1 + V_6, V_2 + V_6 &= V_3 + V_8, V_4 + V_7 + V_8, V_5 + V_7 + V_8
\end{aligned}
$$

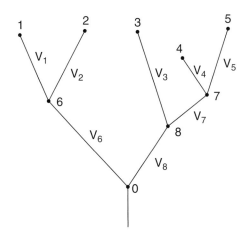

Figure 17.3: Felsenstein (1984) test of clocklike behavior.

A likelihood ratio test $(-2 \cdot (\log likS_0 - \log likS_1))$ can then be performed between the constrained and unconstrained tree with $n - 1$ degrees of freedom (the number of free edges that differ between the two scenarios). This test is less restrictive than that requiring all paths to the root be equal, but will still nearly always reject a clock with real data (*e.g.* Bell and Donoghue, 2005).

17.4   Relaxed Clock Models

Even though many would say the molecular clock was dead with Farris (1981), it lived on until the mid 2000s when statistical methods were developed that allowed dating without a strict clock. The techniques below each allow variation in rates over edges to a greater or lesser extent. In order to keep the variance of internal dating under control, however, there are limits of varying types. At present, there are a large variety of molecular dating techniques (see Rutschman, 2006 for a review); discussed here are three classes into which most methods fall: local clocks, rate smoothing, and Bayesian relaxed clock methods.

17.4.1   Local Clocks

If the global clock represents one (rejected) extreme, and each edge having its own rate the other, local clocks are an intermediate form (Hasegawa et al., 1989; Yoder and Yang, 2000). In this approach, edges are placed into categories with a specific rate parameter. The categories may be aspects of their biology (herbaceous and non-herbaceous plants—Bell and Donoghue, 2005) or groups (subtrees). The key aspect is that these categories are identified *a priori* as constraints on the likelihood analysis. This is also the weak point of the method in that this identification, especially in absence of a tree, can be problematic.

17.4.2   Rate Smoothing

Rate smoothing methods allow all edges to have unique rate parameters in principle, but restrict the abruptness of transitions between adjoining edges. The idea is that there is a correlation in rates between edges: the closer they are, the more similar they are likely to be. This has been implemented in several ways.

Non-Parametric Rate Smoothing

The first of the rate smoothing techniques, NPRS (Sanderson, 1997), estimates local rates for each edge $(\hat{r}_i)$ and minimizes the transitions between descendent rates over the tree, W, (Eq. 17.1).

$$W = \sum_{k \in \text{ internal vertices}} \sum_{j \text{ is descendent edge of } k} |\hat{r}_k - \hat{r}_i| \qquad (17.1)$$

Michael J. Sanderson

Sanderson (2002) pointed out that NPRS is prone to over-fitting rate changes, leading to high variance time estimates. This leads to the idea of more severely penalizing changes in rate (below).

Penalized Likelihood

Penalized likelihood (Sanderson, 2002) combined likelihood and a "roughness" parameter to penalize rapid rate transitions between edges. In this way, the method reduces rate variation over NPRS. The penalty, Φ, is determined by the sum of squares of rate differences between adjacent edges, augmented by the variance in the rate of the edge and root edge (Eq. 17.2).

$$\Phi = \left[\sum_{k\in \text{ internal vertices, not root}} \left(r_k - r_{anc(k)}\right)^2 \right] + Var\left(r_k, r_{\text{root}}\right) \qquad (17.2)$$

Log Φ multiplied by the smoothing factor (λ) is then subtracted from the log likelihood of the tree given the rates. This reduces the optimality of the tree based on the degree of rate variation. This is certainly a penalty, yet the justification seems to me to be unclear, since rate homogeneity hardly seems the default expectation and could well be its opposite.

Heuristic Rate Smoothing (HRS)

HRS Yang (2004) contains elements of local clocks as well as NPRS. HRS examines variation across edges of the tree and across multiple locus-based trees without a clock constraint. These provide variance estimates, allowing the assignment of edges and genes to rate categories as in local clocks. These categories then are combined in a master tree analysis yielding divergence times.

17.4.3 Bayesian Clock

Bayesian rate methods (Thorne et al., 1998; Kishino et al., 2001; Thorne and Kishino, 2002) approach edge rate variation in an appropriately Bayesian manner—with priors. Employing a variety of prior probability distributions on edge rates (*e.g.* log-normal, exponential), these methods are able to evaluate trees with unique rates for each edge. Correlation among rates, as in NPRS, can be tested after the analysis to form the MAP solution. These methods rely on MCMC procedures and can be quite time consuming (see below), but offer freedom from much of the cumbersome and *ad hoc* machinery of other methods.

Jeffrey L. Thorne

17.5   Implementations

Three commonly used implementations are discussed briefly here.

17.5.1   r8s

r8s (Sanderson, 2003, 2004) implements the (Langley and Fitch, 1974) NPRS (Sanderson, 1997) and PL (Sanderson, 2002) methods. The program requires an input tree as well as at least one fixed date (Fig. 17.4).

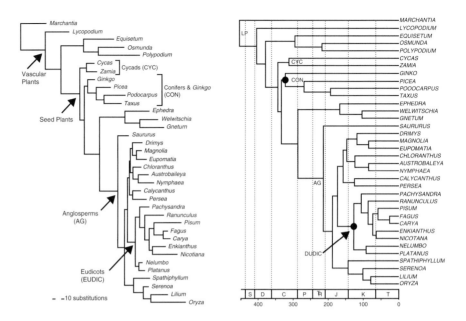

Figure 17.4: r8s (Sanderson, 2003) rate smoothed dating of tree.

17.5.2   MULTIDIVTIME

MULTIDIVTIME is an implementation of the Bayesian rate model and date determinations. Like BEAST (below) it sets priors on edge rates and uses an MCMC approach to determining the posterior probabilities of dates and rates (Thorne et al. (1998); Kishino et al. (2001); Thorne and Kishino (2002)). MULTIDIVTIME operates on a single input tree, but can accommodate multiple genetic data sets (Fig. 17.5).

17.5.3   BEAST

BEAST (Drummon et al., 2006; Drummon and Rambaut, 2007) is a fully Bayesian approach to tree estimation (even from unaligned sequences via Thorne et al., 1992) as well as dating. BEAST uses a Yule prior on tree topology favoring specific tree shapes over others. Given the totality of tasks BEAST attempts to solve, it can be extremely time consuming (Fig. 17.6).

17.6   Criticisms

Graur and Martin (2004) discuss (in enjoyably forceful language) a selection of the many shortcomings of dating procedures, most specifically those of Hedges et al. (1996). They raise four issues with molecular dates:

Daniel Graur

1. Incorrect identification of fossil taxa.

2. Error in age estimation of a particular fossil.

3. Regression error in dates (especially during extrapolation beyond range of actual dates).

4. Recycling of estimated dates as observed dates without error.

Each of these general faults is exemplified by Hedges et al. (1996) in a study of the diversification of birds and mammals. As mentioned above, to assign a date to a clade, the fossil taxon employed must be a member of it. Graur and Martin (2004) point out that the central dating point is attached to an amniote of uncertain affinity. In addition to this, the date attached to the taxon

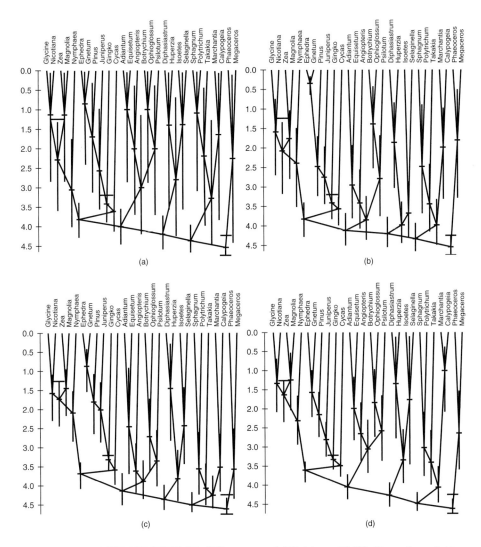

Figure 17.5: MULTIDIVTIME (Thorne and Kishino, 2002) Bayesian dating of tree.

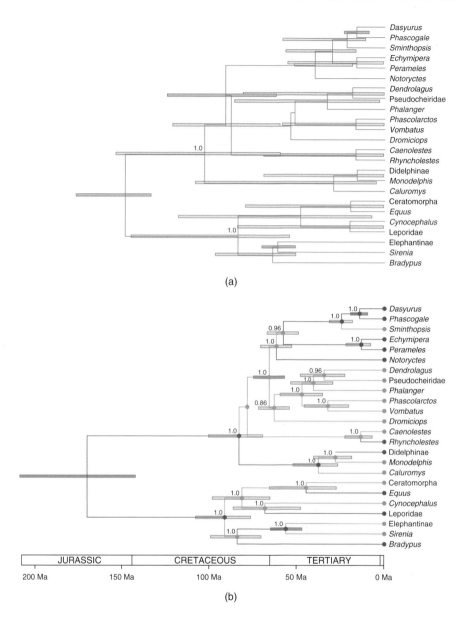

Figure 17.6: BEAST (Drummon and Rambaut, 2007) Bayesian dating of trees with priors on dates (above) and posterior probabilities (below). See Plate 17.6 for the color Figure.

is unsupported by the cited literature. Other literature has a different date from that used by Hedges et al. (1996). Furthermore, the date is used as if it were absolute and without error. As mentioned earlier, fossil dates are often relative, and even radiometric data are based on an interval containing a specimen and are further subject to observational error as are all aspects of experimental science. These errors are exacerbated by incorrect extrapolation variance in the use of dating regressions and (perhaps worst of all) estimated dates used (without error) as if they were observed.

Another salutary case is that of Near et al. (2005), reanalyzed by Parham and Irmis (2008). Of the 17 literature dates cited in their dating of turtle divergences, 15 of the dates are not even mentioned in the literature cited, and two are used with a spurious precision of 1 part in 2000 (when authors used words such as "about" and "approximately" to characterize the dates). This was coupled with an analysis-free and idiosyncratic placement of a key taxon making their dating extremely suspect.

There are two sorts of problems with the above examples: those that are inherent to the methods, and those that come about by misuse, poor scholarship, or over-interpretation of results. Obviously, it is not the fault of methods or their implementations if they are improperly used, but this still leaves the crucial factor of error analysis. Although improving, the incorporation of the myriad sources of experimental, topological, and statistical (*i.e.* regression) error must be incorporated into the determination of dates. When they are, errors of magnitude comparable to or even exceeding the estimated dates can be achieved (Graur and Martin, 2004; Pulquério and Nichols, 2006).

17.7 Molecular Dates?

Recent Bayesian methods have made great progress in allowing edge-specific rate parameters. This does come at the usual Bayesian price of the necessary prior distributions. What is a reasonable prior for evolutionary rates on tree edges? Given our lack of knowledge, sampling, lineage, environmental, and locus specific (among myriad known and unknown other) effects, the uniform distribution would seem as appropriate as log-normal or exponential. Such a prior is likely to lead to large uncertainty in estimated dates, leaving us where we were initially, with rock solid, if limited, minimum ages based on sister taxon relationships.

The central question remains—can hypotheses be falsified with "molecular" dates? Are these estimates sufficiently precise to adjudicate between biogeographic or other scenarios? This does not appear to be the case at present, but time will tell.

17.8 Exercises

1. Should dates be used in tree construction? What would be the optimality criterion?

2. How much variation in a date can be allowed and the date still be useful?

3. What are the pros and cons of different prior distributions for edge rate parameters?

Appendix A

Mathematical Notation

Commonly used mathematical symbols.

$A = \text{argmax}_x f(x)$	A is set to the value of x that maximizes $f(x)$.
$\sum_a^b f(x)$	summation of $f(x)$ from $x = a$ to $x = b$.
$\prod_a^b f(x)$	product of $f(x)$ from $x = a$ to $x = b$.
$\forall x \in X$	for all elements x contained in the set X such as x_0, x_1 etc.
X_i	the ith element in a collection (set or array) of objects X such as x_0, x_1 etc.
$\|X\|$	the size of an object such as the cardinality (number of elements) of a set or the length of an array.
$n!$	n factorial. $\prod_{i=2}^{i=n} i$ or $n \cdot (n-1) \cdot (n-2) \cdot \ldots \cdot 2$.
$X \setminus Y$	the set created by subtracting the elements of set Y from set X.
$\|x\|$	absolute value of x.
$\mathbb{R}^+$	the set of positive real numbers.
$X \subseteq Y$	X is a subset of or equal to Y.
$X \nsubseteq Y$	X is not a subset of or equal to Y.
$X \subsetneq Y$	X is a proper subset of Y.
$\lfloor X \rfloor$	the largest integer $\leq$ than X.
$X \cap Y$	intersection of sets X and Y.
$X \cup Y$	union of sets X and Y.
$\emptyset$	the empty set.
$X \leftarrow Y$	assignment of Y to X.
$\binom{x}{y}$	binomial expansion of x and $y = \frac{x!}{y!(x-y)!}$.
$\log^* x$	iterated logarithm. The number of applications of log until the result is ≤ 1.

Systematics: A Course of Lectures, First Edition. Ward C. Wheeler.
© 2012 Ward C. Wheeler. Published 2012 by Blackwell Publishing Ltd.

A^c	complement of set A $(S \setminus A)$.
$A \therefore B$	A therefore B.
$ab|cd$	unrooted tree $((a,b)(c,d))$.
$a|bc$	rooted tree $(a(b,c))$.
$x \in (a,b)$	values x in the interval $a < x < b$.
$x \in [a,b]$	values x in the interval $a \leq x \leq b$.
$a \propto b$	a is proportional to b.
$\exists$	there exists.
$\nexists$	there does not exist.

Bibliography

Achtman, M., van der Ende, A., Zhu, P., Koroleva, I. S., Kusecek, B., Morelli, G., Schuurman, I. G., Brieske, N., Zurth, K., Kostyukov, N. N., and Platonov, A. E. 2001. Molecular epidemiology of serogroup A meningitis in Moscow, 1969–1997. *Emerging Infectious Disease* 7:420–427.

Adams, E. N. 1972. Consensus techniques and the comparison of taxonomic trees. *Syst. Zool.* 21:390–397.

Aho, A. V., Sagiv, Y., Szymanski, T. G., and Ullman, J. D. 1981. Inferring a tree from lowest common ancestors with an application to the optimization of relational expressions. *SIAM J. Comput.* 10:405–421.

Akaike, H. 1974. A new look at the statistical model identification. *IEEE Trans. Autom. Contr. ACM* 19:716–723.

Allen, B. and Steel, M. 2001. Subtree transfer operations and their induced metrics on evolutionary trees. *Annals of Combinatorics* 5:1–13.

Aristotle 360BCE. History of animals. http://classics.mit.edu/Aristotle. Translated by D'Arcy Wentworth Thompson.

Bachman, P. 1894. Die analytische Zahlentheorie. Teubner, Leipzig.

Bader, D. A., Moret, B. M. E., Warnow, T., Wyman, S. K., Yan, M., Tang, J., Siepel, A. C., and Caprara, A. 2002. GRAPPA, version 2.0. http://www.cs.unm.edu/moret/grappa. Technical report, University of New Mexico.

Baker, R. H. and DeSalle, R. 1997. Multiple sources of character information and the phylogeny of Hawaiian drosophilids. *Syst. Biol.* 46:654–673.

Bandelt, H. J. 1990. Recognition of tree metrics. *SIAM J. Discrete Math.* 3:1–6.

Bandelt, H. J. and Dress, A. W. M. 1986. Reconstructing the shape of a tree from observed dissimilarity data. *Adv. Appl. Math.* 7:309–343.

Bandelt, H. J. and Dress, A. W. M. 1992a. A canonical decomposition theory for metrics on a finite set. *Adv. Math.* 92:47–105.

Systematics: A Course of Lectures, First Edition. Ward C. Wheeler.
© 2012 Ward C. Wheeler. Published 2012 by Blackwell Publishing Ltd.

Bandelt, H. J. and Dress, A. W. M. 1992b. Split decomposition: A new and useful approach to phylogenetic analysis of distance data. *Mol. Phylo. Evol.* 1:242–252.

Barry, D. and Hartigan, J. 1987. Statistical analysis of hominid molecular evolution. *Stat. Sci.* 2:191–210.

Bartley, W. W. 1976. The philosophy of Karl Popper. *Philosophia* 6:3–4.

Batzoglou, S. 2005. The many faces of sequence alignment. *Brief. Bioinform.* 6:6–22.

Baum, B. R. 1992. Combining trees as a way of combining data sets for phylogenetic inference, and the desirability of combining gene trees. *Taxon* 41:3–10.

Baum, L. E., Petrie, T., Soules, G., and Weiss, N. 1970. A maximization technique occurring in the statistical analysis of probabilistic functions of Markov chains. *Ann. Math. Statist.* 41:164–171.

Bayes, T. 1763. An essay towards solving a problem in the doctrine of chances. *Philos. T. Roy. Soc.* 53:370–418.

Bell, C. D. and Donoghue, M. J. 2005. Dating the Dipsicales: comparing models, genes, and evolutionary implications. *Amer. Jour. Botany* 92:284–296.

Bellman, R. E. 1953. An introduction to the theory of dynamic programming. Technical Report R-245, The RAND Corporation.

Belon, P. 1555. L'Histoire de la nature des oyseaux. Chez Guillaume Cauellat, Paris.

Bininda-Emonds, O. R., Beck, R. M., and Purvis, A. 2005. Getting to the roots of matrix representation. *Syst. Biol.* 54:668–672.

Bininda-Emonds, O. R. P. and Bryant, H. N. 1998. Properties of matrix representation with parsimony analyses. *Syst. Biol.* 47:497–508.

Bininda-Emonds, O. R. P., Jeffrey, J. E., Coates, M. I., and Richardson, M. K. 2002. From Haeckel to even-pairing: the evolution of developmental sequences. *Theor. Biol.* 121:297–320.

Bock, W. J. 1974. Philosophical foundations of classical evolutionary classification. *Syst. Zool.* 22:375–392.

Bock, W. J. 1977. Foundations and methods in evolutionary classification, pp. 851–895. *In* M. K. Hecht, P. C. Goody, and B. M. Hecht (eds.), Major Patterns in Vertebrate evolution. Plenum Press, New York.

Bonet, M. L., St. John, K., Mahindru, R., and Amenta, N. 2006. Approximating subtree distances between phylogenies. *J. Comput. Biol.* 13:1419–1434.

Boore, J. L. 2000. The duplication/random loss model for gene rearrangement exemplified by mitochondrial genomes of deuterostome animals, pp. 133–147. *In* D. Sankoff and J. Nadeau (eds.), Comparative genomics, computational biology series, volume 1. Kluwer Academic Publishers, Dordrecht, Netherlands.

Boore, J. L., Collins, T. M., Stanton, D., Daehler, L. L., and Brown, W. M. 1995. Deducing the pattern of arthropod phylogeny from mitochondrial DNA rearrangements. *Nature* 376:163 165.

Boore, J. L., Lavrov, D. V., and Brown, W. M. 1998. Gene translocation links insects and crustaceans. *Nature* 392:667–668.

Bordewich, M., Linz, S., St. John, K., and Semple, C. 2007. A reduction algorithm for computing the hybridization number of two trees. *Evolutionary Bioinformatics* 3:86–98.

Bordewich, M. and Semple, C. 2005. On the computational complexity of the rooted subtree prune and regraft distance. *Ann. Combinatorics* 8:409–423.

Boujenfa, K., Essoussi, N., and Limam, M. 2008. Comparison of phylogenetic trees of multiple protein sequence alignment methods. *World Acad. of Sci., Eng, and Tech.* 44:322–327.

Boyden, A. 1973. Perspectives in Zoology. Pergamon Press, Oxford.

Brady, R. H. 1985. On the independence of systematics. *Cladistics* 1:113–126.

Bremer, K. 1988. The limits of amino acid sequence data in angiosperm phylogenetic reconstruction. *Evolution* 42:795–803.

Bremer, K. 1990. Combinable component consensus. *Cladistics* 6:369–372.

Brent, R. P. 1973. Algorithms for Minimization without Derivatives, chapter 4. Prentice-Hall, Englewood Cliffs, NJ.

Bronn, H. G. 1858. Untersuchungen über die Entwickelungs-Gesetze der organischen Welt. E. Schweizerbart'sche Verlagshandlung, Stuttgart.

Bronn, H. G. 1861. Essai d'une réponse à la question de prix proposé en 1850 par l'académie des sciences pour le concours de 1853, et pius remise pour celui de 1856, savoir: Etudier les lois de la distribution des corps organisés fossiles dans les differents terrains sédimentaires, suivant l'ordre de leur superposition. *Comptes Rendu Hebdomadaire des Séances de l'Académie des Sciences, Supplement* 2:377–918.

Brundin, L. 1966. Transantarctic relationships and their significance, as evidenced by chironomid midges with a monograph of the subfamilies Podonomidae and Aphroteniinae and the austral Heptagyiae. *Kunliga Svenska Vetenskscapsacadamiens Handlingar* 11:1–472.

Bryant, D. and Waddell, P. 1998. Rapid evaluation of least-squares and minimum evolution criteria on phylogenetic trees. *Mol. Biol. Evol.* 15:1346–1359.

Buerki, S., Forest, F., and ans N. Alvarez, N. S. 2011. Comparative performance of supertree algorithms in large data sets using the soapberry family (Sapindaceae) as a case study. *Syst. Biol.* 60:32–44.

Buneman, P. 1971. The recovery of trees from measure of dissimilarity, pp. 387–395. *In* F. R. Hodson, D. G. Kendall, and P. Tautau (eds.), Mathematics and the archeological and historical sciences. Edinburgh University Press, Edinburgh.

Buneman, P. 1974. A note on metric properties on trees. *J. Comb. Theory (B)* 17:48–50.

Camin, J. H. and Sokal, R. R. 1965. A method for deducing branching sequences in phylogeny. *Evolution* 19:311–326.

Caprara, A. 1997. Sorting by reversals is difficult. *In* Proceedings of 1st Conference Computational Molecular Biology, Santa Fe NM, pp. 78–83, New York. ACM Press.

Carpenter, J. M. 1996. Uninformative bootstrapping. *Cladistics* 12:177–181.

Carrillo, H. and Lipman, D. 1988. The multiple sequence alignment problem in biology. *SIAM J. Appl. Math.* 48:1073–1082.

Cassis, G. and Schuh, R. T. 2009. Systematic methods, fossils, and relationships within Heteroptera (Insecta). *Cladistics* 25:1–19.

Cavalli-Sforza, L. L. and Edwards, A. W. F. 1967. Phylogenetic analysis: models and estimation procedures. *Evolution* 21:550–570.

Cerny, V. 1985. A thermodynamical approach to the traveling salesman problem: an efficient simulation algorithm. *J. Optimiz. Theory App.* 45:41–51.

Champion, R. 1985. The purpose of Popper. *Melbourne Age Monthly Review*.

Chang, J. T. 1996. Inconsistency of evolutionary tree topology reconstruction methods when substitution rates vary across characters. *Math. Bio.* 134:189–215.

Chase, M. W., Soltis, D. E., Olmstead, R. G., Morgan, D., Les, D. H., Mishler, B. D., Duvall, M. R., Price, R. A., Hills, H. G., Qiu, Y.-L., Kron, K. A., Rettig, J. H., Conti, E., Palmer, J. D., Manhart, J. R., Sytsma, K. J., Michaels, H. J., Kress, W. J., Karol, K. G., Clark, W. D., Hedren, M., Gaut, B. S., Jansen, R. K., Kim, K.-J., Wimpee, C. F., Smith, J. F., Furnier, G. R., Strauss, S. H., Xiang, Q.-Y., Plunkett, G. M., Soltis, P. S., Swensen, S. M., Willimas, S. E., Gadek, P. A., Quinn, C. J., Eguiarte, L. E., Golenberg, E., Learn, G. H. J., Graham, S. W., Barret, S. C. H., Dayanandan, S., and Albert, V. A. 1993. Phylogenetics of seed plants: An analysis of nucleotide sequences from the plastid gene RBCL. *Ann. Mol. Bot. Gard.* 80:528–580.

Chaudhuri, K., Chen, K., Mihaescu, R., and Rao, S. 2006. On the tandem duplication–random loss model of genome rearrangement. *In* SODA '06: Proceedings of the seventeenth annual ACM-SIAM symposium on Discrete algorithm, pp. 564–570, New York, NY, USA. ACM.

Chen, D., Diao, L., Eulenstein, O., Fenández-Baca, D., and Sanderson, M. J. 2003. Flipping: A supertree construction method, pp. 135–160. *In* M. Janowitz, F.-J. Lapointe, F. R. McMorris, B. Mirkin, and F. S. Roberts (eds.), Bioconsensus, DIMACS series in discrete mathematics and theoretical computer science. American Mathematical Society, Providence, Rhode Island.

Chib, S. and Greenberg, E. 1995. Understanding the Metropolis–Hastings algorithm. *American Statistician* 49:327–335.

Church, A. 1936a. A note on the Entscheidungsproblem. *J. Symbolic Logic* 1:40–41.

Church, A. 1936b. An unsolvable problem of elementary number theory. *Am. J. Math.* 58:345–363.

Church, A. and Rosser, J. B. 1936. Some properties of conversion. *T. Am. Math. Soc.* 39:472–482.

Coddington, J. and Scharff, N. 1994. Problems with zero-length branches. *Cladistics* 10:415–423.

Cormode, G. and Muthukrishnan, S. 2002. The string edit distance matching problem with moves. *In* SODA '02: Proceedings of the thirteenth annual ACM-SIAM symposium on Discrete algorithms, pp. 667–676. Society for Industrial and Applied Mathematics.

Cracraft, J. 1983. Species concepts and speciation analysis. *Current Ornithology* pp. 159–187.

Cracraft, J. 1989. Speciation and its ontology: the empirical consequences of alternative species concepts for understanding patterns and processes of differentiation, pp. 28–59. *In* D. Otte and J. A. Endler (eds.), Speciation and its Consequences. Sinauer Associates, Sunderland, England.

Cracraft, J. 2000. Species concepts in theoretical and applied biology: a systematic debate with consequences, pp. 3–14. *In* Q. D. Wheeler and R. Meier (eds.), Species Concepts and Phylogenetic Theory. Columbia University Press, New York.

Cuvier, G. 1812. Recherches sur les ossemens fossiles de quadrupédes. Déterville, Paris.

Dalevi, D. and Eriksen, N. 2008. Expected gene-order distances and model selection in bacteria. *Bioinformatics* 24:1332–1338.

Darling, A. C., Mau, B., Blattner, F. R., and Perna, N. T. 2004. Mauve: Multiple alignment of conserved genomic sequence with rearrangements. *Genome Research* 14:1394–1403.

Darwin, C. 1859a. Letter to J. D. Hooker.

Darwin, C. R. 1859b. On the Origin of Species by means of natural selection, or the preservation of favoured races in the struggle for life. John Murray, London, 1st edition.

Day, W. H. E. 1987. Computational complexity of inferring phylogenies from dissimilarity matrices. *B. Math. Biol.* 49:461–467.

DeGroot, M. H. and Schervish, M. J. 2006. Probability and Statistics. Addison-Wesley, Boston, 3rd edition.

DePinna, M. G. 1991. Concepts and tests of homology in the cladistic paradigm. *Cladistics* 7:367–394.

Desper, R. and Gascuel, O. 2002. Fast and accurate phylogeny reconstruction algorithms based on the minimum-evolution principle. *J. Comp. Biol.* 9:687–705.

Desper, R. and Gascuel, O. 2007. The minimum evolution distance-based approach to phylogenetic inference, pp. 1–32. *In* O. Gascuel (ed.), Mathematics of Evolution and Phylogeny. Oxford University Press, Oxford.

Dobzhansky, T. 1937. Genetics and the Origin of Species. Columbia University Press, New York.

Donoghue, M. J. 1985. A critique of the biological species concept and reccomendations for a phylogenetic alternative. *Bryologist* 88:172–181.

Donoghue, M. J. and Kadereit, J. W. 1992. Walter Zimmerman and the growth of phylogenetic theory. *Syst. Biol.* 41:74–84.

Donoghue, M. J., Olmstead, R. G., Smith, J. F., and Palmer, J. D. 1992. Phylogenetic relationships of Dipscales based on RBCL sequences. *Ann. Mo. Bot. Gard.* 79:333–345.

Dress, A., Huber, K. T., and Moulton, V. 2007. Some uses of the Farris transform in mathematics and phylogenetics. *Annals of Combinatorics* 11:1–37.

Dress, A. and Krüger, M. 1987. Parsimonious phylogenetic trees in metric spaces and simulated annealing. *Adv. App. Math.* 8:8–37.

Drummon, A. J., Ho, S. Y. W., Phillips, M. J., and Rambaut, A. 2006. Relaxed phylogenetics and dating with confidence. *PLoS Biology* 4:699–710.

Drummon, A. J. and Rambaut, A. 2007. Beast: Bayesian evolutionary analysis by sampling trees. *BMC Evolutionary Biology* 7:214.

Dunn, C. W., Hejnöl, A., Matus, D. Q., Pang, K., Browne, W. E., Smith, S. A., Seaver, E., Rouse, G. W., Obst, M., Edgecombe, G. D., Sørensen, M. V., Haddock, S. H., Schmidt-Rhaesa, A., Okusu, A., Kristensen, R. M., Wheeler, W. C., Martindale, M. Q., and Giribet, G. 2008. Broad phylogenomic sampling improves resolution of the animal tree of life. *Nature* 452:745–749.

Edgar, R. C. 2004a. MUSCLE: a multiple sequence alignment method with reduced time and space complexity. *BMC Bioinformatics* 5:113–132.

Edgar, R. C. 2004b. MUSCLE: multiple sequence alignment with high accuracy and high throughput. *Nucleic Acids Res.* 32:1792–1797.

Edgar, R. C. and Sjölander, K. 2003. SATCHMO: sequence alignment and tree construction using hidden Markov models. *Bioinformatics* 19:1404–1411.

Edwards, A. 1972. Likelihood. Cambridge University Press, Cambridge, England.

Edwards, A. W. F. 1970. Estimation of the branching points of a branching diffusion process. *J. Royal Stat. Soc. B* 32:155–174.

Efron, B. 1979. Bootstrap methods: another look at the jackknife. *Ann. Stat.* 7:1–26.

Efron, B. and Tibshirani, R. 1993. An Introduction to the Bootstrap. Chapman and Hall, Boca Raton, Florida.

Eldredge, N. and Cracraft, J. 1980. Phylogenetic Patterns and the Evolutionary Process. Columbia University Press, New York.

Erdös, P. L., Steel, M. A., Székely, L., and Warnow, T. J. 1997. Constructing big trees from short sequences. *In* P. Degano, R. Gorrier, and A. Marchetti-Spaccalmela (eds.), Automata, Languages and Programming, 24th International Colloquium, ICALP'97, Bologna, Italy, pp. 827–837, Berlin. Springer-Verlag.

Estabrook, G. F. 1978. Some concepts for the estimation of evolutionary relationships in systematic botany. *Syst. Bot.* 3:146–158.

Faith, D. P. 1991. Cladistic permutation tests for monophyly and nonmonophyly. *Syst. Zool.* 40:366–375.

Farris, J. S. 1969. On the cophenetic correlation coefficient. *Syst. Zool.* 18:279–285.

Farris, J. S. 1970. A method for computing Wagner trees. *Syst. Zool.* 19:83–92.

Farris, J. S. 1973a. A probability model for inferring evolutionary trees. *Syst. Zool.* 22:250–256.

Farris, J. S. 1981. Distance data in phylogenetic analysis, pp. 3–23. *In* V. A. Funk and D. R. Brooks (eds.), Advances in Cladistics, volume 1. New York Botanical Garden, New York.

Farris, J. S. 1983. The logical basis of phylogenetic analysis. *In* N. I. Platnick and V. A. Funk (eds.), Advances in cladistics, Proceedings of the second meeting of the Willi Hennig Society, volume 2, pp. 7–36, New York. Columbia University Press.

Farris, J. S. 1988. Hennig86. Program and documentation distributed by the author.

Farris, J. S., Albert, V. A., Källersjö, M., Lipscomb, D., and Kluge, A. G. 1996. Parsimony jackknifing outperforms neighbor-joining. *Cladistics* 12:99–124.

Farris, J. S., Kluge, A. G., and Eckardt, M. J. 1970. A numerical approach to phylogenetic systematics. *Syst. Zool.* 19:172–189.

Farris, S. J. 1972. Estimating phylogenetic trees from distance matrices. *Am. Nat.* 106:645–668.

Farris, S. J. 1973b. On comparing the shapes of taxonomic trees. *Syst. Zool.* 22:50–54.

Farris, S. J. 1974. Formal definitions of paraphyly and polyphyly. *Syst. Zool.* 23:548–554.

Farris, S. J. 1977. Phylogenetic analysis under Dollo's law. *Syst. Zool.* 26:577–88.

Felsenstein, J. 1973. Maximum likelihood and minimum-steps methods for estimating evolutionary trees from data on discrete characters. *Syst. Zool.* 22:240–249.

Felsenstein, J. 1978. Cases in which parsimony or compatibility methods will be positively misleading. *Syst. Zool.* 27:401–410.

Felsenstein, J. 1981. Evolutionary trees from DNA sequences: a maximum likelihood approach. *J. Mol. Evol.* 17:368–376.

Felsenstein, J. 1984. Distance methods for inferring phylogenies: A justification. *Evolution* 38:16–24.

Felsenstein, J. 1985. Confidence limits on phylogenies: An approach using the bootstrap. *Evolution* 39:783–791.

Felsenstein, J. 1993. PHYLIP: Phylogeny Inference Package.

Felsenstein, J. 2004. Inferring Phylogenies. Sinauer Associates, Sunderland, MA.

Feng, D.-F. and Doolittle, R. F. 1987. Progressive sequence alignment as a prerequisite to correct phylogenetic trees. *J. Mol. Evol.* 25:351–360.

Feyerabend, P. K. 1975. Against Method. Verso, London.

Feyerabend, P. K. 1987. Farewell to Reason. Verso, London.

Finden, C. and Gordon, A. 1985. Obtaining common pruned trees. *J. Classification* 2:255–276.

Fisher, R. A. 1912. On an absolute criterion for fitting frequency curves. *Mess. Math.* 41:155–160.

Fisher, R. A. 1950. Statistical methods for research workers. Oliver and Boyd, Edinburgh, 11th edition.

Fitch, W. M. 1970. Distinguishing homologous from analogous proteins. *Syst. Zool.* 19:99–113.

Fitch, W. M. 1971. Toward defining the course of evolution: minimum change for a specific tree topology. *Syst. Zool.* 20:406–416.

Fitch, W. M. and Margoliash, E. 1967. Construction of phylogenetic trees. *Science* 155:279–284.

Fleissner, R., Metzler, D., and von Haeseler, R. 2005. Simultaneous statistical multiple alignment and phylogeny reconstruction. *Syst. Biol.* 54:548–561.

Foulds, L. R. and Graham, R. L. 1982. The Steiner problem in phylogeny is NP-complete. *Adv. Appl. Math.* 3:43–49.

Fraser, A. and Burnell, D. 1970. Computer Models in Genetics. McGraw-Hill, New York.

Freudenstein, J. V., Pickett, K. M., Simmons, M. P., and Wenzel, J. W. 2003. From basepairs to birdsongs: phylogenetic data in the age of genomics. *Cladistics* 19:333–347.

Frost, D. R., Grant, T., Faivovich, J., Haas, A., Haddad, C. F. B., Bain, R., de Sá, R. O., Donnellan, S. C., Raxworthy, C. J., Wilkinson, M., Channing, A., Campbell, J. A., Blotto, B. L., Moler, P., Drewes, R. C., Nussbaum, R. A., Lynch, J. D., Green, D., and Wheeler., W. C. 2006. The amphibian tree of life. *B. Am. Mus. Nat. Hist.* 297:1–371.

Gaffney, E. S. 1979. An introduction to the logic of phylogeny reconstruction, pp. 80–111. *In* J. Cracraft and N. Eldredge (eds.), Phylogenetic analysis and paleontology. Columbia University Press, New York.

Galton, F. 1889. Natural Inheritance. Macmillan, London.

Gardner, P. P. and Giegerich, R. 2004. A comprehensive comparison of comparative RNA structure prediction approaches. *BMC Bioinformatics* 5:140.

Gascuel, O. 1997. Concerning the NJ algorithm and its unweighted version UNJ, pp. 149–170. *In* B. Mirkin, F. S. Roberts, and F. R. McMorris (eds.), Mathematical Hierarchies and Biology. American Mathematical Society, Providence, RI.

Gascuel, O., Bryant, D., and Denis, F. 2001. Strengths and limitations of the minimum evolution principle. *Syst. Biol.* 50:621–627.

Gatesy, J., Baker, R., and Hayashi, C. 2004. Inconsistencies in arguments for the supertree approach: Supermatrices versus supertrees of Crocodylia. *Syst. Biol.* 53:342–355.

Gaut, B. S. and Lewis, P. O. 1995. Success of maximum likelihood phylogeny inference in the four-taxon case. *Mol. Biol. Evol.* 12:152–162.

Gelman, S. and Gelman, G. D. 1984. Stochastic relaxation, Gibbs distributions and the Bayes restoration of images. *IEEE Trans. Pattern Anal. Mach. Intel* 6:721–741.

Georges Louis Leclerc, c. d. B. 1749–1778. Histoire naturelle, générale et particuliére. Paris.

Geyer, C. J. 1991. Markov chain Monte Carlo maximum likelihood, pp. 156–163. *In* E. M. Keramidas (ed.), Computing Science and Statistics: Proc 23rd Symp. Interface. Interface Foundation, Fairfax Station, VA.

Ghiselin, M. T. 1966. On psychologism in the logic of taxonomic controversies. *Syst. Zool.* 15:207–215.

Ghiselin, M. T. 1969. The Triumph of the Darwinian Method. University of California Press, Berkeley and London.

Ghiselin, M. T. 1974. A radical solution to the species problem. *Syst. Zool.* 23:536–544.

Gilbert, E. N. and Pollak, H. O. 1968. Steiner minimal trees. *SIAM J. Applied Math.* 16:1–29.

Gingerich, P. and Schoeninger, M. 1977. The fossil record and primate phylogeny. *J. Human evol.* 6:485–505.

Gingerich, P. D. 1984. Primate evolution: evidence from the fossil record, comparative morphology, and molecular biology. *Yearbook of Physical Anthropology* 27:57–72.

Giribet, G. 2001. Exploring the behavior of POY, a program for direct optimization of molecular data, pp. S60–S70. *In* One day symposium in numerical Cladistics, volume 17. Elsevier.

Giribet, G. 2003. Molecules, development and fossils in the study of metazoan evolution; Articulata versus Ecdysozoa revisited. *Zoology* 106:303–326.

Giribet, G. 2007. Efficient tree searches with available algorithms. *Evolutionary Bioinformatics* 3:341–356.

Giribet, G., DeSalle, R., and Wheeler, W. C. 2002. Pluralism and the aims of phylogenetic research, pp. 141–146. *In* G. Giribet, R. DeSalle, and W. C. Wheeler (eds.), Molecular Systematics and Evolution: Theory and Practice. Birkhäuser Verlag, Basel.

Giribet, G. and Edgecombe, G. D. 2006. Conflict between datasets and phylogeny of centipedes: an analysis based on seven genes and morphology. *Proc. Roy. Soc. B* 273:531–538.

Giribet, G., Edgecombe, G. D., Carpenter, J. M., d'Haese, C., and Wheeler, W. C. 2004. Is Ellipura monophyletic? A combined analysis of basal hexapod relationships with emphasis on the origin of insects. *Organisms, Diversity, and Evolution* 4:319–340.

Giribet, G., Edgecombe, G. D., and Wheeler, W. C. 2001. Arthropod phylogeny based on eight molecular loci and morphology. *Nature* 413:157–161.

Giribet, G. and Wheeler, W. 1999. The position of arthropods in the animal kingdom; Ecdysozoa, islands, trees and the 'parsimony ratchet'. *Mol. Phyl. Evol.* 13:619–623.

Giribet, G. and Wheeler, W. C. 2007. The case for sensitivity: a response to Grant and Kluge. *Cladistics* 23:1–3.

Gödel, K. 1931. über formal unentscheidbare Sätze der Principia Mathematics und verwandter Systeme. *Monatshefte für Mathematik und Physik* 38:173–198.

Goldberg, D. E., Deb, K., and Horn, J. 1989. Genetic algorithms. *In* Search, Optimization, and Machine Learning. Addison-Wesley, Boston, MA.

Goldman, N. 1990. Maximum likelihood inference of phylogenetic trees, with special reference to a poisson process model of DNA substitution and to parsimony analysis. *Syst. Zool.* 39:345–361.

Goloboff, P. 1999a. Analyzing large data sets in reasonable times: solutions for composite optima. *Cladistics* 15:415–428.

Goloboff, P., Farris, S., and Nixon, K. 2003. TNT (Tree analysis using New Technology) version 1.0 ver. beta test v. 0.2. Program and documentation available at http://www.zmuc.dk/public/phylogeny/tnt. Published by the authors. Tucumán, Argentina.

Goloboff, P. A. 1993a. Character optimization and calculation of tree lengths. *Cladistics* 9:433–436.

Goloboff, P. A. 1993b. Estimating character weights during tree search. *Cladistics* 9:83–91.

Goloboff, P. A. 1999b. NONA (No Name) ver. 2. Published by the author. Tucumán, Argentina.

Goloboff, P. A. 2002. Techniques for analysing large data sets, pp. 70–79. *In* R. DeSalle, G. Giribet, and W. Wheeler (eds.), Techniques in Molecular Systematics and Evolution. Birkhäuser Verlag, Basel.

Goloboff, P. A. 2005. Minority rule supertrees? MRP, compatibility, and minimum flip may display the least frequent groups. *Cladistics* 21:282–294.

Goloboff, P. A. 2008. Calculating SPR distances between trees. *Cladistics* 24:591–597.

Goloboff, P. A., Catalano, S. A., Mirande, J. M., Szumik, C. A., Arias, J. S., Källersjö, M., and Farris, J. S. 2009. Phylogenetic analysis of 73 060 taxa corroborates major eukaryotic groups. *Cladistics* 25:211–230.

Goloboff, P. A. and Farris, J. S. 2001. Methods for quick consensus estimation. *Cladistics* 17:S26–S34.

Goloboff, P. A., Mattoni, C. I., and Quinteros, A. S. 2006. Continuous characters analyzed as such. *Cladistics* 22:589–601.

Goloboff, P. A. and Pol, D. 2002. Semi-strict supertrees. *Cladistics* 18:514–525.

Goloboff, P. A. and Pol, D. 2005. Parsimony and Bayesian phylogenetics, pp. 148–159. *In* V. A. Albert (ed.), Parsimony, Phylogeny, and Genomics. Oxford University Press, Oxford.

Goloboff, P. A. and Pol, D. 2007. On divide-and-conquer strategies for parsimony analysis of large data sets: Rec-I-DCM3 versus TNT. *Syst. Biol.* 56:485–495.

Goodman, M., Olson, C. B., Beeber, J. E., and Czelusniak, J. 1982. New perspectives in the molecular biological analysis of mammalian phylogeny. *Acta Zoologica Fennica* 169:19–35.

Gordon, A. D. 1980. On the assessment and comparison of classifications. *In* R. Tomassonne (ed.), Analyse de Donniers et Informatique, pp. 149–160, Le Chesney. INRIA.

Gordon, A. D. 1986. Consensus supertrees: The synthesis of rooted trees containing overlapping sets of labeled leaves. *J. Classification* 3:335–348.

Gornall, R. J. 1997. Practical aspects of the species concept in plants, pp. 171–190. *In* H. A. Dawah and M. R. Wilson (eds.), Species: The units of biodiversity. Chapman and Hall, London.

Gorodkin, J., Heyer, L. J., and Stormo, G. D. 1997. Finding the most significant common sequence and structure motifs in a set of RNA sequences. *Nucl. Acid. Res.* 25:3724–3732.

Gotoh, O. 1982. An improved algorithm for matching biological sequences. *J. Mol. Biol.* 162:705–708.

R. Graham, J. K. Lenstra, and R. R. Tarjan (eds.) 1985. The traveling salesman problem. Wiley Interscience series in discrete mathematics. John Wiley and Sons, New York.

Grant, T. and Kluge, A. G. 2003. Data exploration in phylogenetic inference: Scientific, heuristic, or neither. *Cladistics* 19:379–418.

Grant, T. and Kluge, A. G. 2005. Stability, sensitivity, science, and heurism. *Cladistics* 21:597–604.

Grant, T. and Kluge, A. G. 2007. Ratio of explanatory power (REP): A new measure of group support. *Mol. Phyl. and Evol.* 44:483–487.

Grant, T. and Kluge, A. G. 2008a. Clade support measures and their adequacy. *Cladistics* 24:1051–1064.

Grant, T. and Kluge, A. G. 2008b. Credit where credit is due: The Goodman–Bremer support metric. *Mol. Phyl. and Evol.* 49:405–406.

Graur, D. and Martin, W. 2004. Reading the entrails of chickens: molecular timescales of evolution and the illusion of precision. *TRENDS in Genetics* 20:80–86.

Grimaldi, D. A. and Engel, M. S. 2008. An unusual, primitive Piesmatidae (Insecta: Heteroptera) in Cretaceous amber from Myanmar (Burma). *Amer. Mus. Novit.* 3611.

Gusfield, D. 1997. Algorithms on Strings, Trees, and Sequences: Computer Science and Computational Biology. Cambridge University Press, Cambridge.

Haas, O. and Simpson, G. G. 1946. Analysis of some phylogenetic terms, with attempts at redefinition. *Proc. Amer. Phil. Soc.* 90:319–349.

Haeckel, E. 1866. Generelle Morphologie der Organismen: Allgemeine Grundzüger der orgaischen Formen-Wissenschaft, mechanisch begründet durch die von C. Darwin reformirte Decendenz-Theorie. G. Reimer, Berlin.

Haeckel, E. 1868. Natüliche Schöpsungsgeschichte. Reimer, Berlin.

Haeckel, E. 1876. The history of creation, or, The development of the earth and its inhabitants by the action of natural causes: doctrine of evolution in general, and of that of Darwin. Henry S. King, London, (translation revised by E. Ray Lankester).

Hamming, R. W. 1950. Error detecting and error correcting codes. *Bell System Technical Journal* 26:147–160.

Hannenhalli, S. and Pezvner, P. A. 1995. Transforming men into mice (polynomial algorithms for genomic distance problem). *In* Proceedings of the 36th Annual IEEE Symposium on the Theory of Computing, pp. 581–592, Milwaukee, Wisconsin.

Hardison, R. C. 2008. Globin genes on the move. *J. Biol.* 7:35.

Harper, C. W. 1976. Phylogenetic inference in paleontology. *J. Paleontol.* 50:180–193.

Harper, C. W. 1979. A Bayesian probability view of phylogenetic systematics. *Syst. Zool.* 28:547–553.

Harshman, J. 1994. The effect of irrelevant characters on bootstrap values. *Syst. Biol.* 43:419–424.

Hasegawa, M., Kashina, H., and Yano, T. 1985. Dating the human-ape splitting by a molecular clock of mitochondrial DNA. *J. Mol. Evol.* 22:160–174.

Hasegawa, M., Kishino, H., and Yano, T. 1989. Estimation of branching dates among primates by molecular clocks of nuclear DNA which slowed down in Hominoidea. *J. Hum. Evol.* 18:461–476.

Hastings, W. K. 1970. Monte Carlo sampling methods using Markov chains and their applications. *Biometrika* 57:97–109.

Hecht, M. K. and Edwards, J. L. 1977. The methodology of phylogenetic inference above the species level, pp. 3–51. *In* M. K. Hecht, P. C. Goody, and B. M. Hecht (eds.), Major Patterns in Vertebrate Evolution. Plenum Press, New York.

Hedges, S. B., Parker, P. H., Sibley, C. G., and Kumar, S. 1996. Continental breakup and the ordinal divergence of birds and mammals. *Nature* 381:226–229.

Hein, J. 1989a. A new method that simultaneously aligns and reconstructs ancestral sequences for any number of homologous sequences, when the phylogeny is given. *Mol. Biol. Evol.* 6:649–668.

Hein, J. 1989b. A tree reconstruction method that is economical in the number of pairwise comparisons used. *Mol. Biol. Evol.* 6:669–684.

Hein, J. C., Jensen, J. L., and Pedersen, C. N. S. 2003. Recursions for statistical multiple alignment. *PNAS* 100:14960–14965.

Hejnöl, A., Obst, M., Stamatakis, A., Ott, M., Rouse, G. W., Edgecombe, G. D., Martinez, P., Baguna, J., Bailly, X., Jondelius, U., Wiens, M., Müller, W. E. G., Seaver, E., Wheeler, W. C., Martindale, M. Q., Giribet, G., and Dunn, C. W. 2009. Assessing the root of bilaterian animals with scalable phylogenomic methods. *Proc. R. Soc. B* 22:4261–4270.

Hendy, M. D. and Penny, D. 1982. Branch and bound algorithms to determine minimal evolutionary trees. *Mathematical Biosciences* 60:133–142.

Hennig, W. 1950. Grundzüge einer Theorie der Phylogenetischen Systematik. Deutcher Zentralverlag, Berlin.

Hennig, W. 1965. Phylogenetic systematics. *Ann. Rev. Entomol.* 10:97–116.

Hennig, W. 1966. Phylogenetic Systematics. University of Illinois Press, Urbana.

Hey, J. 1992. Using phylogenetic trees to study speciation and extinction. *Evolution* 46:627–640.

Higgins, D. G. and Sharp, P. M. 1988. Clustal: A package for performing multiple sequence alignment on a microcomputer. *Gene* 73:237–244.

Hillis, D. M., Bull, J. J., White, M. E., Badgett, M. R., and Molineux, I. J. 1992. Experimental phylogenetics: Generation of a known phylogeny. *Science* 255:589–592.

Hillis, D. M. and Bull, J. T. 1993. An empirical test of bootstrapping as a method for assessing confidence in phylogenetic analysis. *Syst. Biol.* 42:182–192.

Hirosawa, M., Totoki, Y., Hoshida, M., and Ishikawa, M. 1995. Comprehensive study on iterative algorithms of multiple sequence alignment. *Comput. Appl. Biosci.* 11:13–18.

Hofacker, I. L., Bernhart, S. H. F., and Stadler, P. F. 2004. Alignment of RNA base pairing probability matrices. *Bioinfo.* 20:2222–2227.

J. H. Holland (ed.) 1975. Adaptation in Natural and Artificial Systems. University of Michigan Press, Ann Arbor, Michigan.

Holmes, I. and Bruno, W. J. 2001. Evolutionary HMMs: a Bayesian approach to multiple alignment. *Bioinfo.* 17:803–820.

N. Houser, D. D. Roberts, and J. V. Evra (eds.) 1997. Studies in the Logic of Charles Sanders Peirce. Indiana University Press, Bloomington, IN.

Hromkovic, J. 2004. Theoretical Computer Science, Introduction to Automata, Computability, Complexity, Algorithmics, Randomization, Communication, and Cryptography. Springer-Verlag, Berlin.

Huelsenbeck, J. P. and Ronquist, F. 2003. MrBayes: Bayesian inference of phylogeny, 3.0 edition. Program and documentation available at http://morphbank.uuse/mrbayes/.

Hughey, R. and Krogh, A. 1996. Hidden Markov models for sequence analysis: extensions and analysis of the basic method. *Comput. Appl. Biosci.* 12:95–107.

Hull, D. L. 1976. Are species really individuals? *Syst. Zool.* 25:174–191.

Hull, D. L. 1978. A matter of individuality. *Philosophy of Science* 45:335–360.

Hull, D. L. 1988. Science as a Process: An Evolutionary Account of the Social and Conceptual Development of Science. University of Chicago Press, Chicago.

Hull, D. L. 1997. The ideal species concept–and why we can't get it, pp. 357–380. *In* M. F. Claridge, H. A. Dawah, and M. R. Wilson (eds.), Species: The units of biodiversity. Chapman and Hall, London.

Hume, D. 1748. An Enquiry concerning Human Understanding, T. Cadell, London.

Huson, D., Nettles, S., and Warnow, T. 1999. Disk-covering, a fast converging method for phylogenetic tree reconstruction. *J. Comput. Biol.* 6:368–386.

Huson, D. H. and Steel, M. 2004. Distances that perfectly mislead. *Syst. Biol.* 53:327–332.

J. Huxley (ed.) 1940. The new systematics. Oxford University Press, London. 583 pp.

Huxley, J. S. 1959. Grades and clades, pp. 21–22. *In* A. J. Cain (ed.), Function and taxonomic importance, number 2. Systematics Association, London.

Huxley, T. H. 1863. The structure and classification of the Mammalia. Hunterian Lectures. *Medical Times and Gazette*, Volume 1, pp. 607.

Jardin, N. 1970. The observational and theoretical components of homology: a study based on the morphology of the dermal skull-roofs of rhipidistian fishes. *Biol. J. Linn. Soc.* 1:327–361.

Jensen, J. L. and Hein, J. 2005. Gibbs sampler for statistical multiple alignment. *Statistica Sinica* 15:889–907.

Jin, G., Nakhleh, L., Snir, S., and Tuller, T. 2006. Maximum likelihood of phylogenetic networks. *Bioinformatics* 22:2604–2611.

Jukes, T. H. and Cantor, C. R. 1969. Evolution of protein molecules, pp. 21–132. *In* N. H. Munro (ed.), Mammalian Protein Metabolism. Academic Press, New York.

Karp, R. 1972. Reducibility among combinatorial problems, pp. 85–104. *In* R. Miller (ed.), Complexity of Computer Computation. Plenum Press, New York.

Kass, R. E. and Raftery, A. E. 1995. Bayes factors. *Journal of the American Statistical Association* 90:773–795.

Katoh, K., Misawa, K., Kuma, K., and Miyata, T. 2002. MAFFT: a novel method for rapid multiple sequence alignment based on fast fourier transform. *Nucl. Acid. Res.* 30:3059–3066.

Kendall, M. and Stuart, A. 1973. The advanced theory of statistics, volume 2nd. Haffner, New York, 3rd edition.

Kidd, K. K. and Sgaramella-Zonta, L. A. 1971. Phylogenetic analysis: concepts and methods. *Am. J. Hum. Genet.* 23:235–252.

Kim, J. 1998. Large-scale phylogenies and measuring the performance of phylogenetic estimators. *Syst. Biol.* 47:43–60.

Kimura, M. 1983. The neutral theory of molecular evolution. Cambridge University Press, Cambridge.

Kirkpatrick, S., Gelatt, C. D., and Vecchi, M. P. 1983. Optimization by simulated annealing. *Science* 220:671–680.

Kishino, H. and Hasegawa, M. 1989. Evaluation of the maximum likelihood estimate of the evolutionary tree topologies from DNA sequence data, and the branching order in Hominoidea. *J. Mol. Evol.* 29:170–179.

Kishino, H., Thorne, J. L., and Bruno, W. J. 2001. Performance of divergence time estimation method under a probablistic model of rate evolution. *Mol. Biol. Evol.* 18:352–361.

Kjer, K. M. 2004. Aligned 18S and insect phylogeny. *Syst. Biol.* 53:506–514.

Kleene, S. C. 1936. General recursive functions of natural numbers. *Mathematische Annalen* 112:727–742.

Kluge, A. G. 1989. A concern for evidence and a phylogenetic hypothesis of relationships among Epicrates (Boidae, Serpentes). *Syst. Zool.* 38:7–25.

Kluge, A. G. 1997. Testability and the refutation and corroboration of cladistic hypotheses. *Cladistics* 13:81–96.

Kluge, A. G. 1998. Sophisticated falsification and research cycles: Consequences for differential character weighting in phylogenetic systematics. *Zoologica Scripta* 26:349–360.

Kluge, A. G. 2005. What is the rationale for 'Ockham's razor' (a.k.a. parsimony) in phylogenetic analysis, pp. 15–42. *In* V. Albert (ed.), Parsimony, Phylogeny, and Genomics. Oxford University Press, Great Britain.

Kluge, A. G. 2009. Explanation and falsification in phylogenetic inference: Exercises in Popperian philosophy. *Acta Biotheoretica* 57:171–186.

Knuth, D. 1973. The Art of Computer Programming. Addison-Wesley, Reading, MA.

Kolaczkowski, B. and Thornton, J. W. 2004. Performance of maximum parsimony and likelihood phylogenetics when evolution is heterogeneous. *Nature* 431:980–984.

Kolaczkowski, B. and Thornton, J. W. 2009. Long-branch attraction bias and inconsistency in Bayesian phylogenetics. *PLOS One* 4:e7891.

Kornet, D. J. 1993. Permanent splits as speciation events: a formal reconstruction of the internodal species concept. *J. Theor. Biol.* 164:407–435.

Krivánek, M. 1986. Computing the nearest neighbor interchange metric for unlabeled binary trees is NP-Complete. *J. Classification* 3:55–60.

Krogh, A., Brown, M., Mian, I. S., Sjölander, K., and Haussler, D. 1994. Hidden Markov models in computational biology: applications to protein modeling. *J. Mol. Biol.* 235:1501–1531.

Kruskal, J. B. 1956. On the shortest spanning subtree of a graph and the traveling salesman problem. *Proc. Amer. Math. Soc.* 7:48–50.

Kuhn, T. S. 1962. The Structure of Scientific Revolutions. University of Chicago Press, Chicago.

Kupczok, A., Schmidt, H. A., and von Haeseler, A. 2010. Accuracy of phylogeny reconstruction methods combining overlapping gene data sets. *Algorithms for Molecular Biology* 5:37.

Lakatos, I. 1970. Criticism and the Growth of Knowledge. Cambridge University Press, Cambridge.

Lamarck, J.-B. 1809. Philosophie zoologique ou exposition des considérations relatives á l'histoire naturelle des animaux.

Lanave, C., Preparata, G., Saccone, C., and Serio, G. 1984. A new method for calculating evolutionary substitution rates. *J. Mol. Evol.* 20:86–93.

Land, A. H. and Doig, A. G. 1960. An automatic method of solving discrete programming problems. *Econometrica* 28:497–520.

Langley, C. H. and Fitch, W. M. 1974. An examination of the constancy of the rate of molecular evolution. *J. Mol. Evol.* 3:161–177.

Lankester, E. R. 1870a. On the use of the term "homology". *Annals and Magazine of Natural History, Zoology, Botany, and Geology* 6:342.

Lankester, E. R. 1870b. On the use of the term homology in modern zoology, and the distinction between homogenetic and homoplastic agreements. *Annals and Magazine of Natural History, Zoology, Botany, and Geology* 6:34–43.

Lapointe, F.-J. and Cucumel, G. 1997. The average consensus procedure: Combination of weighted trees containing identical or overlapping sets of taxa. *Syst. Biol.* 46:306–312.

Larget, B., Kadane, J. B., and Simon, D. L. 2005. A Bayesian approach to the estimation of ancestral genome arrangements. *Mol. Phyl. Evol.* 36:214–223.

Larget, B. and Simon, D. L. 1999. Markov chain Monte Carlo algorithms for the Bayesian analysis of phylogenetic trees. *Mol. Biol. Evol.* 16:750–759.

Larget, B., Simon, D. L., Kadane, J. B., and Sweet, D. 2004. A Bayesian analysis of metazoan mitochondrial genome arrangements. *Mol. Biol. Evol.* 22:486–495.

Lavine, M. and Schervish, M. J. 1999. Bayes factors: What they are and what they are not. *Am. Stat.* 53:119–122.

Lee, M. S. Y. and Hugall, A. F. 2003. Partitioned likelihood support and the evaluation of data set conflict. *Syst. Biol.* 52:15–22.

LeSy, V., Varón, A., Janies, D., and Wheeler, W. C. 2007. Towards phyloge-
nomic reconstruction. *In* Proceedings of the 2007 International Conference on
Bioinformatics and Computational Biology, pp. 98–104, Las Vegas Nevada,
USA.

LeSy, V., Varón, A., and Wheeler, W. C. 2006. Pairwise alignment with rearrangement. *Genome Informatics* 17:141–151.

Levin, D. A. 1979. The nature of plant species. *Science* 204:381–384.

Lewis, P. O., Holder, M. T., and Holsinger, K. E. 2005. Polytomies and Bayesian
phylogenetic inference. *Sys. Biol.* 54:241–253.

Lherminera, P. and Solignac, M. 2000. L'espèce: définitions d'auteurs.
C. R. Acad. Sci. Paris, Sciences de la vie 323:153–165.

Linnaeus, C. 1753. Species Plantarum. Holmiae, Stockholm.

Linnaeus, C. 1758. Systema Naturae. Holmiae, Stockholm. 10th edition.

Lipscomb, D. L., Farris, J. S., Kallersjo, M., and Theler, A. 1998. Support,
ribosomal sequences and the phylogeny of the eukaryotes. *Cladistics* 14:
303–338.

Little, A. T. D. P. and Farris, J. S. 2003. The full-length phylogenetic tree from
1551 ribosomal sequences of chitinous fungi, Fungi. *Mycological Research*
107:901–916.

Liu, K., Nelesen, S., Raghavan, S., Linder, C. R., and Warnow, T. 2009. Barking
up the wrong treelength: The impact of gap penalty on alignment and tree
accuracy. *IEEE Trans. Comput. Biol. Bioinf.* 6:7–20.

Löytynoja, A. and Milinkovitch, M. C. 2003. ProAlign, a probabilistic multiple
alignment program. *Bioinformatics* 19:1505–1513.

Lundberg, J. G. 1972. Wagner networks and ancestors. *Syst. Zool.* 21:398–413.

Lundy, M. 1985. Applications of the annealing algorithm to combinatorial problems in statistics. *Biometrika* 72:191–198.

Lunter, G., Drummond, A. J., Miklós, I., and Hein, J. 2005. Statistical align-
ment: Recent progress, new applications, and challenges, pp. 375–406. *In*
R. Nielsen (ed.), Statistical Methods in Molecular Evolution. Springer, New
York.

Macleay, W. S. 1819. Horae Entomologicae. London edition.

Maddison, D. R. 1991. The discovery and importance of multiple islands of
most parsimonious trees. *Syst. Zool.* 40:315–328.

Maddison, W. 1995. Calculating the probability distributions of ancestral states
reconstructed by parsimony on phylogenetic trees. *Syst. Biol.* 44:474–481.

Margush, T. and McMorris, F. R. 1981. Consensus n-trees. *Bull. Math. Biol.* 43:239–244.

Martiz, J. S. and Lwin, T. 1989. Empirical Bayes Methods. Chapman and Hall, London. 2nd Edition.

Mathews, D. H. and Turner, D. H. 2002. Dynalign: An algorithm for finding the secondary structure common to two RNA sequences. *J. Mol. Biol.* 317:191–203.

Matsuda, H. 1996. Protein phylogenetic inference using maximum likelihood with a genetic algorithm. *In* L. Hunter and T. E. Klein (eds.), Pacific Symposium on Biocomputing '96, pp. 512–523, London. World Scientific.

Mau, B., Newton, M. A., and Larget, B. 1999. Bayesian phylogenetic inference via Markov chain Monte Carlo methods. *Biometrics* 55:1–12.

Mayden, R. L. 1997. A hierarchy of species concepts: the denouement in the saga of the species problem, pp. 381–423. *In* H. A. Dawah and M. R. Wilson (eds.), Species: The units of biodiversity. Chapman and Hall, London.

Mayr, E. 1940. Speciation phenomena in birds. *Am. Nat.* 74:249–278.

Mayr, E. 1942. Systematics and the origin of species. Cambridge University Press, Cambridge. 334 pp.

Mayr, E. 1963. Animal species and evolution. Harvard University Press, Cambridge, MA.

Mayr, E. 1965. Numerical phenetics and taxonomic theory. *Syst. Zool.* 14:73–97.

Mayr, E. 1969. Principles of systematic biology. McGraw-Hill, New York.

Mayr, E. 1982. The growth of biological thought. Harvard University Press, Cambridge, MA.

Mayr, E. and Ashlock, P. D. 1991. Principles of Systematic Zoology. McGraw-Hill, New York.

McGuire, G., Denham, M. C., and Balding, D. J. 2001. Models of sequence evolution for DNA sequences containing gaps. *Mol. Biol. Evol.* 18:481–490.

McKenna, M. and Bell, S. K. 1997. Classification of mammals above the species level. Columbia University Press, New York, NY. 535 pp.

Meier, R. 1997. Test and review of the empirical performance of the ontogenetic criterion. *Syst. Zool.* 46:699–721.

Meier, R. and Willmann, R. 2000. The Hennigian species concept, pp. 30–43. *In* Q. D. Wheeler and R. Meier (eds.), Species Concepts and Phylogenetic Theory. Columbia University Press, New York.

Metropolis, N. and Ulam, S. 1949. The Monte Carlo method. *J. Am. Stat. Assoc.* 44:335–341.

Metropolis, N. A., Rosenbluth, A., Rosenbluth, M., Teller, A., and Teller, E. 1953. Equation of state calculations by fast computing machine. *J. Chem. Phys.* 21:1087–1092.

Michener, C. D. 1977. Discordant evolution and the classification of allopodine bees. *Syst. Zool.* 26:32–56.

Michener, C. D. and Sokal, R. R. 1957. A quantitative approach to a problem in classification. *Evolution* 11:130–162.

Mickevich, M. F. and Farris, J. S. 1980. PHYSYS: Phylogenetic analysis system.

Mishler, B. D. 1985. The morphological, developmental, and phylogenetic basis of species concepts in bryophytes. *Bryologist* 88:207–214.

Mishler, B. D. and Brandon, R. N. 1987. Individuality, pluralism, and the phylogenetic species concept. *Biol. Philos.* 2:397–414.

Mishler, B. D. and Donoghue, M. J. 1982. A case for pluralism. *Syst. Biol.* 31:491–503.

Mishler, B. D. and Theriot, E. C. 2000. The phylogenetic species concept (*sensu* Mishler and Theriot): Monophyly, apomorphy, and phylogenetic species concepts, pp. 44–54. *In* Q. D. Wheeler and R. Meier (eds.), Species Concepts and Phylogenetic Theory. Columbia University Press, New York.

Moilanen, A. 1999. Searching for most parsimonious trees with simulated evolutionary optimization. *Cladistics* 15:39–50.

Moilanen, A. 2001. Simulated evolutionary optimization and local search: Introduction and application to tree search. *Cladistics* 17:S12–S25.

Moores, A. O., Harmon, L. J., Blum, M. G. B., Wong, D. H. J., and Heard, S. B. 2007. Some models of phylogenetic tree shape, pp. 149–170. *In* O. Gascuel and M. Steel (eds.), Reconstructing evolution: New mathematical and computational advances. Oxford University Press, New York.

Moret, B. M. E., Tang, J., Wang, L. S., and Warnow, T. 2002. Steps toward accurate reconstruction of phylogenies from gene-order data. *J. Comput. Syst. Sci.(special issue on computational biology)*.

Morgan, G. J. 1998. Emile Zuckerkandl, Linus Pauling, and the molecular evolutionary clock. *J. Hist. Biol.* 31:155–178.

Morgenstern, B. 1999. DIALIGN 2: Improvement of the segment-to-segment approach to multiple sequence alignment. *Bioinformatics* 15:211–218.

Morgenstern, B., Dress, A., and Werner, T. 1996. Multiple DNA and protein sequence alignment based on segment-to-segment comparison. *PNAS* 93:12098–12103.

Mossel, E., Roch, S., and Steel, M. 2009. Shrinkage effect in ancestral maximum likelihood. *IEEE/ACM Transactions in computational biology and bioinformatics* 6:126–133.

Mosteller, F. and Wallace, D. L. 1984. Applied Bayesian and classical inference: The case of the Federalist papers. Springer-Verlag, New York, 2nd edition.

Murienne, J., Harvey, M. S., and Giribet, G. 2008. First molecular phylogeny of the major clades of Pseudoscorpiones (Arthropoda: Chelicerata). *Mol. Phyl. Evol.* 49:170–184.

Nadeau, J. H. and Taylor, B. A. 1984. Lengths of chromosomal segments conserved since divergence of man and mouse. *PNAS* 81:814–818.

Near, T. J., Meylan, P. A., and Shaffer, H. B. 2005. Assessing concordance of fossil calibration points in molecular clock studies: an example using turtles. *Am. Nat.* 165:137–146.

Needleman, S. B. and Wunsch, C. D. 1970. A general method applicable to the search for similarities in the amino acid sequences of two proteins. *J. Mol. Biol.* 48:443–453.

Nei, M. and W.-H.Li 1979. Mathematical model for studying genetic variation in terms of restriction endonucleases. *PNAS* 76:5269–5273.

Nelson, G. 1978. Ontogeny, phylogeny, paleontology, and the biogenetic law. *Syst. Zool.* 27:324–345.

Nelson, G. 1979. Cladistic analysis and synthesis: Principles and definitions, with a historical note on Adanson's Familles des plantes (1763–1764). *Syst. Zool.* 28:1–21.

Nelson, G. and Ladiges, P. Y. 1994. Three-item consensus: Empirical test of fractional weighting, pp. 193–209. *In* R. Scotland, D. J. Siebert, and D. M. Williams (eds.), Models in phylogeny reconstruction, The Systematics Association Special Volume No. 52. Oxford University Press, New York.

Nelson, G. J. 1972. Review of Die Rekonstruktion der Phylogenese mit Hennig's Prinzip by D. Schlee (1971). *Syst. Zool.* 21:350–352.

Nelson, G. J. and Platnick, N. I. 1981. Systematics and Biogeography: Cladistics and Vicariance. Columbia University Press, New York.

Nelson, G. J. and Platnick, N. I. 1991. Three-taxon statements: A more precise use of parsimony? *Cladistics* 7:351–366.

Neyman, J. 1952. Lectures and conferences on mathematical statistics and probability. Washington graduate school, USDA, Washington, DC, 2nd edition.

Neyman, J. 1971. Molecular studies in evolution: a source of novel statistical problems. *In* S. S. Gupta and J. Yackel (eds.), Statistical Decision Theory and Related Topics, pp. 1–27.

Nixon, K. and Wheeler, Q. D. 1990. An amplification of the phylogenetic species concept. *Cladistics* 6:211–223.

Nixon, K. C. 1999. The parsimony ratchet, a new method for rapid parsimony analysis. *Cladistics* 15:407–414.

Nixon, K. C. and Carpenter, J. M. 1993. On outgroups. *Cladistics* 9:413–426.

Nixon, K. C. and Carpenter, J. M. 1996a. On consensus, collapsability, and clade concordance. *Cladistics* 12:305–321.

Nixon, K. C. and Carpenter, J. M. 1996b. On simultaneous analysis. *Cladistics* 12:221–241.

Nixon, K. C. and Davis, J. I. 1991. Polymorphic taxa, missing values and cladistic analysis. *Cladistics* 7:23–241.

Nixon, K. C. and Wheeler, Q. D. 1992. Extinction and the origin of species, pp. 119–143. *In* M. J. Novacek and Q. D. Wheeler (eds.), Extinction and phylogeny. Columbia University Press, New York, NY.

Norell, M. A. 1987. The phylogenetic determination of taxonomic diversity: implications for terrestrial vertebrates at the K-T boundary. *J. Vertebr. Paleontol.* 7 (Suppl.).

Norell, M. A. 1992. Taxic origin and temporal diversity: the effect of phylogeny, pp. 88–118. *In* M. J. Novacek and Q. D. Wheeler (eds.), Extinction and phylogeny. Columbia University Press, New York, NY.

Norell, M. A. and Makovicky, P. J. 2004. Dromaeosauridae, pp. 196–210. *In* D. B. Weishampel, P. Dodson, and H. Osmólska (eds.), The Dinosauria. University of California Press, Berkeley, 2nd edition.

Notredame, C. 2002. Recent progresses in multiple sequence alignment: a survey. *Pharmacogenomics* 3:131–144.

Notredame, C., Higgins, D., and Heringa, J. 2000. T-Coffee: A novel method for multiple sequence alignments. *J. Mol. Biol.* 302:205–217.

Notredame, C. and Higgins, D. G. 1996. Saga: sequence alignment by genetic algorithm. *Nucleic Acids Res.* 24:1515–1524.

Notredame, C., Holm, L., and Higgins, D. G. 1998. Coffee: an objective function for multiple sequence alignments. *Bioinformatics* 14:407–422.

Ockham, W. 1323. Summa Logicae.

Ogden, T. H. and Rosenberg, M. S. 2007. Alignment and topological accuracy of the direct optimization approach via POY and traditional phylogenetics via ClustalW + PAUP*. *Syst. Biol.* 56:182–193.

Oken, L. 1802. Grundriss der Naturphilosophie, der Theorie der Sinne, mit der darauf gegründeten Classification der Thiere.

Owen, R. 1843. Lectures on comparative anatomy. Longman, Brown, Green, and Longmans, London.

Owen, R. 1847. Report on the archetype and homologies of the vertebrate skeleton, pp. 169–340. Report on the 16th meeting of the British Association for the advancement of science. Murray, London.

Owen, R. 1848. On the archetype and homologies of the vertebrate skeleton. John van Voorst, London.

Owen, R. 1849. On the nature of limbs, a discourse. John van Voorst, London.

Parham, J. F. and Irmis, R. B. 2008. Caveats on the use of fossil calibrations for molecular dating: a comment on Near et al. *Am. Nat.* 171:132–136.

Patterson, C. 1982. Morphological characters and homology, pp. 21–74. *In* K. A. Joysey and A. E. Friday (eds.), Problems of Phylogenetic Reconstruction. Academic Press, New York.

C. Patterson (ed.) 1987. Molecules and Morphology in Evolution: Conflict or Compromise?, number 3. International Congress of Systematic and Evolutionary Biology, Cambridge University Press, Cambridge.

Patterson, C. 1988. Homology in classical and molecular biology. *Mol. Biol. Evol.* 5:603–625.

Patterson, H. E. H. 1980. A comment on "mate recognition systems". *Evolution* 34:330–331.

Patterson, H. E. H. 1985. The recognition concept of species, pp. 21–29. *In* E. S. Vrba (ed.), Species and speciation. Transvaal Museum, Pretoria, South Africa.

Pauplin, Y. 2000. Direct calculation of a tree length using a distance matrix. *J. Mol. Evol.* 51:41–47.

Penny, D. 1982. Towards a basis for classification: the incompleteness of distance measures, incompatibility analysis and phenetic classification. *J. Theoretical Biology* 96:129–142.

Penny, D., Hendy, M. D., Lockhart, P. J., and Steel, M. A. 1996. Corrected parsimony, minimum evolution, and Hadamard conjugations. *Syst. Biol.* 45:596–606.

Penny, D., Hendy, M. D., and Steel, M. A. 1991. Testing the theory of descent, pp. 155–183. *In* M. M. Miyamoto and J. Cracraft (eds.), Phylogenetic Analysis of DNA Sequences. Oxford University Press, London.

Philippe, H., Zhou, Y., Brinkmenn, H., Rodrigue, N., and Delsuc, F. 2005. Heterotachy and long-branch attraction in phylogenetics. *BMC Evolutionary Biology* 5:50.

Phillips, A., Janies, D., and Wheeler, W. C. 2000. Multiple sequence alignment in phylogenetic analysis. *Mol. Phyl. Evol.* 16:317–330.

Pickett, K. M. and Randle, C. P. 2005. Strange Bayes indeed: uniform topological priors. *Mol. Phyl. Evol.* 34:203–211.

Pickett, K. M., Tolman, G. L., Wheeler, W. C., and Wenzel, J. W. 2005. Parsimony overcomes statistical inconsistency with the addition of more data from the same gene. *Cladistics* 21:438–445.

Pimentel, R. A. and Riggins, R. 1987. The nature of cladistic data. *Cladistics* 3:301–209.

Platnick, N. I. 1989. Cladistics and phylogenetic analysis today. *In* B. Fernholm, K. Bremer, and H. Jörnvall (eds.), The Hierarchy of Life, pp. 17–24. Elsevier Press, Amsterdam.

Platnick, N. I. and Gaffney, E. S. 1977. Systematics: A Popperian perspective (Reviews of "The Logic of Scientific Discovery," and "Conjectures and Refutations." by Karl R. Popper). *Syst. Zool.* 26:360–365.

Platnick, N. I., Griswold, C. E., and Coddington, J. A. 1991. On missing entries in cladistic analysis. *Cladistics* 7:337–343.

Pleijel, F. 1995. On character coding for phylogenetic reconstruction. *Cladistics* 11:309–315.

Pol, D. and Siddall, M. E. 2001. Biases in maximum likelihood and parsimony: A simulation approach to a 10-taxon case. *Cladistics* 17:266–281.

Popper, K. 1934. Logik der Forschung. Mohr Siebeck, Tübingen.

Popper, K. 1959. The Logic of Scientific Discovery. Routledge, London.

Popper, K. 1963. Conjectures and Refutations. Routledge, London, 2002 edition.

Popper, K. 1972. Objective Knowledge: An Evolutionary Approach. Oxford University Press, Oxford.

Popper, K. 1983. Realism and the Aim of Science. Routledge, London and New York.

Posada, D. and Buckley, T. R. 2004. Model selection and model averaging in phylogenetics: Advantages of Akaike information criterion and Bayesian approaches over likelihood ratio tests. *Syst. Biol.* 53:793–808.

Posada, D. and Crandall, K. A. 1998. MODELTEST: Testing the model of DNA substitution. *Bioinformatics* 14:817–818.

Post, E. 1936. A variant of a recursively unsolvable problem. *J. Symbolic Logic* 1:103–105.

Prendini, L., Weygoldt, P., and Wheeler, W. C. 2005. Systematics of the Damon variegatus group of African whip spiders (Chelicerata: Amblypygi): Evidence from behaviour, morphology and DNA. *Organisms, Diversity and Evolution* 5:203–236.

Prim, R. C. 1957. Shortest connection networks and some generalizations. *Bell Systems Techn.* 36:1389–1401.

Provine, W. B. 1986. Sewall Wright and evolutionary biology. University of Chicago Press, Chicago.

Pulquério, M. J. and Nichols, R. A. 2006. Dates from the molecular clock: how wrong can we be? *Trends Ecol. Evol.* 22:180–184.

Purvis, A. 1995a. A composite estimate of primate phylogeny. *Phil. Trans. R. Soc. Lond. B* 348:405–421.

Purvis, A. 1995b. A modification to Baum and Ragan's method for combining phylogenetic trees. *Syst. Biol.* 44:251–255.

Quenouille, M. H. 1949. Approximate tests of correlation in time-series. *J. R. Statist. Soc. B* 11:68–84.

Quenouille, M. H. 1956. Notes on bias estimation. *Biometrika* 43:353–360.

Ramírez, M. J. 2007. Homology as a parsimony problem: a dynamic homology approach for morphological data. *Cladistics* 23:1–25.

Rannala, B. and Yang, Z. 1996. Probability distribution of molecular evolutionary trees: A new method of phylogenetic inference. *J. Mol. Evol.* 43:304–311.

Raup, D. M., Gould, S. J., Schopf, T. J., and Simberloff, D. S. 1973. Stochastic models of phylogeny and the evolution of diversity. *J. Geol.* 81:525–542.

Ray, J. 1686. Historia Plantarum.

Redelings, B. D. and Suchard, M. A. 2005. Joint Bayesian estimation of alignment and phylogeny. *Syst. Biol.* 54:401–418.

Reeder, J. and Giegerich, R. 2005. Consensus shapes: an alternative to the Sankoff algorithm for RNA consensus structure prediction. *Bioinformatics* 21:3516–3523.

Regan, C. T. 1926. Organic evolution. *Rept. British Assoc. Advmt. Sci.* 1925:75–86.

Reidl, R. 1978. Order in living organisms: a systems analysis of evolution. Wiley, Chichester.

Rieppel, O. C. 1988. Fundamentals of comparative biology. Birkhauser Verlag, Basel, Boston, Berlin.

Remane, A. 1952. Die Grundlagen des Naturlichen Systems der Vergleichenden Anatomie und der Phylogenetik. Geest and Portig, Leipzig, Germany.

Rice, K. A., Donoghue, M. J., and Olmstead, R. G. 1997. Analyzing large data sets: RBCL 500 revisited. *Syst. Biol.* 46:554–563.

Richards, R. 2005. The aesthetic and morphological foundations of Ernst Haeckel's evolutionary project. *In* M. Kemperink and P. Dassen (eds.), The Many Faces of Evolution in Europe, 1860–1914. Peeters, Amsterdam.

Robillard, T., Legendre, F., Desutter-Grandcolas, L., and Grandcolas, P. 2006. Phylogenetic analysis and alignment of behavioral sequences by direct optimization. *Cladistics* 22:602–633.

Robinson, D. F. 1971. Comparison of labelled trees with valency three. *J. Combinatorial Theory* 11:105–119.

Robinson, D. F. and Foulds, L. R. 1981. Comparison of phylogenetic trees. *Mathematical Biosciences* 53:131–147.

Robinson, J. B. 1949. On the Hamiltonian game (a traveling-salesman problem). *Rand Reports* pp. 1–10.

Roch, S. 2006. A short proof that phylogenetic tree reconstruction by maximum likelihood is hard. *Computational Biology and Bioinformatics, IEEE/ACM Transactions on* 3:92–94.

Rosa, D. 1918. Ologenesi. R. Bemporad, Florence.

Rosen, D. E. 1978. Vicariant patterns and historical explanation in biogeography. *Syst. Zool.* 27:159–188.

Rosen, D. E. 1979. Fishes from the uplands and intermontane basin of Guatemala: Revisionary studies and comparative geography. *Bull. Am. Mus. Nat. Hist.* 162:267–376.

Roshan, U. 2004. Algorithmic techniques for improving the speed and accuracy of phylogenetic methods. PhD thesis, The University of Texas at Austin.

Roshan, U., Moret, B., Williams, T., and Warnow, T. 2004. Rec-I-DCM3: A fast algorithmic technique for reconstructing large phylogenetic tree. *In* Proc. IEEE Computer Society Bioinformatics Conference CSB 2004, Stanford U.

Russell, E. S. 1916. Form and Function a Contribution to the History of Animal Morphology. John Murray, London.

Rutschman, F. 2006. Molecular dating of phylogenetic trees: A brief review of current methods that estimate divergence times. *Diversity Distrib.* 12:35–48.

Rzhetsky, A. and Nei, M. 1993. Theoretical foundation of the minimum-evolution method of phylogenetic inference. *Mol. Biol. Evol.* 4:406–425.

Saint-Hilaire, E. G. 1818. Philosophie Anatomique. Vol. 1, Des Organes Respiratoires sous le Rapport de la Détermination et de l'Identité de leurs Pièces Osseuses. Baillière, J.-B., Paris.

Saint-Hilaire, E. G. 1830. Principes de Philosophie Zoologique, discutés en Mars 1830, au Sein de l'Académie Royale des Sciences. Pichon et Dider, Paris.

Saint-Hilaire, E. G. 1833. Le degré d'influence du monde ambiant pour modifier les formes animales; question intéressant l'origine des espéces téléosariens et successivement cele des animaux de l'époque actuelle. *Mém. Acad. R. Sci. Inst. France* 12:63–92.

Saitou, N. and Nei, M. 1987. The neighbor-joining method: A new method for reconstructing phylogenetic trees. *Mol. Biol. Evol.* 4:406–425.

Salter, L. A. and Pearl, D. K. 2001. Stochastic search strategy for estimation of maximum likelihood phylogenetic trees. *Syst. Biol.* 50:7–17.

Sanderson, M. 2004. r8s software package. Available from http://ginger.ucdavis.edu/r8s/.

Sanderson, M. J. 1997. A nonparametric approach to estimating divergence times in the absence of rate constancy. *Mol. Biol. Evol.* 14:1218–1231.

Sanderson, M. J. 2002. Estimating absolute rates of molecular evolution and divergence times: A penalized likelihood approach. *Mol. Biol. Evol.* 19:101–109.

Sanderson, M. J. 2003. r8s: inferring absolute rates of molecular evolution and divergence times in the absence of a molecular clock. *Bioinformatics* 19:301–302.

Sanderson, M. J. and Kim, J. 2000. Parametric phylogenetics? *Syst. Biol.* 49:817–829.

Sankoff, D. 1985. Simultaneous solution of the RNA folding, alignment and protosequence problems. *SIAM J. Appl. Math.* 45:810–825.

Sankoff, D., Abel, Y., and Hein, J. 1994. A tree, a window, a hill; generalization of nearest-neighbor interchange in phylogenetic optimization. *J. Classif.* 11:209–232.

Sankoff, D., Sundaram, G., and Kececioglu, J. 1996. Steiner points in the space of genome rearrangements. *Internat. J. Found. Comp. Sci.* 7:1–9.

Sankoff, D. M. 1975. Minimal mutation trees of sequences. *SIAM J. Appl. Math.* 28:35–42.

Sankoff, D. M. and Cedergren, R. J. 1983. Simultaneous comparison of three or more sequences related by a tree, pp. 253–263. *In* D. M. Sankoff and J. B. Kruskal (eds.), Time warps, string edits, and macromolecules: the theory and practice of sequence comparison, chapter 9. Addison Wesley, Reading, MA.

Sankoff, D. M. and Rousseau, P. 1975. Locating the vertices of a Steiner tree in arbitrary space. *Math. Program.* 9:240–246.

Sarich, V. M. and Wilson, A. C. 1967. Immunological time scale for hominid evolution. *Science* 158:1200–1203.

Sastry, K., Goldberg, D., and Kendall, G. 2005. Genetic Algorithms, chapter Search Methodologies: Introductory Tutorials in Optimization and Decision Support Methodologies. Springer. eds. Burke E. K. and Kendall G.

Schaeffer, B., Hecht, M. K., and Eldredge, N. 1972. Phylogeny and paleontology. *Evol. Biol.* 6:31–46.

Schleicher, A. 1869. Darwinism Tested by the Science of Language. John Camden and Hotten, London.

Schröder 1870. Vier combinatoriche probleme. *Zeitschrift für Matemmatik und Physik* 15:489–503.

Schuh, R. T. and Brower, A. V. Z. 2009. Biological Systematics. Cornell University Press, 2nd edition.

Schuh, R. T. and Polhemus, J. T. 1980. Analysis of taxonomic congruence among morphological, ecological, and biogeographic data sets for the Leptopodomorpha (Hemiptera). *Syst. Zool.* 29:1–26.

Schuh, R. T., Weirauch, C., and Wheeler, W. 2009. Phylogenetic analysis of family-group relationships in the Cimicomorpha (Hemiptera). *Syst. Entomol.* 34:15–48.

Schulmeister, S. and Wheeler, W. C. 2004. Comparative and phylogenetic analysis of developmental sequences. *Evolution and Development* 6:50–57.

Schulmeister, S., Wheeler, W. C., and Carpenter, J. M. 2002. Simultaneous analysis of the basal lineages of Hymenoptera (Insecta) using sensitivity analysis. *Cladistics* 18:455–484.

Schwaz, G. 1978. Estimating the dimension of a model. *Ann. Statist.* 6:461–464.

Semple, C. and Steel, M. 2000. A supertree method for rooted trees. *Disc. Appl. Math* 105:147–158.

Semple, C. and Steel, M. 2003. Phylogenetics. Oxford University Press, Oxford.

Shannon, C. E. 1950. Prediction and entropy of printed english. *The Bell System Technical Journal* 30:50–64.

Shimodaira, H. and Hasegawa, M. 1999. Multiple comparisons of log-likelihoods with applications to phylogenetic inference. *Mol. Biol. Evol.* 16:1114–1116.

Sibley, C. and Ahlquist, J. 1984. The phylogeny of the hominoid primates, as indicated by DNA-DNA hybridization. *J. Mol. Evol.* 20:2–15.

Sibley, C. G. and Ahlquist, J. E. 1990. Phylogeny and classification of birds. Yale University Press, New Haven, CT.

Siddall, M. E. 1998. Success of parsimony in the four-taxon case: Long-branch repulsion by likelihood in the Farris zone. *Cladistics* 14:209–220.

Simpson, G. G. 1944. Tempo and mode in evolution. Columbia University Press, New York. 237 pp.

Simpson, G. G. 1961. Principles of animal taxonomy. Columbia University Press, New York.

Slowinski, J. B. 1998. The number of multiple alignments. *Mol. Phylogen. Evol* 10:264–266.

Smets, B. F. and Barkay, T. 2005. Horizontal gene transfer: perspectives at a crossroads of scientific disciplines. *Nature Reviews Microbiology* 3:675–678.

Smouse, P. E. and Li, W.-H. 1987. Likelihood analysis of mitochondrial restriction-cleavage patterns for the human-chimpanzee-gorilla trichotomy. *Evolution* 41:1162–1176.

Sneath, P. H. A. and Sokal, R. R. 1973. Numerical Taxonomy. Freeman, San Francisco. Revision of 1963 edition.

Sober, E. 1980. Evolution, population thinking, and essentialism. *Philos. Sci.* 47:350–383.

Sober, E. 1988. Reconstructing the past; parsimony evolution, and inference. The MIT Press, Cambridge, MA.

Sober, E. 2004. The contrast between parsimony and likelihood. *Syst. Biol.* 53:644–653.

Sober, E. 2005. Parsimony and its presuppositions, pp. 43–53. *In* V. Albert (ed.), Parsimony, Phylogeny, and Genomics. Oxford University Press, Oxford.

Sokal, R. R. and Michener, C. D. 1958. A statistical method for evaluating systematic relationships. *University of Kansas Scientific Bulletin* 28:1409–1438.

Sokal, R. R. and Rohlf, F. J. 1962. The comparison of dendrograms by objective means. *Taxon* 11:33–40.

Sokal, R. R. and Rohlf, F. J. 1981. Taxonomic congruence in the Leptopodomorpha re-examined. *Syst. Zool.* 30:309–325.

Sokal, R. R. and Sneath, P. H. A. 1963. Numerical Taxonomy. Freeman, San Francisco.

Solomonoff, R. 1964. A formal theory of inductive inference. *Information Control* 7:1–22, 224–254. parts 1 and 2.

Soltis, D. E., Soltis, P. S., Chase, M. W., Mort, M. E., Albach, D. C., Zanis, M., Savolainen, V., Hahn, W. H., Hoot, S. B., Fay, M. F., Axtell, M., Swensen, S. M., Prince, L. M., Kress, W. J., Nixon, K. C., and Farris, J. S. 2000. Angiosperm phylogeny inferred from 18S rDNA, RBCL, and atpB sequences. *Bot. J. Linnean Soc.* 133:381–461.

Sota, T. and Vogler, A. P. 2001. Incongruence of mitochondrial and nuclear gene trees in the carabid beetles *Ohomopterus*. *Syst. Biol.* 50:39–59.

Stamos, D. N. 2003. The species problem: biological species, ontology, and the metaphysics of biology. Lexington Books, Lanhan, MD.

Steel, M. 1992. The complexity of reconstructing trees from qualitative characters. *J. Classif.* 9:91–116.

Steel, M. 2000. Sufficient conditions for two tree reconstruction techniques to succeed on sufficiently long sequences. *SIAM J. Discrete Amth.* 14:36–48.

Steel, M. 2009. A basic limit on inferring phylogenies by pairwise sequence comparisons. *J. Theor. Biol.* 256:467–472.

Steel, M. 2011a. Consistency of Bayesian inference of resolved phylogenetic trees. Technical report, University of Canterbury, Christchurch, New Zealand.

Steel, M., Dress, A. W. M., and Bocker, S. 2000. Simple but fundamental limitations on supertree and consensus tree methods. *Syst. Biol.* 49:363–368.

Steel, M. and Penny, D. 2000. Parsimony, likelihood, and the role of models in molecular phylogenetics. *Mol. Biol. Evol.* 17:839–850.

Steel, M. and Penny, D. 2004. Two further links between MP and ML under the Poisson model. *Applied Mathematics Letters* 17:785–790.

Steel, M. and Pickett, K. M. 2006. On the impossibility of uniform priors on clades. *Mol. Phyl. Evol.* 39:585–586.

Steel, M. A. 2011b. Can we avoid "SIN" in the house of "No Common Mechanism? *Syst. Biol.* 60:96–109.

Steel, M. A., Hendy, M. D., and Penny, D. 1993. Parsimony can be consistent. *Syst. Biol.* 42:581–587.

Steel, M. A., Szekely, L. A., and Hendy, M. D. 1994. Reconstructing trees from sequences whose sites evolve at variable rates. *J. Comput. Biol.* 1:153–163.

Stocsits, R. R., Letsch, H., Hertel, J., Misof, B., and Stadler, P. F. 2009. Accurate and efficient reconstruction of deep phylogenies from structured rnas. *Nucl. Acid. Res.* 37:6184–6193.

Strang, G. 2006. Linear Algebra and Its Applications. Wellesley-Cambridge Press, Wellesley, MA, 4th edition.

Strimmer, K. and Moulton, V. 2000. Likelihood analysis of phylogenetic networks using directed graphical models. *Mol. Biol. Evol.* 17:875–881.

Suchard, M. A. and Redelings, B. D. 2006. BAli-Phy: simultaneous Bayesian inference of alignment and phylogeny. *Bioinformatics* 22:2047–2048.

Swofford, D. L. 1990. PAUP: Phylogenetic Analysis Using Parsimony.

Swofford, D. L. 1993. PAUP: Phylogenetic Analysis Using Parsimony.

Swofford, D. L. 2002. PAUP*: Phylogenetic analysis using parsimony (*and other methods), version 4.0b 10. Sinauer Associates, Sunderland, Massachusetts.

Swofford, D. L., Olsen, G. L., Waddell, P. J., and Hillis, D. M. 1996. Phylogenetic inference. Molecular Systematics. Sinauer, Sunderland, 2nd edition.

Swofford, D. L., Waddell, P. J., Huelsenbeck, J. P., Foster, P. G., Lewis, P. O., and Rogers, J. S. 2001. Bias in phylogenetic estimation and its relevance to the choice between parsimony and likelihood methods. *Syst. Biol.* 50:525–539.

Szalay, F. S. 1977. Ancestors, descendants, sister-groups and testing of phylogenetic hypotheses. *Syst. Zool.* 26:12–18.

Székely, L. and Steel, M. A. 1999. Inverting random functions. *Annals of Combinatorics* 3:103–113.

Tavaré, S. 1986. Some probabilistic and statistical problems on the analysis of DNA sequences. *Lec. Math. Life Sci.* 17:57–86.

Thompson, J. D., Higgins, D. G., and Gibson, T. J. 1994. CLUSTAL W: improving the sensitivity of progressive multiple sequence alignment through sequence weighting, position specific gap penalties and weight matrix choice. *Nucleic Acids Res.* 22:4673–4680.

Thompson, J. D., Koehl, P., Ripp, R., and Poch, O. 2005. BAliBASE 3.0: latest developments of the multiple sequence alignment benchmark. *Proteins* 61:127–136.

Thorne, J. L. and Kishino, H. 2002. Divergence time and evolutionary rate estimation with multilocus data. *Syst. Biol.* 51:689–702.

Thorne, J. L., Kishino, H., and Felsenstein, J. 1991. An evolutionary model for maximum likelihood alignment of DNA sequences. *J. Mol. Evol.* 33:114–124.

Thorne, J. L., Kishino, H., and Felsenstein, J. 1992. Inching toward reality: an improved likelihood model of sequence evolution. *J. Mol. Evol.* 34:3–16.

Thorne, J. L., Kishino, H., and Painter, I. S. 1998. Estimating the rate of molecular evolution of the rate of molecular evolution. *Mol. Biol. Evol.* 15:1647–1657.

Torres, A., Cabada, A., and Nieto, J. J. 2003. An exact formula for the number of alignments between two DNA sequences. *DNA Sequence* 19:427–430.

Tuffley, C. and Steel, M. 1997. Links between maximum likelihood and maximum parsimony under a simple model of site substitution. *Bull. Of Math. Biol.* 59:581–607.

Tukey, J. W. 1958. Bias and confidence in not-quite large samples (abstract). *Ann. Math. Statist.* 29:614.

Turing, A. 1936. On computable numbers, with an application to the Entscheidungsproblem. *P. Lond. Math. Soc., 2* 42:230–265.

Turing, A. 1937. On computable numbers, with an application to the Entscheidungsproblem: A correction. *P. Lond. Math. Soc., 2* 43:544–546.

Ukkonen, E. 1985. Finding approximate patterns in strings. *J. Algorithm* 6:132–137.

Vácsy, C. 1971. Les origens et les principes du développement de la nomenclature binaire en botanique. *Taxon* 20:573–590.

Varón, A., Vinh, L. S., Bomash, I., and Wheeler, W. C. 2008. POY 4.0. American Museum of Natural History. http://research.amnh.org/scicomp/projects/poy.php.

Varón, A., Vinh, L. S., and Wheeler, W. C. 2010. POY version 4: Phylogenetic analysis using dynamic homologies. *Cladistics* 26:72–85.

Velhagen, W. A. 1997. Analyzing developmental sequences using sequence units. *Syst. Biol.* 46:204–210.

Viterbi, A. J. 1967. Error bounds for convolutional codes and an asymptotically optimum decoding algorithm. *IEEE T. Inform. Theory* 13:260–269.

von Goethe, J. W. 1790. Die Metamorphose der Pflanzen. Versuch die Metamorphose der Pflanzen zu erklären. C. W. Ettinger, Gotha.

Voss, R. A. 2003. Accelerated likelihood surface exploration: The likelihood ratchet. *Syst. Biol.* 52:368–373.

Vrana, P. and Wheeler, W. C. 1992. Individual organisms as terminal entities: Laying the species problem to rest. *Cladistics* 8:67–72.

Wagner, W. H. 1961. Problems in the classification of ferns. *In* Recent Advances in Botany, pp. 841–844. University of Toronto Press, Toronto.

Wald, A. 1949. Note on the consistency of the maximum likelihood estimate. *Annals of Mathematical Statistics* 20:595–601.

Wang, L. and Gusfield, D. 1997. Improved approximation algorithms for tree alignment. *J. Algorithm* 25:255–273.

Wang, L. and Jiang, T. 1994. On the complexity of multiple sequence alignment. *J. Comput. Biol.* 1:337–348.

Wang, L., Jiang, T., and Gusfield, D. 2000. A more efficient approximation scheme for tree alignment. *SIAM J. Comput.* 30:283–299.

Wang, L., Jiang, T., and Lawler, E. L. 1996. Approximation algorithms for tree alignment with a given phylogeny. *Algorithmica* 16:302–315.

Wang, L.-S. and Warnow, T. 2001. Estimating true evolutionary distances between genomes. *In* Proceedings of 33rd Annual ACM Symposium on Theory of Computing, pp. 637–646. ACM Press, New York.

Wang, L.-S. and Warnow, T. 2005. Distance-based genome rearrangement phylogeny, pp. 453–383. *In* O. Gascuel (ed.), Mathematics of Evolution and Phylogeny. Oxford University Press, Oxford.

Waterman, M. S., Smith, T., and Beyer, W. A. 1976. Some biological sequence metrics. *Adv. Math.* 20:367–387.

Watrous, L. E. and Wheeler, Q. D. 1983. The out-group comparison method of character analysis. *Syst. Zool.* 30:1–11.

Wheeler, Q. D. 1990. Ontogeny and character phylogeny. *Cladistics* 6:225–268.

Q. D. Wheeler and R. Meier (eds.) 2000. Species Concepts and Phylogenetic Theory. Columbia University Press, New York.

Wheeler, Q. D. and Platnick, N. I. 2000. The phylogenetic species concept (*sensu* Wheeler and Platnick), pp. 55–69. *In* Q. D. Wheeler and R. Meier (eds.), Species Concepts and Phylogenetic Theory. Columbia University Press, New York.

Wheeler, W. C. 1991. Congruence among data sets: a Bayesian approach, pp. 334–346. *In* M. M. Miyamoto and J. Cracraft (eds.), Phylogenetic Analysis of DNA Sequences. Oxford University Press, London.

Wheeler, W. C. 1993. The triangle inequality and character analysis. *Mol. Biol. Evol.* 10:707–712.

Wheeler, W. C. 1994. Sources of ambiguity in nucleic acid sequence alignment, pp. 323–352. *In* G. W. B. Schierwater, B. Streit and R. DeSalle (eds.), Molecular Ecology and Evolution: Approaches and Applications. Birkhäuser Verlag, Basel, Switzerland.

Wheeler, W. C. 1995. Sequence alignment, parameter sensitivity, and the phylogenetic analysis of molecular data. *Syst. Biol.* 44:321–331.

Wheeler, W. C. 1996. Optimization alignment: The end of multiple sequence alignment in phylogenetics? *Cladistics* 12:1–9.

Wheeler, W. C. 1999. Fixed character states and the optimization of molecular sequence data. *Cladistics* 15:379–385.

Wheeler, W. C. 2001a. Homology and DNA sequence data, pp. 303–318. *In* G. Wagner (ed.), The Character Concept in Evolutionary Biology. Academic Press, New York.

Wheeler, W. C. 2001b. Homology and the optimization of DNA sequence data. *Cladistics* 17:S3–S11.

Wheeler, W. C. 2003a. Implied alignment. *Cladistics* 19:261–268.

Wheeler, W. C. 2003b. Iterative pass optimization. *Cladistics* 19:254–260.

Wheeler, W. C. 2003c. Search-based character optimization. *Cladistics* 19:348–355.

Wheeler, W. C. 2006. Dynamic homology and the likelihood criterion. *Cladistics* 22:157–170.

Wheeler, W. C. 2007a. The analysis of molecular sequences in large data sets: where should we put our effort? pp. 113–128. *In* T. R. Hodkinson and J. A. N. Parnell (eds.), Reconstructing the Tree of Life: Taxonomy and Systematics of Species Rich Taxa. Systematics Association, Oxford University Press, Oxford.

Wheeler, W. C. 2007b. Chromosomal character optimization. *Mol. Phyl. Evol.* 44:1130–1140.

Wheeler, W. C. 2010. Distinctions between optimal and expected support. *Cladistics* 26:657–663.

Wheeler, W. C. 2011. Trivial minimization of extra-steps under dynamic homology. *Cladistics* 27:1–2.

Wheeler, W. C., Aagesen, L., Arango, C. P., Faivovich, J., Grant, T., D'Haese, C., Janies, D., Smith, W. L., Varón, A., and Giribet, G. 2006a. Dynamic Homology and Systematics: A Unified Approach. American Museum of Natural History, New York.

Wheeler, W. C. and Giribet, G. 2009. Phylogenetic hypotheses and the utility of multiple sequence alignment, pp. 95–104. *In* M. S. Rosenberg (ed.), Perspectives on Biological Sequence Alignment. University of California Press, Berkeley, CA.

Wheeler, W. C., Giribet, G., and Edgecombe, G. D. 2004. Arthropod systematics: The comparative study of genomic, anatomical, and paleontological information, pp. 281–295. *In* J. Cracraft and M. J. Donoghue (eds.), Assembling the Tree of Life. Oxford University Press, Oxford.

Wheeler, W. C. and Gladstein, D. S. 1991–1998. MALIGN. Program and documentation available at http://research.amnh.org/scicomp/projects/malign.php. New York, NY. Documentation by Daniel Janies and W. C. Wheeler.

Wheeler, W. C., Gladstein, D. S., and De Laet, J. 1996–2005. POY version 3.0. Program and documentation available at http://research.amnh.org/scicomp/projects/poy.php (current version 3.0.11). Documentation by D. Janies and W. C. Wheeler. Commandline documentation by J. De Laet and W. C. Wheeler. American Museum of Natural History, New York.

Wheeler, W. C. and Pickett, K. M. 2008. Topology-Bayes versus clade-Bayes in phylogenetic analysis. *Mol. Biol. Evol.* 25:447–453.

Wheeler, W. C., Ramírez, M. J., Aagesen, L., and Schulmeister, S. 2006b. Partition-free congruence analysis: implications for sensitivity analysis. *Cladistics* 22:256–263.

Wheeler, W. C., Schuh, R. T., and Bang, R. 1993. Cladistic relationships among higher groups of Heteroptera: congruence between morphological and molecular data sets. *Ent. Scand.* 24:121–138.

Wheeler, W. C., Whiting, M. F., Wheeler, Q. D., and Carpenter, J. C. 2001. The phylogeny of the extant hexapod orders. *Cladistics* 17:113–169.

Whewell, W. 1847. The philosophy of the inductive sciences. John W. Parker, London.

Whiting, A. S., Sites Jr., J. W., Pellegrino, K. C., and Rodrigues, M. T. 2006. Comparing alignment methods for inferring the history of the new world lizard genus Mabuya (Squamata: Scincidae). *Mol. Phyl. Evol.* 38:719–730.

Wiley, E. O. 1975. Karl R. Popper, systematics and classification: A reply to Walter Bock and other evolutionary taxonomists. *Syst. Zool.* 24:233–243.

Wiley, E. O. 1978. The evolutionary species concept reconsidered. *Syst. Zool.* 27:88–92.

Wiley, E. O. and Lieberman, B. S. 2011. Phylogenetics. Wiley-Blackwell, second edition.

Wiley, E. O. and Mayden, R. L. 2000. The evolutionary species concept, pp. 70–89. *In* Q. D. Wheeler and R. Meier (eds.), Species Concepts and Phylogenetic Theory. Columbia University Press, New York.

Wilkinson, M., Cotton, J. A., Creevey, C., Eulenstein, O., Harris, S. R., Lapointe, F.-J., Levasseur, C., McInerney, J. O., Pisani, D., and Thorley, J. L. 2005. The shape of supertrees to come: Tree shape related properties of fourteen supertree methods. *Syst. Biol.* 54:419–431.

Wilkinson, M., Thorley, J., Littlewood, D. T. J., and Bray, R. A. 2001. Towards a phylogenetic supertree for the Platyhelminthes?, pp. 292–301. *In* D. T. J. Littlewood and R. A. Bray (eds.), Interrelationships of the Platyhelminthes. Chapman-Hall, London.

Williams, D. M. and Ebach, M. C. 2008. Foundations of Systematics and Biogeography. Springer, New York.

Wright, S. 1931. Evolution in Mendelian populations. *Genetics* 16:97–100, 155–159.

Wright, S. 1940. The statistical consequences of Mendelian heredity in relation to speciation, pp. 161–184. *In* J. Huxley (ed.), The new systematics. Oxford University Press, London. 583 pp.

Yancopoulos, S., Attie, O., and Friedberg, R. 2005. Efficient sorting of genomic permutations by translocation, inversion and block interchange. *Bioinformatics* 21:3340–3346.

Yang, Z. 1994a. Estimating the pattern of nucleotide substitution. *J. Mol. Evol.* 39:105–111.

Yang, Z. 1994b. Statistical properties of the maximum likelihood method of phylogenetic estimation and comparison with distance matrix methods. *Syst. Biol.* 43:329–342.

Yang, Z. 1996. Phylogenetic methods using parsimony and likelihood methods. *J. Mol. Evol.* 42:294–307.

Yang, Z. 1997. How often do wrong models produce better phylogenies? *Mol. Biol. Evol.* 14:105–108.

Yang, Z. 2004. A heuristic rate smoothing procedure for maximum likelihood estimation of species divergence times. *Acta Zool. Sinica* 50:645–656.

Yang, Z. 2006. Computational Molecular Evolution. Oxford University Press, Oxford.

Yang, Z., Goldman, N., and Friday, A. E. 1995. Maximum likelihood trees from DNA sequences: a peculiar statistical estimation problem. *Syst. Biol.* 44:384–399.

Yang, Z. and Rannala, B. 1997. Bayesian inference using DNA sequences: a Markov chain Monte Carlo method. *Mol. Biol. Evol.* 14:717–724.

Yang, Z. and Rannala, B. 2005. Branch-length prior influences Bayesian posterior probability of phylogeny. *Syst. Biol.* 54:455–470.

Yoder, A. D. and Yang, Z. 2000. Estimation of primate speciation dates using local molecular clocks. *Mol. Biol. Evol.* 17:1081–1090.

Ypma, T. J. 1995. Historical development of the Newton-Raphson method. *SIAM Review* 37:531–551.

Yule, G. U. 1925. A mathematical theory of evolution based on the conclusions of Dr. J. C. Willis, F. R. S. *Phil. Trans. Royal Soc. London Biol.* 213:21–87.

Zhang, G., Miyamoto, M. M., and Cohn, M. J. 2006. Lamprey type II collagen and Sox9 reveal an ancient origin of the vertebrate collagenous skeleton. *PNAS* 103:3180–3185.

Zhang, H. and Gu, X. 2004. Maximum likelihood for genome phylogeny on gene content. *Statistical Applications in Genetics and Molecular Biology* 3:1–16. Article 31.

Zimmermann, W. 1931. Arbeitsweise der botanischen Phylogenetik und anderer Gruppierungswissenschaften, pp. 941–1053. *In* E. Abderhalden (ed.), Handbuch der biologischen Arbeitsmethoden, volume 9. Urban und Swartzenberg, Berlin.

Zuckerkandl, E. and Pauling, L. 1965. Evolutionary divergence and convergence in proteins, pp. 97–116. *In* V. Bryson and H. J. Vogel (eds.), Evolving Genes and Proteins. Academic Press, New York.

Zuckerkandl, E. and Pauling, L. B. 1962. Molecular disease, evolution, and genetic heterogeneity, pp. 189–225. *In* M. Kasha and B. Pullman (eds.), Horizons in Biochemistry. Academic Press, New York.

Index

Systematics: A Course of Lectures, First Edition. Ward C. Wheeler.
© 2012 Ward C. Wheeler. Published 2012 by Blackwell Publishing Ltd.

Church (1936b), 77, 380

Coddington and Scharff (1994), 188, 211, 380

Cormode and Muthukrishnan (2002), 205, 380

Cracraft (1983), 58, 380

Cracraft (1989), 55, 380

Cracraft (2000), 63, 380

Cuvier (1812), 8, 380

Dalevi and Eriksen (2008), 233, 380, 434

Darling et al. (2004), 207, 381

Darwin (1859a), 12, 381

Darwin (1859b), 9, 64, 112–114, 117, 381

Day (1987), 83, 156, 162, 290, 381

DeGroot and Schervish (2006), 89, 235, 273, 333, 381

DePinna (1991), 20, 117, 119, 189, 381

Desper and Gascuel (2002), 164, 167, 381

Desper and Gascuel (2007), 162, 163, 381

Dobzhansky (1937), 14, 54, 55, 381

Donoghue and Kadereit (1992), 16, 381

Donoghue et al. (1992), 332, 381

Donoghue (1985), 55, 56, 64, 381

Dress and Krüger (1987), 314, 381

Dress et al. (2007), 153, 381

Drummon and Rambaut (2007), 370, 372, 381, 437

Drummon et al. (2006), 370, 381

Dunn et al. (2008), 25, 275, 329, 339, 382

Edgar and Sjölander (2003), 265, 382

Edgar (2004a), 132, 138, 382

Edgar (2004b), 132, 138, 382

Edwards (1970), 240, 382

Edwards (1972), 216, 280, 382

Efron and Tibshirani (1993), 325, 382

Efron (1979), 325, 382

Eldredge and Cracraft (1980), 23, 28, 43, 58, 382

Erdös et al. (1997), 149, 382

Estabrook (1978), 38, 39, 50, 382

Faith (1991), 332, 382

Farris et al. (1970), 153, 154, 383

Farris et al. (1996), 328, 383

Farris (1969), 160, 382

Farris (1970), 25, 28, 164, 175, 176, 178, 180, 296, 297, 382

Farris (1972), 164, 166, 276, 383

Farris (1973a), 213, 217, 224, 240, 272, 279, 285, 382

Farris (1973b), 355, 356, 383

Farris (1974), 15, 35, 38, 383

Farris (1977), 383

Farris (1981), 366, 368, 382

Farris (1983), 74, 271, 272, 383

Farris (1988), 299, 383

Felsenstein (1973), 213, 215, 219, 383

Felsenstein (1978), 213–215, 274, 277, 278, 280, 383

Felsenstein (1981), 222, 279, 333, 334, 367, 383

Felsenstein (1984), 367, 383

Felsenstein (1985), 325, 383

Felsenstein (1993), 314, 383

Felsenstein (2004), 162, 235, 240, 367, 383

Feng and Doolittle (1987), 134, 136, 265, 383

Feyerabend (1975), 73, 383

Feyerabend (1987), 73, 383

Finden and Gordon (1985), 349, 384

Fisher (1912), 101, 384

Fisher (1950), 280, 384

Fitch and Margoliash (1967), 136, 160–164, 276, 384

Fitch (1970), 41, 43, 384

Fitch (1971), 25, 179, 180, 182, 365, 384

Fleissner et al. (2005), 230, 231, 264, 384

Foulds and Graham (1982), 87, 290, 384

Fraser and Burnell (1970), 85, 384

Freudenstein et al. (2003), 21, 22, 384

Frost et al. (2006), 198, 319–321, 323, 384

Gaffney (1979), 271, 384

Galton (1889), 21, 384, 428